Simulation and Wargaming

Simulation and Wargaming

Edited by

Charles Turnitsa
Regent University
Virginia Beach, USA

Curtis Blais
Naval Postgraduate School MOVES Institute
Monterey, USA

Andreas Tolk
The MITRE Corporation
Charlottesville, USA

This edition first published 2022
© 2022 John Wiley & Sons, Inc.

The right of Charles Turnitsa, Curtis Blais, and Andreas Tolk to be identified as the author(s) of the editorial material in this work has been asserted in accordance with law.

Registered Office
John Wiley & Sons, Inc., 111 River Street, Hoboken, NJ 07030, USA

Editorial Office
111 River Street, Hoboken, NJ 07030, USA

For details of our global editorial offices, customer services, and more information about Wiley products visit us at www.wiley.com.

Wiley also publishes its books in a variety of electronic formats and by print-on-demand. Some content that appears in standard print versions of this book may not be available in other formats.

Library of Congress Cataloging-in-Publication Data

Names: Turnitsa, Charles, editor. | Blais, Curtis, editor. | Tolk, Andreas, editor.
Title: Simulation and wargaming / edited by Charles Turnitsa, Regent University, Virginia Beach, USA, Curtis Blais, Naval Postgraduate School MOVES Institute, Monterey, USA, Andreas Tolk, The MITRE Corporation, Charlottesville, USA.
Description: Hoboken, NJ : Wiley, 2021. | Includes bibliographical references and index.
Identifiers: LCCN 2021008520 (print) | LCCN 2021008521 (ebook) | ISBN 9781119604785 (hardback) | ISBN 9781119604792 (adobe pdf) | ISBN 9781119604808 (epub) | ISBN 9781119604815 (obook)
Subjects: LCSH: War games. | Simulation methods.
Classification: LCC U310 .S487 2021 (print) | LCC U310 (ebook) | DDC 658.4/0352-dc23
LC record available at https://lccn.loc.gov/2021008520
LC ebook record available at https://lccn.loc.gov/2021008521

Cover Design: Wiley
Cover Image: © Unsplash

Set in 9.5/12.5pt STIXTwoText by Straive, Pondicherry, India

SKY10032262_123021

This text is dedicated to **Dr. Stuart H. Starr** *(1942 to 2018), a mentor, colleague, and friend. He devoted his life to military operations research and analysis, including pushing wargaming towards new horizons. His work made him a Fellow of the Military Operations Research Society and was recognized by the Clayton J. Thomas Award as well as the Vance R. Wanner Memorial Award, but far more worth and longer lasting than his many recognitions are the impressions that his personality left on all of us: his openness and willingness to step in and educate and support everyone on the team to create new insights. He truly will be missed.*

Contents

Foreword

Reiner K. Huber

I was pleased to receive, and gladly accepted, the invitation to contribute the foreword to the timely book "Wargaming and Simulation" dedicated to Stuart Starr. I have known Stuart since the 1970s when we met at many professional and project meetings and discussions on transatlantic defense issues related, among others, to modeling and simulations in the context of assessment studies to support military and political decision-makers during the Cold War and the decade thereafter. The product of the last project in which both of us participated actively, from 2000 to 2003, was a revised version of the Code of Best Practice for C2 Assessment[1] that NATO had laid out in a technical report in 1999. Considering this report "a framework for thinking about the changing nature of war gaming," Stuart developed a highly interesting paper on how the sophisticated tools of collaboration technology emerging may revolutionize wargaming.[2]

Wargaming and simulation accompanied my professional life as a military OR/ SA analyst and an academic teacher in one way or another. It began with an episode of what wargamers want from OR models, which are on the heart of simulation. I was a junior OR analyst and captain of the German Air Force (GAF) Reserve when, in the mid-1960s, I was called up for a wargame by the Air Staff in the Defense Ministry. It was the first time I participated in a two-sided map display manual wargame for estimating success and losses to be expected in counter-air operations against well-defended Warsaw Pact (WP) airbases. Most of the players were fighter bomber pilots, some of them with WW 2 combat experience, and GAF air defense-officers familiar with the WP's air defense capabilities. OR

---

1 It was a project of NATO's Research and Technology Organization sponsoring the research task group SAS-026 chaired by Dr. David S. Alberts. It was published first by the Pentagon's C2 Research Program in 2002.
2 Starr, Stuart H. (2001) ""Good Games" – Challenges for the War-Gaming Community," *Naval War College Review*: Vol. 54: no.2, Article 9.

analysts of the Air OR Group of IABG[3] followed the players' mission plans calculating the attack aircraft lost and the damage caused to the targets using the respective mathematical models taken from IABG's air war model.[4] Never have I forgotten the disputes between players and OR analysts. The blue players considered the target damage calculated as too low and the loss of their air sorties as too high. In the midst of the game they suggested that the analysts manipulate the "critical" inputs of the assessment models so that the outputs would be closer to their judgment. Not surprisingly, the analysts rejected the suggestion arguing that rather than manipulating the game halfway, to improve its results for Blue, it would make more sense to end the game and, thereafter, revisit its data and the assumptions underlying the assessment models. That is what we did.

On the basis of the air war model the Air OR group then developed, together with the military advisory group associated with IABG's Study Division, an interactive wargame that was successfully tested within a high-level planning exercise of the GAF in 1970. In 1972 I became head of IABG's System Studies Division that included OR/SA support for all three service branches of the Bundeswehr. In a discussion with STC's Andreas Mortensen[5], about the upcoming issue of modeling in support of overall force capability assessment, we agreed that the most difficult and least documented aspect of military OR/SA seems to be the land war. Thus we felt that a scientific conference on land battle systems modeling would not only contribute to a better understanding of the implications of different modeling approaches, but also help preclude undesirable redundancy through better familiarity with models available elsewhere.

The Special Program Panel of Systems Science of the NATO Scientific Affairs Division[6] accepted and funded, together with the German Ministry of Defense,

---

3 Founded in 1962, IABG became, in addition to its engineering facilities for testing military hardware designs, the principal Operations Research and Defense Analysis Institution of West-Germany's Defense Ministry.

4 The mathematical air war model had been developed, together with analysts of the RAND Corporation and the USAF Systems Command, by IABG's Air OR group for the assessment of combat aircraft designs proposed by industry.

5 At the time, Mortensen, a senior member of the Norwegian Defence Research Establishment, was head of STC's OR division.

6 The Special Program Panel on Systems Science (SPOSS) was the follow-on of the Advisory Panel on Operations Research (APOR), which was established by the NATO Science Committee in 1958 tasked to organize conferences and symposia, develop educational programs, and award scholarships for study visits to spread methods and applications of operational research in WW 2 and thereafter, thus facilitating the buildup of both NATO and national military institutions to support defense planners and militaries in sustaining a NATO force structure capable of deterring a Soviet aggression. The initial Chairman of APOR was Prof. Phillip Morse, a pioneer of OR research in WW 2 and chairman of the US Navy's Anti-Submarine Warfare Operations Research Group (ASWORG). In 1973, the Science Committee replaced APOR by SPOSS arguing that, after 15 years, APOR has accomplished its mission and a reorientation of its effort was necessary toward applying previously developed techniques und theories to deal with real large-scale systems continuing, however, along the high scientific standards of APOR and its chairmen.

my proposal to organize, together with my co-chairmen, Lynn F. Jones of the UK's Royal Armaments Research and Development Establishment and Egil Reine of the Norwegian Defence Research Establishment, the scientific conference on "Modeling Land Battle Systems for Military Planning" held at the War Gaming Center of IABG[7] in Ottobrunn, Germany, 26–30 August 1974. The keynote address of the Deputy Under Secretary of the US Army for Operations Research Dr. Wilbur Payne began with the following statement:

> *As we are less and less able to rely on historical European combat data and as we see more and more the necessity of evaluating issues in large contexts,* **gaming and simulation** *emerge perhaps as the only tools able to organize large quantities of information and discipline our thinking and communication about them.*[8]

While not having strong opinions on some methodological aspects of the problem of modeling land battle systems, Dr. Payne expressed strong opinions on certain problems that require more attention than we have given in the past. From the point of view of a senior member of the profession[9], and a bureaucrat involved in trying to use the results of research to generate defense programs and convince others that the programs are worthy of support, he pointed out some of these problems such as (1) identifying radical changes in the general structures of combat; (2) interactions between weapon system development and tactics development; (3) the issue of "quality versus quantity" in the weapon systems design and selection, and (4) developing estimates of combat losses. He believed that all of these imply basic and extensive improvements in both modeling techniques and how models are used. Thus, he proposed to discuss these key problems "to get a better idea of the direction we should take to improve the ability of our models to handle these four problems."

In the context of the problems listed by Payne, a constructive assessment technique in form of force on force models would be appropriate such as, for example,

---

7 The War Gaming Center evolved from a classical manual "Kriegspiel" group set up by retired German Army officers in the mid-1960s, and the Air OR Group mentioned above, to eventually develop air/land games and simulations for investigating force structure and theater-wide defense operations.

8 Reiner K. Huber, Lynn F. Jones, Egil Reine (Eds.): *Military Strategy and Tactics – Computer Modeling of Land War Problems.* 1974 Plenum Press, New York, pp. 9–12.

9 Dr. Payne was a pioneer in army operations research and pre-eminent leader in the field for more than three decades. He began his career in 1955 as an analyst at the Army Operation Research Office at Johns Hopkins University to eventually become the first deputy under secretary of the Army Operations Research from 1968 to 1975 and the first director of the systems analysis activity of the US Army Training and Doctrine Command (TRADOC) from 1975 to 1986. In 1990, the Department of the Army Systems Analysis Award became the Dr. Wilbur B. Payne Memorial Award for Excellence in Analysis in order to honor the memory and contributions of Dr. Payne to the Operations Research field.

changing the structure and tactics of defense forces to improve their deterrent capability. This was exactly the idea of a group of German political scientists and retired military officers who proposed, in the late 1970s and early 1980s, that Germany's all-active armored defense forces be supplemented by reactive forces thus improving the chance that an eventual attack by WP forces could be stopped at the demarcation line, between East and West, without NATO having to employ nuclear weapons.[10] The sometimes bitter debates between the retired military members of the group and their active peers in the Ministry of Defense, on the pros and cons of the group's proposals, were characterized by arguments based mostly on military judgement rather than analysis.

Therefore, together with my colleague Prof. Hans Hofmann of the Institute of Applied Systems and Operations Research (IASFOR) at the Bundeswehr University in Munich[11], we initiated a project to take a look at the arguments of both sides on the basis of the outcomes of battle simulations involving 12 reactive defense options of four categories[12] using the Monte Carlo-type model BASIS. In cooperation with the military authors of the options, this model was developed by Hofmann and his research assistants over a period of three years, accounting, in great detail, all essential interactions affecting the dynamics and outcome of ground battles, for the simulation of battles between battalion-sized German ground forces defending against a sequence of regimental-sized Soviet attacking

---

10 The terms "reactive" and "active" were used as defined by Saadia Amiel to characterize two complementary land force elements within the framework of a defensive strategy. In contrast to the all armor general-purpose forces of the Cold War designed to fight in both, the reactive and active mode, Amiel proposed a land force structure ". . .*of two components: one consisting of defensive combined arms teams committed to reactive defense where precise and high firepower is at a premium, and the second of offensive combined arms formations where maneuverability is at a premium.*" [Amiel, Sadia: "Deterrence by Conventional Forces". Survival March, April 1978. pp. 56–64].

11 Initially, the Bundeswehr Universities in Munich and Hamburg were founded in 1973 primarily as academic institutions to meet the significant deficit of officer candidates at the time by preparing them for professional carriers after leaving the military. As one of the institutes of the computer science departments in Munich, IASFOR's teaching and research emphasized defense and security issues as applications for computer science.

12 The categories of reactive defense include: 1) Static Area Defenses characterized by small combat teams fighting from a network of prepared positions and field fortification; 2) Dynamic Area Defenses fighting a mobile attrition-oriented delaying battle by falling back through a series of prepared and partially reinforced positions; 3) Continuous Fire Barrier Defenses feature a fire belt along the demarcation line acting as a barrier that the enemy has to penetrate; 4) Selective Barrier Defenses are implemented depending on the tactical situation by "*barrier brigade*" composed of thee infantry battalions and one engineer battalion deployed to fight the initial forward defense battle from prefabricated field fortifications set up prior for the opponents expected assault. The brigades are manned largely by reservists living in the vicinity of their battalions.

forces supported by organic and higher level fire support on both sides. More than 500 battle simulation experiments were conducted in different type of terrain and visibility to generate sufficient data for a detailed analysis of each of the 12 reactive options.[13]

The results were discussed at a Workshop with international experts organized by the German Strategy Forum on "Long-Term Development of NATO's Forward Defense," held 2–4 December, 1984, in Bad Godesberg/Bonn. The overall conclusion of the analysis suggested that properly equipped and trained reactive defense forces being available on short notice might be an effective and efficient tool to absorb the initial attack by fighting, at the demarcation line, an attrition-oriented delaying battle thus providing the time for the active defenses to deploy at the points of the enemy's main thrusts and for counterattacks into the enemy's exposed flanks. The main reason why the Bundeswehr and its NATO partners did not consider following up the options investigated by IASFOR was that restructuring the all-active forces, deployed at the time, in the Central Region close to the demarcation line, would involve some time of conventional weakness and strategic risk considering the strategic situation in the 1970s and 1980s. However, given today's strategic situation between NATO and Russia, it seems that NATO partners in the East might well revisit some of the reactive options investigated by IASFOR for their territorial defense forces.

Comparing the proceedings[14] of the follow-on conference held in Brussels under the aegis of Panel 7 of the Defense Research Group (DRG)[15] – eight years later – with the proceedings of the conference in Ottobrunn (see footnote 8), both published by Plenum Press, the number of papers that addressed inter-active and computer simulation models had increased by 36% and their average length of papers by 60% suggesting that battle simulation modeling had expanded and

---

13 For a detailed description of BASIS the reader is referred to Hans W. Hofmann, Reiner K. Huber and Karl Steiger: "On Reactive Defense Options – A Comparative Systems Analysis for the Initial Defense Against the first Strategic Echelon of the Warsaw Pact in Central Europe," *Modeling and Analysis of Conventional Defense in Europe* (Huber, Ed.). 1986 Plenum Press, New York. pp. 97–139.
14 Reiner K. Huber (Ed.): *Systems Analysis and Modeling in Defense*. 1984 Plenum Press, New York.
15 In 1976, after the mandate of SPOSS was widened to cover scientific and environmental affairs, the NATO Science Committee decided to terminate supporting System Science and OR/SA altogether feeling that military applications should fall within the realms of other more endowed NATO divisions and directorates concerned with armaments and defense research such as the DRG and its Panel 7 (on Defense Applications of Operations Research) that adopted the members of SPOSS. DRG had been established in 1967 by the NATO Military Committee with the mandate to specific problems that NATO militaries may be faced with.

intensified significantly in the eight years between the two conferences.[16] And this expansion went on with the reorganization of NATO in 1998 when the former DRG panel 7 established, parallel to its System Analysis and Studies Group (SAS), the Modeling and Simulation Group (MSG).[17] While the mandate of SAS did not change, MSG's primary mission areas include standardization of modeling and simulation (M&S), and education and associated science and technology "to promote co-operation among Alliance Bodies, NATO Member and Partner Nations to maximize the efficiency with which M&S is used."

Concerning wargaming at the Brussels conference, I well remember G. G. Armstrong of the Directorate of Land Operations Research and Analysis Establishment (DLOR) in Ottawa, who discussed Canada's evolutionary army wargaming approach that is in fact a package of simulations in either training game or research game formats together with an experience in playing wargames.[18] The research game was a manual game with the purpose to produce data from which the sponsor's questions may be answered involving a great deal of computer assistance required to carry out the assessments and record the data that the games produce. The training wargames were carried out by the same people who create the research wargames with the purpose to exercise staff colleges and formations thus exposing DLOR's methods and findings throughout the Canadian army. It seems that the Canadian approach was a good fertilizer for today's renaissance of manual wargaming in most NATO nations.

Armstrong ended his talk by emphasizing that civilian scientist and military officers must work together. "An all-scientist war game can easily become a 'black box mathematician's delight' which is tactically ridiculous. Conversely, an all-military wargame can very easily become an exercise carried out without regard to its purpose."

---

16 Therefore, many of the participants of the Brussels conference suggested that, in order to keep up with state of the art of scientific military modeling and analysis, a time interval between meetings of five years would be more appropriate than the eight years since the Ottobrunn conference in 1974.

17 As part of the reorganization of NATO in 1998, NATO's Research and Technology Organization (RTO) was created through the merger of DRG and AGARD (Advisory Group for Aerospace Research and Development founded in 1952 as an agency of the NATO Military Committee). Among others, RTO reorganized, not without some controversy, DRG's Panel 7 by establishing two panels, the new "Modeling and Simulation Group" (MSG) and the panel "System Analysis and Studies" (SAS). Both, SAS and MSG build on the methodological legacy of the Science Committee's APOR/SPOSS and DRG's Panel 7, albeit pursuing different objectives. Finally, in 2012, RTO was replaced by the NATO Science and Technology Organization (STO) "with a view to meeting, to the best advantage, the collective needs of NATO, NATO Nations and partner Nations in the fields of [military] Science and Technology."

18 G.G. Armstrong: "Canadian Land Wargaming." *Systems Analysis and Modeling in Defense (Huber, Ed.).* 1984 Plenum Press, New York, pp. 171–179.

At the end of this Foreword, let me come back to Stuart Starr's initially mentioned paper on the changing nature of wargaming by the emergence of sophisticated collaboration tools allowing geographically dispersed individuals to participate fully in the deliberation and decisions in a wargame. Contrary to delegating the participation to subordinates because of short of time, this would allow actual decision-makers (commanders, heads of agencies, senior executives) to participate and play personally in wargames, thus increasing both the fidelity of the games and the real value of the entire activity by educating the decision-makers directly about the intricacies and nuances of the problems being considered. In the long term, Stuart was convinced that the state of the art of collaboration technology had already advanced to the point of integrating the available standalone collaboration tools into "virtual buildings" in which participants interact "face to face" in real time. In fact, decision-makers would, in a crisis, be able to play relevant wargames practically ad hoc provided, however, the military wargaming community had generated, together with battle simulation institutions, the respective data and gaming rules.

Thus, Stuart Starr's foreseen challenges are facing both of the communities, the War-Gaming and the Simulation.

Reiner K. Huber
Emeritus Professor
University of the German Armed Forces

Munich, Germany, September 2019

Preface

Several parallel events did lead to the development of this compendium on wargaming and simulation.

In December 2017, the United States Marines Corps (USMC) conducted a two-day workshop on the future of wargaming in the light of the plans to set up a new USMC Wargaming Center. The rapid change of technology accessible by opponents as well as useful to improve own forces' and allies' capabilities requires a better support for experts when developing new concepts or new structures to fight in such a complex environment, with the objective of enabling the best USMC planning process. This "5th generation USMC Wargaming Center" shall not break this focus, but add more capabilities to it, using the latest technologies, including but not limited to simulation, artificial intelligence, data science, and visualization. Out of question was that the human mind is and must remain the center of wargaming. Reducing wargaming to computational analysis would lead to results that are too limited, as human innovation and creativity are pivotal components of the process. Nonetheless, models, methods, and tools are needed to refine conditions, inform decisions, and clarify factors to identify issues, substantiate findings, and indicate directions. Decision tools, scenario tools, adjudication tools, and synthesis tools must support the human mind in the wargaming process. Industry and academic partners presented methods and tools pushing the limit of what has only been theoretically possible toward what is now practically feasible. These discussions contributed to initiate this book: how can such new methods and tools enable the next generation of wargaming?

Unfortunately, one of the titans of military operations research, Dr. Stuart Howard Starr, born on 29 January 1942, passed away on 17 March 2018. Many authors have been influenced by his research results and works, some as colleagues, others as scholars or even students, but all as friends. Stuart was often ahead of his time and introduced innovative ideas and concepts early, any many of them would need years to get fully understood to finally make it into the mainstream approaches. Examples are his early works on cyber warfare when computers were just being introduced to military headquarters to support command and control with digital means, and the command and control assessments for the new operations of the armed forces after the end of the Cold War, asking for command structures that just today are seriously evaluated to be introduced in the armed forces. His vision on a more prevalent role of simulation solutions enabling rigid wargaming was driving us, the editors, as well as many of our contributors. We therefore decided in agreement with all participants to dedicate this volume to his memory.

Finally, the recent findings of complexity theory showed the need for better support of decision-makers. They need training and evaluation environments presenting the same complexity observed in reality, and they need decision support that helps to detect, understand, and govern complexity, in particular emerging behaviors typical for complexity. Organizations such as the International Counsel of Systems Engineering (INCOSE) identified many simulation related methods and tools in their primer for complexity, and these insights should drive military decision-making as well: We need simulation to represent complex environment as well as to detect and manage it in our command and control processes. Using the latest simulation methods in support of wargaming is therefore a must.

The result of these parallel events is a compendium that shows the state of the art of wargaming and tools that allow the implementation of the vision of a fifth-generation wargaming center. Although the original ideas are rooted in the USMC initiative, the insights presented here are neither limited to the Marines nor the United States. Methods and tools are generally applicable to support decision-making in complex environments, applied to defense challenges. However, the focus of our description lies on the methods, not the application domain. In addition, some chapters provide the historical context to better understand where we are going, and some give application examples to showcase the complexity of the decisions and the power of available solutions. Overall, we hope to contribute to the discussion about decision-making in complex environments, and how simulation and related other computational support can realize the vision of a wargaming center for the challenges of our times.

Andreas Tolk, Hampton, Virginia
Curtis L. Blais, Monterey, California
Chuck Turnitsa, Newport News, Virginia
December 2019

List of Contributors

Steven Aguiar
Naval Undersea Warfare Center
Newport, Rhode Island
USA

Ryszard Antkiewicz
Military University of Technology
Warsaw
Poland

Jeffrey Appleget
Naval Postgraduate School
Montrey, California
USA

Curtis L. Blais
Naval Postgraduate School
Monterey, California
USA

Karsten Brathen
FFI – Norwegian Defence Research
Establishment
Kjeller, Norway

Paul K. Davis
RAND Corporation
Santa Monica, California
USA

Armin Fügenschuh
Brandenburg Technical University
Cottbus–Senftenberg
Germany

Richard J. Haberlin
The MITRE Corporation
McLean, Virginia
USA

Dean S. Hartley III
Hartley Consulting
Oak Ridge, Tennessee
USA

Alejandro S. Hernandez
Naval Postgraduate School
Monterey, California
USA

M. Fatih Hocaoğlu
Istanbul Medeniyet University
Istanbul
Turkey

Jan Hodicky
NATO HQ SACT
Norfolk, Virginia
USA

Reiner K. Huber
University of the Federal
Armed Forces
Munich
Germany

Leonie Marguerite Johannsmann
German Air Force
Wunstorf
Germany

William Lademan
USMC Wargaming Division
Quantico, Virginia
USA

Sönke Marahrens
German Institute for Defense and
Strategic Studies, Hamburg
Germany

Sandra Matuszewski
German Air Force
Kalkar
Germany

Ole Martin Mevassvik
FFI – Norwegian Defence Research
Establishment
Kjeller, Norway

Daniel Müllenstedt
German Air Force Command
Cologne
Germany

Andrzej Najgebauer
Military University of Technology
Warsaw
Poland

Ernest H. Page
The MITRE Corporation
McLean, Virginia
USA

Dariusz Pierzchała
Military University of Technology
Warsaw
Poland

Phillip E. Pournelle
United States Navy, Retired
Fairfax, Virginia
USA

Joseph M. Saur
Taurus TeleSYS
Yorktown, Virgnia
USA

Johannes Schmidt
Brandenburg Technical University
Cottbus–Senftenberg
Germany

Rikke Amilde Seehuus
FFI – Norwegian Defence Research
Establishment
Kjeller, Norway

Mark Sisson
United States Strategic Command
Omaha, Nebraska
USA

Andreas Tolk
The MITRE Corporation
Hampton, Virginia
USA

Charles Turnitsa
Regent University
Newport News, Virginia
USA

Paul Vebber
Naval Undersea Warfare Center
Newport, Rhode Island
USA

Jorit Wintjes
Julius-Maximilians-Universität
Würzburg
Germany

Sławomir Wojciechowski
Multinational Corps North East
Szczecin
Poland

Author Biography

Steven Aguiar is a senior engineer in the Undersea Warfare (USW) Combat Systems department and program manager for the Virtual Undersea Battlespace program at the Naval Sea Systems Command Warfare Centers. He holds several leadership roles across the Department of the Navy's Modeling & Simulation (M&S) community including M&S lead for Visualization, Virtual Reality and Augmented Reality. Over the last five years he has focused his efforts on the role of M&S technologies in wargaming, working with various US Fleet commands to prototype and deploy innovative wargaming environments. In 2008 he also pioneered the exploration and subsequent adoption of Virtual World technologies across the US Navy enterprise. He has an MS and a BS in electrical engineering from the University of Massachusetts Dartmouth.

Ryszard Antkiewicz is an associate professor at the Cybernetics Faculty of Military University of Technology and director of Operations Research and Decision Support Department. He holds a PhD and MSc in computer science from the Military University of Technology in Warsaw, and habilitation from Systems Research Institute Polish Academy of Sciences. His scientific interest is focused on modeling and performance evaluation of computer systems and computer networks, application of quantitative methods in computer security, combat modeling and simulation, mathematical methods of decision support. He has taken part in many scientific projects connected with combat simulation, artificial intelligence application in combat decision support system, capability-based planning of armed forces development, cyber warfare modeling, terrorist threat prediction, crisis management. He published more than 100 peer-reviewed journal articles, book chapters, and conference papers.

Jeff Appleget is a retired army colonel who served 20 of his 30 years on active duty as an operations research analyst. His first analysis tour was at the U.S. Army Concepts Analysis Agency (CAA; now Center for Army Analysis), working on the validation of the artillery module of the FORCEM simulation and using CEM to model Operation Desert Storm, which earned him the CAA Director's Award for Excellence and his first Dr. Wilbur B. Payne Memorial Award for Excellence in Analysis. He served as the director of the TRADOC Analysis Center-Monterey, supervising the creation of advanced simulation techniques and models, to include leading the Land Warrior Training Initiative project that converted a COTS first-person shooter (Delta Force 2) into a training simulation for the US Army's Land Warrior program. He served at the TRADOC Analysis Center-WSMR, supervising the use of the combat simulation CASTFOREM and the human-in-the-loop simulation JANUS for the Future Combat Systems (FCS) Analysis of Alternatives (AoA), earning his second Wilbur Payne Award. He served at TRAC-FLVN, where he continued supervising FCS AoA update analyses using Vector-In-Command (VIC) at FLVN and CASTFOREM and JANUS at WSMR. He then finished his army career as the deputy to the TRAC director where he was one of the lead architects of the TRAC Irregular Warfare Tactical Wargame, earning the 2011 Army Modeling and Simulation Team Award (Analysis). He has been a senior lecturer in the NPS Operations Research Faculty since 2009. He teaches the Wargaming Applications, Combat Modeling, and Advanced Wargaming Applications resident courses at NPS. He also teaches week-long Basic Analytic Wargaming Mobile Education Team (MET) courses for US and international sponsors around the world. In 2016, he earned the Richard W. Hamming Faculty Award for Interdisciplinary Achievement. Along with Dr. Rob Burks, Jeff directs the activities of the NPS Wargaming Activity Hub.

Curtis L. Blais is on the research faculty in the Naval Postgraduate School's Modeling, Virtual Environments, and Simulation (MOVES) Institute, Monterey, California. He holds a PhD in MOVES from the Naval Postgraduate School and bachelor and master of science degrees in mathematics from the University of Notre Dame. Dr. Blais has over 45 years of experience in all phases of modeling and simulation development, from requirements definition through test and employment of simulation systems for training and analysis. He has served in various levels of software engineering management and develops and delivers modeling and simulation education. His research interests include agent-based simulation, interoperability across command and control systems, simulation systems, and unmanned systems, and semantic web technologies for knowledge representation. Dr. Blais serves in various positions in the Simulation Interoperability Standards Organization (SISO) and the Military Operations Research Society (MORS), and is a member of the Institute of Electrical and Electronics Engineers (IEEE) and the Society for Modeling and Simulation International (SCS).

Karsten Brathen is a chief scientist at FFI, Norwegian Defence Research Establishment, Kjeller, Norway. He holds a siv. ing. degree in engineering cybernetics from the Norwegian Institute of Technology, Trondheim, Norway. His research interests include modeling and simulation methods applied in support of training, operations, systems engineering, and concept development and experimentation. He has published more than 100 journal articles, book chapters, conference papers, and scientific-technical reports. He is a senior member of the Institute of Electrical and Electronics Engineers (IEEE) and a Member of the Association for Computing Machinery (ACM), the Society for Modeling and Simulation (SCS) and the International Council on Systems Engineering (INCOSE).

Paul K. Davis is a retired principal researcher at RAND (still active as an adjunct) and a professor in the Pardee RAND Graduate School. His research has included strategic planning (particularly defense planning), strategy, deterrence theory, arms control, and advanced methods of modeling and analysis, notably multiresolution modeling, pioneering work on exploratory analysis under uncertainty, and semi-qualitative methods of modeling social-behavioral phenomena such as terrorism and insurgency, and heterogeneous information fusion. Dr. Davis received a BS in chemistry from the University of Michigan and a PhD in theoretical chemical physics from the Massachusetts Institute of Technology. He has worked at the Institute for Defense Analyses, the State Department, the Department of Defense (as a senior executive), and – since 1981 – for the RAND Corporation. He reviews for or is associate editor of multiple scholarly journals and has served on numerous national panels.

Armin Fügenschuh is full professor for engineering mathematics and numerics of optimization at the Brandenburg University of Technology in Cottbus (BTU). He studied mathematics from 1995 to 2000 in Oldenburg, Germany, and at the Jagiellonian University in Cracow, Poland. In 2000, he became a research associate at the Darmstadt University of Technology where he received a doctorate degree in 2005. After that he held postdoc positions in Darmstadt, Berlin, Atlanta (Georgia, USA), and Erlangen. Between 2013 and 2017 he was an associate professor at the Helmut Schmidt University/University of the Federal Armed Forces (HSU) in Hamburg. His main research interests are linear and nonlinear mixed-integer programming and their applications, with a focus on operations research problems from engineering, transportation, and logistics. His further interest is in wargaming. He gave several courses at the HSU on board game conflict simulations as well as computer-based conflict simulation tools. He is a member of several academic societies (GOR, SIAM, DMV, EMS, VDI). His work was awarded with several academic prizes, such as the EURO Excellence in Practice Award (2016) or a Dissertation Award of the German OR Society.

Richard Haberlin is a senior principal computer scientist and chief engineer of the Modeling, Simulation, Experimentation and Analysis Technical Center at The

MITRE Corporation in McLean, VA. He holds a PhD and MS in operations research from George Mason University and a BS in ocean engineering from the US Naval Academy. His research interests include inferential reasoning and decision support aided through interactive visualization. He is also evaluating application of reusable frameworks with combat simulations to support rapid integration of multiple algorithmic solutions including mixed-integer linear programming and artificial intelligence. He leverages 20 years of navy operational and staff experience to produce tailored, relevant, and defensible analyses informing executive-level decisions across a wide breadth of government organizations. He serves on the editorial board of ASCE Infrastructure Systems and holds membership with the MITRE Veteran's Council and the Military Officers Association of America (MOAA).

Dean S. Hartley III is the principal of Hartley Consulting. Previously he was a senior member of the Research Staff at the Oak Ridge National Laboratory. Hartley graduated summa cum laude, Phi Beta Kappa, from Wofford College in 1968, majoring in mathematics and foreign languages. He received his PhD in piecewise linear topology from the University of Georgia in 1973. Dr. Hartley is a director of the Military Operations Research Society (MORS), a past vice president of the Institute for Operations Research and Management Science (INFORMS), and a past president of the Military Applications Society (MAS). Dr. Hartley has published *An Ontology for Unconventional Conflict, Unconventional Conflict: A Modeling Perspective, Predicting Combat Effects*, contributed 10 chapters to eight other books, and written more than 150 articles and technical documents. Hartley received the Koopman Prize for best publication in military operations research in 1994 and the Steinhardt Prize for lifetime achievement in operations research in 2013.

Alejandro (Andy) Hernandez is an associate professor in the Systems Engineering Department at the Naval Postgraduate School (NPS), Monterey, California. He holds a BS in civil engineering from the U.S. Military Academy, a MS and PhD in operations research from NPS, and a master's in strategic studies from the Army War College. He is a retired army officer whose assignments include Director of Analysis & Assessment – Iraq, and Chief of the Warfighting Analysis Division in the Department of the Army Programs and Resources Directorate. Dr. Hernandez teaches courses in capability engineering, fundamentals of systems engineering, system suitability, probability and statistics, and research methods. He serves as the deputy director for the Simulation Experiments and Efficient Designs Center and has focused some of his most recent studies on codifying the application of modeling and simulation techniques in mission engineering. His continued research efforts combine scenarios, computerized simulation experiments, systems analysis, and systems engineering methodologies to improve decision-making for the design, development, operations, and management of complex systems.

M. Fatih Hocaoğlu is an associate professor at Istanbul Medeniyet (Civilization) University in Turkey and a scientist in simulation, artificial intelligence, and mathematical programming scientist at Agena Information and Defense Technologies Ltd. that he is also the founder of. He holds his PhD in industrial engineering in the area of simulation and qualitative reasoning. His research interests include modeling and simulation, reasoning technologies, operational research (optimization theory), simulation and agent programming languages. He is the designer and developer of a simulation and agent programming language called AdSiF (Agent driven Simulation Framework). He developed several simulation projects in defense domain. Some of these are land-based air defense simulations, simulation for C4ISR systems, marine warfare simulation, and simulations for civil sectors. He is a member of the Operational Research Society in Turkey. He received multiple awards from the Scientific and Technological Research Council of Turkey.

Jan Hodicky is a modeling and simulation advisor at the NATO Headquarter Supreme Allied Commander Transformation in Norfolk, Virginia. He earned his PhD in informatics and computer science with a special focus on modeling and simulation of autonomous systems. He is an author of around 100 papers in international journals/conferences, co-author of the patent of Virtual Reality in Command Control Systems with 23 years of service in the Czech Armed Forces. His research efforts focus on applied modeling and simulation to military problem domains. He has been the head of the Aviation Technology Department at the University of Defense in Brno in 2019 and a member of strategic management in Defense Department at the Centre for the Security and Military Strategic Studies University of Defense in 2018. From 2013 to 2017 he was chairing the Doctrine Education & Training Branch at the NATO Modelling and simulation Centre of Excellence in Rome.

Reiner K. Huber is emeritus professor at the German Armed Forces University Munich (UniBwM) where he held the chair of Operations Research and Systems Analysis (with emphasis on defense and security issues) from 1975 to 2000. Prior to this appointment he was head of the systems studies division of IABG which he had joined in 1964 as an analyst after three years of military service in the German Air Force. IABG was the German Defense Ministry's modeling and analysis institution founded in the early 1960s to support weapon systems and operations assessment, and defense planning. His research at UniBwM included, among others, exploratory analysis for improving military stability in Central Europe between NATO and WP and, after the disintegration of the WP and USSR, the stability of multipolar defense arrangements on a regional and global scale. His recent work is focused on command and control as an invited expert to a NATO project. Dr. Huber received his academic education at the Technical University Munich (TUM) and, as a Fulbright scholar, at the University of Texas in Austin. He was awarded a doctorate (equivalent to a PhD) in 1970 from TUM in aerospace engineering.

Leonie Johannsmann is a lieutenant in the German Air Force. With a successfully completed officer training she started her study of industrial engineering at the Helmut Schmidt University/University of the Federal Armed Forces in Hamburg in 2013. She wrote her master's thesis, during a trimester abroad, at the Naval Postgraduate School in Monterey, USA. In her master's thesis she optimized the spare parts inventory for a military deployment with methods of operations research. Her work was awarded with two academic prizes: the DWT Student Prize of the German Society for Defense Technology and a Best Paper Award of the 12th NATO OR&A Conference 2018. In 2017, she started her pilot training for cargo airplanes for the air force.

Ambrose Kam is a fellow in Cyber at Lockheed Martin and chief engineer in Cyber Operations Analysis, Rotary & Mission Systems (RMS) Cyber Innovations in Moorestown, New Jersey. He has MSc degrees from the Massachusetts Institute of Technology, the MIT Sloan School of Management, and Cornell University, and a BSc from the University at Buffalo. He is a specialist in cyber risk assessment and agile methods and pioneered the application of modeling and simulation techniques to cybersecurity. He collaborated with MIT (School of Engineering), Georgia Tech, Air Force Academy and West Point to develop Cyber Risk Assessment methodology. In 2017, he won the Asian American Engineer of the Year (AAEOY) award for his technical contributions and leadership.

William J. Lademan is a professional wargamer with extensive experience in the field that includes more than four decades of practice, research, and participation. Graduating from the United States Naval Academy, he received a commission in the United States Marine Corps. During his service, he held various command and staff positions, which included high-level planning positions. He also attended the Naval War College and the Fletcher School of Law and Diplomacy. After service, he obtained a PhD in chemistry from Lehigh University and spent over a decade in academia and the chemical industry before joining a consulting firm as a wargame designer. Currently, he is the technical director of the Wargaming Division, Marine Corps Warfighting Laboratory, charged with the execution of the Wargaming Program in support of examining service concepts, combat development, and operational plans. He is also involved in the planning for the construction of the Marine Corps' purpose-built wargaming center and the development of the Next Generation Wargame it will facilitate. In support of realizing the Next Generation Wargame concept, he has formulated two principles of action allowing for the representation of warfighting functions and the manipulation of wargame information and is conducting research into the metrics necessary to define the efficient application of these principles in wargame design and execution.

Sönke Marahrens is the program director of the German Institute for Defense and Strategic Studies, a cooperation between the Joint Forces Staff College and the University of the Armed Forces Hamburg. He is Colonel (GS) of the German Air Force and holds an MSc in computer science from the University of the Federal Armed Forces Munich and a master of public administration from the Royal Canadian College in Kingston. His research interests include artificial intelligence, the Prussian Wargame as well Military Command & Control and leadership for the 21st Century including M&S decision support. He is an expert on the application of NATO Modelling and Simulation, OR and NATO and National CD&E. He has received multiple awards, including the Clausewitz Medal as well as the Artur K Cebrowski Award.

Sandra Matuszewski is a general staff officer in the German Air Force. Her job specialization in the air force is information technology. She studied social sciences and administrative law from 2005 to 2009 in Munich, Germany, at the University of the Federal Armed Forces. Between October 2009 and March 2010, she took part in the ISAF Mission in Afghanistan. From 2015 to 2017 she completed her master's in military leadership and international security during the General Staff Officers Course at the Leadership Academy of the Federal Armed Forces in Hamburg, where she studied wargaming and created the board game "Enhanced Luna Warrior." She works at the Air Operations Center in Kalkar, Germany, where she is responsible for the CIS mission planning for the German Air Force.

Ole Martin Mevassvik is a principal scientist at FFI and project manager for M&S research. He received a siv. ing. degree in cybernetics from the Norwegian Institute of Technology, Trondheim, Norway in 1995. His main research interests are systems architecture and simulation interoperability with the focus on Command and Control to Simulation (C2SIM) interoperability. Ole Martin has participated in several national research projects and international activities on defense modeling and simulation. He has contributed to more than 80 peer-reviewed conference and journal papers and scientific reports. Ole Martin Mevassvik has also acted as a consultant for the Norwegian Armed Forces in several simulator acquisition projects.

Daniel Müllenstedt studied mechanical engineering from 2011 to 2015 at the Helmut Schmidt University in Hamburg, Germany. He received the DWT-Student

Prize of the German Society for Defense Technology for his Master Thesis, and the Böttcher Prize for the Best Student of the Year 2016. In 2016 he became a maintenance officer at the Tactical Air Wing 71 "Richthofen" in Wittmund, Germany. After his training as a systems engineer for Eurofighter at AIRBUS Defence and Space in Manching, Germany in 2018, he was deployed as maintenance operations officer in the technical group of the Tactical Air Wing 71 "Richthofen." Since 2019 he is a weapon system officer for the Eurofighter at the Air Force Forces Command in Cologne, Germany.

Andrzej Najgebauer is professor of computer and information systems and the chair of the Modelling and Simulation for Decision Support in Conflict and Crisis Situations Team at the Military University of Technology in Warsaw, Poland. He is also Polish member of STO/NATO Modelling and Simulation Group. He held the position of Dean of Cybernetics Faculty and vice president of the University for scientific affairs. He holds MSc and PhD in computer science from the Military University of Technology of Poland. He also holds a doctor of science in computer science, decision support systems from Warsaw University of Technology. His scientific, professional, and educational activities are mainly focused on artificial intelligence, modeling and simulation, designing of military decision support systems, threat prediction, wargames designing, cybersecurity and cyberwar. He was project leader of Polish Army Simulation System for CAXes and many Polish or international projects on DSS in the area of security and defense. He is the member of IFORS and member of Polish Society of Operations Research and Systems Analysis, vice president of Polish Society of Computer Simulation, the supervisor of 10 doctorates and general chair or co-chair of many international scientific conferences in the area of MCIS and AI, and author of 5 books and over than 130 publications. He is a member of special group of analysts, who participated in the evaluation of possible results of international war games for eastern Europe. He is an expert in the Strategic Defense Review of Polish Armed Forces.

Ernest H. Page is the DARPA portfolio manager at The MITRE Corporation. Previously he served as chief engineer within the Modeling, Simulation, Experimentation and Analytics Technical Center, and founding Director of MITRE's Simulation Experimentation and Analytics Lab (SEAL). He holds a PhD in computer science from Virginia Tech. With a research interest in simulation modeling methodology, and distributed systems, he has served as principal investigator on numerous government-funded and Independent Research and Development (IR&D) projects. He has held a variety of senior advisory roles, including: technical advisor for the U.S. Army Model and Simulation Office, chief scientist for the U.S. Army Future Combat Systems (FCS) Modeling Architecture for Research and Experimentation (MATREX), and member of the Defense Science Board Task Force on Gaming, Exercising and Modeling and Simulation.

Dr. Page has published over 50 peer-reviewed articles in the areas of simulation modeling methodology, and parallel and distributed systems. He served as the chair of the Association for Computing Machinery (ACM) Special Interest Group on Simulation (SIGSIM), the Board of Directors of the Winter Simulation Conference (WSC), and serves on the editorial boards of Transactions of the Society for Modeling and Simulation International, Journal of Defense Modeling and Simulation, and Journal of Simulation.

Dariusz Pierzchała is an assistant professor at the Faculty of Cybernetics at the Military University of Technology (MUT) in Warsaw, Poland. He graduated from MUT with a MSc degree in information systems. In 2002 he obtained PhD in simulation and decision support. He is also Polish member of STO/NATO Modelling and Simulation Group. He has been teaching a variety of subjects over the last 15 years, from computer engineering to modeling and simulation and knowledge management. With the beginning of 2013, on retiring from the Polish Armed Forces as Colonel, he assumed the position of civilian assistant professor and deputy director at the Institute of Computer and Information Systems. His scientific interests concern decision support systems, machine learning, and computer simulation in the domain of national security, defense and crisis management. He received multiple awards, individually and as a team member, including the NATO STO Scientific Achievement Award (2015).

Phillip E. Pournelle retired as commander from the US Navy after 26 years of service as a surface warfare officer. He served on cruisers, destroyers, amphibious ships, and an experimental high-speed vessel. He served on the Navy Staff doing campaign analysis, at the Office of Secretary of Defense Cost Assessment and Program Evaluation, and at the Office of Net Assessment. He is now the senior director for wargaming and analysis at the Long-Term Strategy Group. He has a master of science degree from the Naval Postgraduate School in Monterey, California. He lives in Fairfax, Virginia.

Joseph Saur is currently working as a software engineer for Taurus TeleSYS on an R&D project supporting Newport News Shipbuilding in Virginia. He holds an MS/CS and a BS/CS from Old Dominion University, graduated w/distinction from the Naval War College, and a BS in biology from St. John's University. A wargamer for over 50 years, he has studied both wargaming and combat modeling academically, has taught multiple classes in computer science at a variety on institutions, including Georgia Tech, Regent University, and ECPI. He also taught courses in wargaming at the Marine Corps University and Marine Corps Warfighting Lab. He served as the government's assessment lead for the DARPA/JFCOM "Integrated Battle Command" project, which attempted to create an integrated set of models representing the entire Political, Military, Economic, Social, Infrastructure and

Information (PMESII) spectrum, and the ability to postulate Diplomatic, Information, Military and Economic (DIME) Courses of Action (COA) in an attempt to predict (short-term) potential multi-dimensional outcomes.

Johannes Schmidt studied business mathematics from 2013 to 2018 at the Freiberg University of Mining and Technology, Germany. Since 2018 he is a research associate at the Brandenburg University of Technology Cottbus-Senftenberg. He works in the field of mixed-integer optimization with differential equation-based constraints. He is particularly interested in the mission planning of unmanned aerial vehicles.

Rikke Amilde Seehuus is a senior scientist at FFI, Kjeller Norway. She holds a PhD in computer science and a MSc in mathematics from the Norwegian University of Science and Technology, Trondheim, Norway. Her research interests include artificial intelligence, behavior modeling, and autonomous systems.

Mark Sisson is currently an operations research analyst with over 10 years' experience at United States Strategic Command (USSTRATCOM). As a graduate of the Air Force Institute of Technology (AFIT), a USSTRATCOM fellow and distinguished graduate from the University of Foreign Military and Cultural Studies Red Teaming School, he is engaged in USSTRATCOM plans analysis. In his previous life, he was an aviator specializing in electronic warfare (EW) with over 4500 hours (including combat) in bombers, reconnaissance, and foreign military sales. He is currently working on his doctorate in strategic security, where he is exploring how to combine wargames with other analytical tools.

Andreas Tolk is a senior principal chief scientist at the MITRE Corporation in Charlottesville Virginia, and adjunct full professor at Old Dominion University in Norfolk, Virginia. He holds a PhD and MSc in computer science from the University of the Federal Armed Forces of Germany. His research interests include computational and epistemological foundations and constraints of model-based solutions in computational sciences and their application in support of model-based systems engineering, including the integration of simulation methods and tools into the systems engineering education and best practices. This includes the application of simulation methods in support of command and control, wargaming, and training domains. He published more than 250 peer-reviewed journal articles, book chapters, and conference papers, and edited 14 textbooks and compendia on systems engineering and modeling and simulation topics. He is a fellow of the Society for Modeling and Simulation (SCS) and senior member of the Institute of Electrical and Electronics Engineers (IEEE) and the Association for Computing Machinery (ACM). He received multiple awards, including distinguished contribution awards from SCS and ACM.

Charles Turnitsa is assistant professor in the Engineering and Computer Science Department at Regent University. He is the lead of the Computer Engineering program. He continues to do research work for the US Government, as a senior research scientist with Georgia Tech Research Institute. He holds a PhD in modeling and simulation, and MSc in electrical and computer engineering from Old Dominion University. He has been a wargamer, both as a hobbyist and a professional, for four decades. His research interests are in the application of combat modeling, the representation of knowledge within computer simulation systems, and the modeling of complex information. He has participated in numerous research projects related to the above areas, and other related areas, and continues to be active in both teaching and research. He is a member of the Association for Computing Machinery (ACM) and the Society for Modeling and Simulation (SCS). He has received numerous awards for his research contributions and remains a very active participant in the hobby wargaming field, currently serving as the president of the Old Dominion Military Society (the historical wargaming club of Southeast Virginia).

Paul Vebber, CDR, USNR (ret), leads wargaming efforts at the Naval Sea Systems Command Warfare Centers. He has an MS in applied science – undersea warfare from the Naval Postgraduate School and a BS in history of science from the University of Wisconsin-Madison. He is active in the wargaming communities of practice associated with the US Navy and the Military Operations Research Society. He is one of the instructors for the wargaming certificate course affiliated with MORS and Virginia Tech University, with nearly 40 years of wargaming experience in the military, contractor, government and hobby sectors. He was one of the founders of www.matrixgames.com and part of the team that won the 2000 Charles S. Roberts award for Best 20th Century Computer wargame for "Steel Panthers: World at War." He is active in the Connections wargaming conference organizing committee and has participated as a player, umpire, or analyst in wargames sponsored by a variety of US Navy and DoD organizations for over 20 years.

Jorit Wintjes is senior lecturer in the History Department at Julius-Maximilians-Universität Würzburg, teaching in both the university's History and Digital Humanities programs. He received a doctor's degree and qualified as a professor in history. He studied classics and history and has published several books on ancient and nineteenth-century military history. His current research interests include Roman naval history as well as the history of professional wargaming.

Sławomir Wojciechowski is lieutenant general in the Polish Army and has been the commander of Multinational Corps Northeast since September 2018. During his 35 years in the military, he served in a variety of Polish Army units and formations. He commanded the Air Defence regiment and later the Infantry brigade. Additionally, he served in key positions in Iraq and Afghanistan, while in parallel preparing and commanding the first Polish European Battlegroup. He has been assigned to several high-level positions in the General Staff, including being department director responsible for strategy and defense planning in the Ministry of Defence. Before becoming Corps Commander, he served successfully on joint level as the deputy, and later as the commander of the Operational Command of the Polish Armed Forces. Educated in the Academy of the National Defence in Warsaw, the UK Joint Services and Command Staff College in Shrivenham, and the US Army War College, General Wojciechowski developed his scientific interests in the area of strategic thinking as well as the state security, defense and development strategies, successfully pursuing a PhD in this domain. He participates in the seminars and conferences covering issues pertaining to problems of geopolitics, military security and NATO or EU military activities.

Prologue to Wargaming and Simulation – An Introduction to the Viewpoints and Challenges

Andreas Tolk and Bill Lademan

Introduction

Since the introduction of the "Kriegsspiel" (wargame) to the Prussian General Staff by Baron von Reisswitz in 1811, which was improved by his son in 1824 by introducing paper maps, unit markers, and well-documented rule books, wargaming has had a place in military education and planning. From this beginning, General von Muffling, the Prussian Chief of Staff, ordered the use of wargames throughout the Prussian Army, and many allied and visiting armies copied these ideas. Wargames help to think through options, investigate new ideas for operations, and prepare military decision-makers by confronting them with surprises requiring a quick response. Following disruptive events requiring a reorientation, like the end of the Cold War in the nineties, or the emerging of new nuclear armed rogue nations in our day, wargames help to set the stage by providing dynamic context including the necessary complexity of the challenge for decision-making.

Wargames are no longer limited to military planning. Domain-specific tabletop games are conducted today in various other domains, from preparing local administration and government for conducting large events, like the Olympics or a sports world championship, or for responding to natural or man-made disasters, like earthquakes, hurricanes, tsunamis, or terror attacks. Even in business, wargames are conducted to evaluate different options, strategies, and possibilities.

The rise of computer simulation changed the role of wargames. Computer simulation systems driving computer-assisted exercises now play the dominant role, especially in the training and education domain. Simulation systems are used to plan and optimize procurement and development as well of a wide array of physical systems. Even operational testing and evaluation is heavily supported by simulation systems, offering a high degree of fidelity in physics-based computations. With the increasing capabilities of artificial intelligence methods, simulations also are becoming more realistic in their representation of command and control.

However, wargames are on the rise again. After years of placing trust into the power of computation, using human creativity and intuition in wargames is becoming increasingly important in the search for new doctrines or concepts of operations. The power of our simulation systems rests on our representation of systems; capturing human ingenuity requires us to look beyond our simulated representations.

What I want to show within this chapter is that wargaming and computer simulation are not competing methods, but that with the advances in both domains a new approach is possible that will enable deeper insights into the complex domain of modern operations, in which we take full advantage of both technologies. New wargaming centers will have to take more advantage of the computational power of simulation systems, while the creativity of wargamers will guide the activities. The following sections will provide several domains that will benefit from such a symbiosis.

This introduction presents two viewpoints on the challenges: those of a simulation expert with more than 20 years' experience in the development and application of simulation systems on many scale and in many domains, and those of a wargaming expert, preparing, conducting, and evaluating wargame events of highest interest in the defense domain.

A Simulationist's Perspective

Modern wargaming centers provide at least three components to support the wargame, namely the operational components, analysis component, and the simulation component. The operations components prepare, conduct, and evaluate the wargame. Since wargames have been conducted, this group has been the important counterpart to the subject matter experts who participate in the wargame itself. To make a wargame successful, it needs to be defined, planned, designed, developed, rehearsed, and finally conducted. The operations group is responsible for all these tasks, from the first ideas to the detailed game plan. During and after the game, they must analyze the results and prepare evaluation reports, outbrief presentations, etc. Some of them may be given as interim reports to the subject matter experts, others are collected to provide the overarching insights captured in the final reports about the wargame. In the earlier days of wargaming, the experts analyzed the situation by themselves, very much like they would do in headquarters. With the increasing complexity of the situation on the battlefield and a more complex solution space, more professional support needed to be provided. Within the defense domain, this analysis group is referred to as Operations Research & System Analysis (ORSA). ORSA experts assist decision-makers in solving complex problems by producing the analysis and logical reasoning necessary to inform and underpin those critical decisions. They are as much part of modern headquarters as they are part of wargaming support components. The simulation component provides numerical insight into the dynamic behavior of the complex battlefield. This is the youngest component, as only with the rise of computational capabilities was it possible to develop simulation systems that implement the theory of war, movement, attrition, and other relevant effects through the computational representation of entities, relations, activities, and effects. While traditionally rooted in the domain of physics-based modeling of mostly kinetic phenomena, recent developments in human and organizational behavior modeling research address such elements of modern warfare as well. As such, simulations did not only replace the rulebooks and result tables of traditional wargames, but also support the ORSA group with the evaluation of decision spaces. Furthermore, modern simulation systems provide powerful interfaces that allow not only the immersive displays of combat simulations, they also provide analytic tools to capture and display wargame metrics interactively.

This section evaluates the role of simulation in more detail to show that simulation is a powerful tool in support of wargaming in many phases, from early design to the generation of after-action reviews and reports. However, there are also several modeling challenges that must be addressed to ensure the best use of these powerful methods.

In order to better understand the potential as well as the pitfalls and dangers of simulation in wargaming, it is necessary to clearly understand modeling and simulation. Every simulation is based on a model, if this model has been documented explicitly, or if it is just implicitly captured in the form of the concepts, properties, relations, and processes implemented in the simulation. A model is a task-driven, purposeful simplification and abstraction of a perception of reality. As so often, the modeling process starts with the problem of the sponsor, which can be a research question to be answered, or a training task to be conducted, or ideas for a new system or new tactics, techniques, and procedures – or even doctrine – to be evaluated. The task usually drives the required abstraction level: is the sponsor's problem on the level of system components, or the entity level of weapon systems, or are units and organizations the topic of concern? Once the abstraction level is clear, not everything on this level is important. Just as scientists plan their experiments with focus on the research question, so too modelers focus on only the concepts of interest, simplifying their perception to the essential components. This perception is shaped by physical, cognitive, and even moral constraints: It reflects the understanding of the modeler, and is shaped by knowledge, experience, and other factors. Therefore, two models from modelers with different background can be quite different, even if they start with the same problem and the same references.

A simulation implements such a model. Simulations are often understood as the execution of models over time, and in the scope of this chapter, the focus is on those using computers to execute a programmed version of the model to do so. The implementation is characterized by numerical challenges, computational complexity, and use of heuristics. Different programming languages, compilers, and platforms add more challenges. Even the same model can therefore result in various and quite different simulations. Even the change of the hardware can lead to surprising changes in predicted outcomes, in particular in complex, nonlinear systems with a high dependency on the initial conditions. When NATO upgraded their hardware, some of the important analysis results had to be revisited, as some battle outcomes changed significantly using the new hardware. This is not a mistake of any programmer, it is just the nature of highly complex, nonlinear systems that become discretized and solved numerically.

Despite such obstacles, modeling and simulation is a powerful tool that helps to reproduce well-known effects, predominantly of physical and kinetic nature, under diverse circumstances and constraints. The next subsection will evaluate where within the process of wargaming simulation can be of help.

During the execution of the wargame, the focus will be on the human players. They provide the creativity allowing for operational agility in planning and decision-making. They have the insights to support new ideas, such as multidomain operations planning in the national and international context. They understand how to explore human decision-making and how to react to unanticipated decisions in the operation. In summary, they are the main players in the wargame, providing creativity and the vision for the big picture. However, the role of simulation is similarly important. It falls to the simulation to compute the mission thread by unbiased execution of decisions in the virtual battle space. This includes computing the mission thread effects as well as the effects of the wargamer's human decisions. Simulations compute all orders of effects, and some higher order effects in complex, nonlinear environments can be surprising. It is not only possible but highly likely that some of these effects will lead to emergent behavior in the scenario, properties that the complex systems expose, but that cannot be exposed by the individual systems themselves. The immersive visualization of results, including detection and visualization of new emergent behaviors, is a pivotal role for the simulation. In other words, simulation can take over the role of secondary players, opposing forces, and supporting roles in planning and preparation, while also supporting computational and visualization requirements in execution and evaluation.

However, the use of modeling and simulation is not limited to the execution phase but can be applied by the operations group in support of many of their tasks. During the design and the development of the wargame, tools can help to visualize ideas and support the composition and reuse of services and rules developed in earlier wargames. Quick consistency checks can make sure that all entities necessary to evaluate a new idea are represented, and all of them have rules assigned that can help with the necessary scenario generation process. For rehearsing, artificial intelligence can be used to calibrate software agents and rules to play the role of human gamers to look for inconsistencies, opportunities to cheat, and other optimization of actions. It may be premature to think about artificial gamers as subject matter experts, but for the pure testing of the limits of the game, current technologies are sufficient. Communicating the results of the wargame is also a task well known by the simulation community, as in the domains of analysis and training providing after action reviews (AAR) has been required for many years. If done correctly, results and lessons learned can inform the development of new rules for the next events. A picture says more than a thousand words, but an executable simulation can say more than a thousand pictures. Scenario generation tools and other support software should also be utilized in this context.

There are several challenges that simulations for defense operations must overcome to be fully supportive of today's requirements. Nonetheless, the increasing complexity of the highly nonlinear operational environment needs such computational support. The recently published primer on complexity for systems engineers, developed and published by the International Council on Systems Engineering (INCOSE), explicitly mentions simulation and artificial intelligence methods as necessary tools for decision-makers in such environments, as complexity requires a new operational agility from the decision-makers, which means to rapidly compose high-performance teams out of the available systems to react quickly and precisely to often unforeseen, maybe even emergent challenges.

Simulation solutions provided for defense are reasonably effective in the modeling of physical and kinetic effects, such as needed for attrition-focused force-on-force modeling. However, the structure of the opposing forces is changing rapidly, driven by increased use of robotic systems and other autonomous systems that lead to new tactics and procedures. New weapon systems, such as the 5th-and 6th-generation systems, provide a new set of capabilities. Opposing systems and future systems are hard to capture, as the parameters – or even the underlying architecture – are unknown or uncertain. Many lessons learned are no longer applicable.

How and where we will have to solve future conflicts is not only challenged by more than such technical uncertainties. With more and more people moving toward mega-cities, many of them in coastal regions, the likelihood of an armed conflict in these regions increases. This will require high-resolution modeling of this environment with high fidelity on a big scale. This will require sensor and weapon system models with equal resolution and fidelity, and the adaptation of rule sets on how to apply these systems, for all participating organizations. International multidomain operations require a new level of coordination between the systems of various services and nations as well as the local commanders utilizing their capabilities. These new kinds of operations are more than joint and combined activities, they are the creative mix of several mission threads optimally creating mutually supportive effects in all domains.

Furthermore, human, cultural, and social behavior modeling will be needed, which implies the use of computational social science models for both opposing and friendly forces. With the advancement of combat medicine saving more soldiers, new challenges emerged, like having to deal with post-traumatic stress syndrome (PTSD). The use of information, including social media, to influence opponents, disseminate information in support of the objective of the organization, and other nontraditional intelligence operations may influence future warfare as well.

The traditional use of predictive simulations used for point optimizations in a well-defined context, possibly supported by some sensitivity analysis, does not meet these emerging requirements. Composable simulation services provided as smart components are needed. The resulting compositions need to be applied to conduct exploratory modeling and analysis addressing the deep uncertainties of these complex environments by allowing a broad evaluation of the solution space. By combining the power of computer simulation-based generation of data with technology of big data allows for a new application of simulation.

In summary, modeling, simulation, analysis, and visualization methods can and should enrich wargaming activities. No other methods allow the exploitation of options within a complex, nonlinear environment, such as the modern battlefield presents. Several chapters in this book provide examples of how these methods and derived tools help in the decision-making process. Not utilizing these methods and tools to the full extent possible would be a mistake.

A Wargamer's Perspective

Recent DoD level interest has highlighted the importance and value of wargaming as a vital and neglected element in the comprehensive understanding of operational environments, force design, and operating concepts. There is both clarity and confusion in this interest. Clarity in that it recognizes a problem and gives momentum to a solution; confusion in that it has not distinguished between the natures of computational analysis and wargaming. It has even suggested that "reinvigorating" wargaming is merely a matter of incorporating analytical techniques (methods, models, and tools – MMT) into wargame designs. However, the effort to optimize this relationship entails more than the simple incorporation of

MMT into wargame design. This incorporation is not new. The struggle is to synchronize the process of sophisticated analytical methodologies with the action of the human intellect such that the potential of both are integrated and optimized by using the computational result as a substrate for human decision.

Analysis is based upon mathematical process; wargaming is based upon human judgement. Both are powerful and are compatible. But, they are not different expressions of the same thing. Computational analysis relies for its manipulation of data and its precision of results upon a methodology involving the quantification of variables and the specification of their interactions. In analysis, exact conclusions emerge from the connection of method to a specific problem. However, analysis is limited by the very tenants of its science to what is measurable. It cannot go beyond statements of trends and precision (accuracy is another matter) because it cannot substantiate what it cannot measure. Further, a particular resulting measurement does not necessarily imply a universal pattern.

Wargaming rests upon what cannot be measured. This stands in contrast to but not in opposition to the computational analytical approach. A wargame does this by embracing, assembling, and organizing many variables without an attempt to assign values or calculate interactions. These variables, which reside in the situation, the individual, and emerge in the dynamic friction of play, are impossible to measure separately or in assembly. The action of the wargame generates interactions and relationships that could not have been anticipated and relies upon the emergence of results not subject to prediction. All of this is synthesized and organized in the human imagination and no science is capable of quantifying the path, dynamic, or chance that transforms this complexity into a comprehensible and coherent whole. And yet this is what both drives a game and defines its results.

Thus, wargames explore the interlocking coherence of the whole while computational analysis produces precision in isolation. The question is: How to associate the two to mutual benefit? The problem is one of relating processed facts and human imagination. The analyst and the wargame designer must combine the two realms without losing the essential strength of either in the midst of the constant dynamic and change in game play. The answer to this dilemma involves the recognition of the distinct natures of the two approaches and the effort to forge complimentary methods. Wargaming permits judgment to be influenced in a dynamic context by emerging evidence as a precursor to decision. Analysis can aid this process by injecting "points of precision" into play, which then merge with and act as an informing substrate for decision . . . the universal requirement in any wargame. In other words, analytical methods can inform imagination with a precision designed to influence but not direct decision in game play.

The benefit of a wargame supported by analytical methods that provide points of departure and situational precision as the basis for decision is the production of informed and defensible insights that can shape and direct subsequent efforts in a concept or combat development sequence. There is no analytical methodology by which the outcomes of the inherently human activity of play can be transitioned into a rigid accuracy. But then war is an inherently human activity that only rarely adheres to the requirements of scientific law and rigidity in war rarely produces brilliance or success. The key to understanding the benefit of the incorporation of analytical methods into wargaming is that, while sharper and more focused insight can be expected as outcomes, one learns that knowledge does not have to be quantifiable in order to be defensible. This informed combination of the analytic science of the necessary with the wargaming art of the possible promises to provide a foundation for the objective substantiation and justification for the resources and programs required for future military success.

Conclusion

These viewpoints show that wargaming and simulation both play important roles in the development and evaluation of new concepts, tactics, techniques, and procedures in complex defense situations. The support seems to be more complementary than competitive.

As a rule, the role of the simulation focuses mainly in the quantitative sphere, reliable presentation of computable effects in a situated synthetic environment for the wargamers and provide the results in immersive form to them. The immersion can be in the form of virtual or augmented reality presentation of the battle sphere, but also in form of intuitive representation of results. The latter may also help the wargamer to evaluate alternative courses of action in their decision cycles. The role of the wargamer is more in the qualitative realm. Humans provide the creativity needed to come up with truly innovative solutions when confronted with complexity, uncertainty, and vagueness of new situations. They make the decision in multidimensional, multi-scope, and multi-resolution solution space. It seems to be obvious that a tighter connection will likely provide a better support. What conceptual and technical methods are useful to support this is a topic of ongoing research.

In this prologue we raise questions about what simulations can provide to enable better wargames, what wargamers need from simulationists to help them be more creative and innovative, and what is needed to generate better aligned compositions and tools. The authors of the various book chapters, which are making up this compendium, address these challenges as well as ideas how to better cope with them. There are many facets and viewpoints reflected in the contributions, providing the basis for more discussions, research, and hopefully many practical applications of solution contributions in the future.

Part I

Introduction

1

An Introduction to Wargaming and Modeling and Simulation

Jeffrey Appleget

Naval Postgraduate School, Montrey, CA, USA

Introduction

Nations have long utilized simulations of combat to help understand how to better man, train, equip, and employ forces in preparation for future combat operations. These force generation, force structure, and force design decisions are often informed by simulating combat against potential adversaries in projected future scenarios, and then analyzing the simulation results to determine the necessary future investments to ensure the force is prepared to meet these potential adversaries. This book will discuss the current practitioner use of both wargames that investigate the human decision-making processes and computer simulations that investigate the quantifiable aspects of combat. Our goal is to provide the reader a better understanding of how each tool brings unique qualities and attributes to bear on the assessment of the phenomenology of combat that allows our senior leaders to make better informed decisions.

Terminology

There are many different types of combat simulations that exist today, and in order to have a fruitful discussion we will need to adopt a standard lexicon. The first step in developing this lexicon is to define the terms "model," "simulation," and "wargaming." A model is "a physical, mathematical, or otherwise logical

Simulation and Wargaming, First Edition. Edited by Charles Turnitsa, Curtis Blais, and Andreas Tolk.
© 2022 John Wiley & Sons, Inc. Published 2022 by John Wiley & Sons, Inc.

representation of a system, entity, phenomenon, or process."[1] A simulation is "a method for implementing a model over time."[2] A wargame is "a representation of conflict or competition in which people make decisions and respond to the consequences of those decisions."[3] Often in today's Department of Defense (DoD), the term "simulation" implies that all the models that comprise the simulation are instantiated in executable computer code, and because of that, most wargames are not thought of or referred to as "simulations." From this point forward, when we use the term "simulation" it will refer exclusively to computer-hosted closed-loop combat simulations, and thus will not include wargames.

Combat simulations are categorized by the amount of human interaction required, the use of probabilistic processes and the level of war they represent. Combat simulations that require periodic human decisions are called "Human-in-the-Loop" or *H-I-T-L simulations*, and these are often used as computer-hosted wargames, with human commanders or command and staff teams making the necessary decisions. *Closed-loop simulations* have totally automated the human decision-making processes in computer code and can simulate hours, days, weeks, or months of combat without any human intervention during the simulation's execution. Simulations that will produce the same output for a fixed set of input parameters are *deterministic*, while simulations that have one or more probabilistic parameters whose value will be determined during the simulation's execution using a random number seed are *stochastic*, which are sometimes referred to as "Monte Carlo" simulations. Simulations are also segregated by the level of war that they represent. A strategic simulation will represent an entire campaign, such as the European or Pacific theaters of war during World War II. An operational simulation will represent a specific operation that is part of a campaign, and a tactical simulation will represent some portion of an operation. In most cases, the higher the level of war, the more abstract the models of the simulation are. Most tactical simulations represent each weapon system and soldier of a unit and are called *entity* simulations. Many strategic simulations represent entire units, such as a company, battalion, or brigade, by aggregating the weapons systems and soldiers of a unit, and treating the unit as a single object with attributes derived by combining the attributes of the unit's entities into a single value that represents some combat capability of the unit. In ground combat simulations, the single value assigned to such a unit is often called the "combat power" of the unit. These simulations are predictably called *aggregate* simulations.

---

1 DoDI 5000.61 "DoD Modeling and Simulation (M&S) Verification, Validation, and Accreditation (VV&A)" December 9, 2009 w/Change 1, October 15, 2018, p.10.
2 DoDI 5000.61 "DoD Modeling and Simulation (M&S) Verification, Validation, and Accreditation (VV&A)" December 9, 2009 w/Change 1, October 15, 2018, p.10.
3 US DoD Joint Publication 5-0, "Joint Planning" 16 June 2017, p. V-31.

Wargames are categorized by the purpose of the wargame, the manner in which players are engaged and the amount of information provided to players. There are three widely recognized purposes for defense wargames: *analytic, educational, and experiential.* Educational and experiential wargames seek to impart knowledge or experience, respectively, to its players. In other words, these wargames produce better educated or more experienced players. Analytic wargames are designed to address an objective and a set of issues that the wargame's sponsor provides. The products of an analytic wargame are findings that address the sponsor's objective and issues. Wargames are also categorized by the method that the players interact. Players are engaged directly in a *seminar* wargame and the facilitator of the wargame usually adjudicates any player interactions. Conversely, players engage each other indirectly in *system* wargames. In system wargames, there are usually models that are designed to adjudicate player interactions. Finally, wargames must provide players with information that they will then use to make their next decision. A wargame that only provides the players the information that would actually be available to them is called a *closed* wargame. In closed wargames, each side (typically red or blue) is sequestered in its own cell, and the wargame's white cell makes the determination of what information each side's collection assets would produce, and then communicates only that information to the appropriate cell. Wargames that allow all the players to have all available information are called *open* wargames.

And, as for the model or models that are used in simulations and wargames, users must heed the statistician George Box's warning: "All models are wrong, some are useful."[4]

An Abbreviated History of Wargames and Simulations

Before the rise of the computer, the primary form of combat simulation was wargaming. Wargaming has a rich history and has been used by many cultures and in many different forms. The ancient games of chess and Go are but a few of the games that were believed to have usefulness in training and testing military commanders' decision-making capabilities, and this belief led to the designing of games focused on modeling combat for the development of military leaders. In the nineteenth century, the Prussians developed Free and Rigid Kriegspiel as methods to educate their officers. Rigid Kriegspiel focused on the calculations of combat, with the hypothesis that good combat leaders had to be able to employ a type of "combat calculus" to mathematically understand what decisions should be made on the battlefield.[5] Free

---

4 Box, G. E. P. (1979), "Robustness in the strategy of scientific model building", in *Launer, R. L.*; Wilkinson, G. N. (eds.), Robustness in Statistics, Academic Press, pp. 201–236
5 Vego, "German War Gaming," p. 108.

Kriegspiel used battle-tested Prussian officers to assess junior officers' responses as they were presented with possible combat situations to which they had to react.[6]

In the early part of the twentieth century, F.W. Lanchester proposed two sets of differential equations that could be used to simulate combat, referred to as Lanchester's linear and square laws. Although he proposed these laws to simulate aerial combat, they gained traction for use in modeling ground combat. The square law rewards a combatant's ability to concentrate forces and was seen as relevant to modern warfare, and the linear law has been accepted as a model of ancient warfare where combatants were unable to mass fires.[7] Lanchester came to realize, through studying the Battle of Trafalgar, that if a battle can be decomposed into a series of concurrent and consecutive sub-battles, separated by space and time, then it is more appropriate to apply the square law to each sub-battle, and sum all sub-battle losses to gain a more accurate accounting of the entire battle than it is to apply the square law a single time to the entire battle.[8] As computer-based combat simulations were developed in the latter half of the century, many ground aggregate simulations used some adaptation of Lanchester's laws to model attrition.[9]

In the first half of the twentieth century, the US Navy made great use of wargaming to examine a potential war with Japan, beginning over two decades of focused wargaming in 1919 at the US Naval War College.[10] This detailed examination of war in the Pacific proved to be so successful that, after the conclusion of World War II, Admiral Chester Nimitz said "...nothing that happened during the war was a surprise – absolutely nothing except the Kamikaze..."[11]

Wargames and Computer-Based Combat Simulations: From the Cold War to Today

As the North Atlantic Treaty Organization (NATO)–Warsaw Pact "Cold War" consumed the US defense establishment for the latter half of the twentieth century, two events greatly impacted the use and characterization of combat simulations in the US DoD. One was the development of the closed-loop combat simulation. The

---

6 Vego, "German War Gaming," p. 110.

7 Lanchester F.W., *Mathematics in Warfare* in *The World of Mathematics,* Vol. 4 (1956) Ed. Newman J.R., Simon and Schuster, 2138–2157

8 Perry, Nigel. "Verification and Validation of the Fractal Attrition Equation," *Defence Systems Analysis Division, Defence Science and Technology Organisation, DSTO-TR-1822, PO Box 1500 Edinburgh South Australia* 5111, Australia, January 2006.

9 Appleget, J. "The combat simulation of Desert Storm with applications for contingency operations" *Naval Research Logistics (NRL),* Volume 42, Issue 4, Pages 710–711, 1995.

10 Edward Miller. War Plan Orange: the U.S. Strategy to Defeat Japan 1897–1945 (Annapolis. MD: Naval Institute Press, 1991), 156.

11 Donald C. Winter, remarks at Naval War College Current Strategy Forum, Newport, RI, 13 June 2006.

earliest computer simulation of ground combat is believed to be CARMONETTE, developed by the US Army's Operations Research Office in 1953, and used to inform defense decisions from 1956 to 1970.[12] The other was the US DoD's embracing of Defense Secretary Robert McNamara's "Systems Analysis" philosophy (McNamara served as SECDEF from 1961 to 1968), which led to the thinking that every important defense concept or procurement program required quantifiable justification.[13]

In the late 1950s, the use of CARMONETTE, a stochastic tactical-level closed-loop combat simulation was still referred to as a "wargaming" technique. It was used for research purposes such as "testing the value of new weapons, fighting techniques, or war plans."[14] The developers of CARMONETTE realized that the concept of repeatability combined with modeling some of the processes of combat probabilistically could be used to remedy one of the identified shortcomings of wargaming: the results of a wargame are only a single realization of a range of potential outcomes. They determined it would be necessary to "repeat the battle calculations allowing nothing but the play of chance to vary and so identify the spectrum of the possible outcomes and the associated frequency distribution." They also realized that the scientific method would be useful if they were to attempt to compare battle outcomes of forces equipped with two different weapons systems; in particular, "we must be able to repeat the battle simulation many times while holding fixed all parameters except that one under investigation."[15]

However, these same developers also came to the realization that simulations such as CARMONETTE could not reproduce the complex decision process a military commander uses in combat:

The design of the simulation is such that it can create a realistic representation of close combat during the brief intense engagement phase lasting for approximately 1 hour or so. Continuation of the simulated combat beyond 1 hour becomes unrealistic because a decision fundamental to the execution of the maneuver would undoubtedly be made at that point. CARMONETTE has no capability to reproduce a military commander's mind, and thus the simulation must be terminated, and the results reviewed. The simulation may then continue with an appropriate order from a commander if desired.[16]

---

12 "A History of Serious Games" powerpoint presentation, Roger Smith, Chief Technology Officer, US Army PEO STRI.

13 John Hanley, Changing DoD's Analysis Paradigm," Naval War College Review, 70, 1 (2017): Article 5. pp. 65-66,

14 CARMONETTE, A Concept of Tactical War Games, Operations Research Office TACSPIEL Group, Department of the Army, *Staff Paper ORO-SP-33*, November 1957, p.1.

15 CARMONETTE, A Concept of Tactical War Games, Operations Research Office TACSPIEL Group, Department of the Army, *Staff Paper ORO-SP-33*, November 1957, p.1.

16 CARMONETTE Volume I General Description Prepared By General Research Corporation Operations Analysis Division Westgate Research Park Mclean, Virginia 22101 Under Contract DAAG 39-74-C-0128 For US Army Concepts Analysis Agency 8120 Woodmont Avenue Bethesda, *Maryland* 20014, November 1974 p 1.

The implication here is that CARMONETTE could be integrated into a larger iterative process where a commander's human decision-making process would shape the approximately hour-long CARMONETTE engagement, and then, after the commander reviewed the engagement results, another human decision would be made that would shape the next CARMONETTE engagement.

In the 1970s, independent of the budding combat simulation community, military wargamers realized that some of the book-keeping functions that wargames required could be better performed by computers. In 1976, the Training and Doctrine Command Systems Analysis Activity integrated a Wang 2200 minicomputer into the Dunn–Kempf manual wargame to create BATTLE, the Battalion Analyzer and Tactical Trainer for Local Engagements. They used the computer to calculate the results of simultaneous direct and indirect fire engagements, and to perform "book-keeping functions" such as tracking movements and ammunition expenditures. They realized that the computer could be leveraged to lessen the burden on wargaming staff to free them up to focus on more vital aspects of wargaming such as the tactical decision-making process.[17]

In the 1980s, the use of computerized combat simulations became ubiquitous in the defense analysis community. The dominant scenario that the US DoD used to underpin acquisition decisions was the NATO–Warsaw Pact fight for Western Europe. This projected conflict had been analyzed continually for decades and both sides' intelligence had been so well developed that, by the mid-1980s, nearly the entire world understood how the battle on the north German plain would unfold: attack corridors, force compositions, and equipment, even opposing commanders were all known. Tom Clancy's novel "Red Storm Rising" (Putnam, 1986) provided a realistic description of what that encounter would have looked like and demonstrated the amount of information commonly available about that potential conflict.[18] Quantifying the kinetic combat capabilities of forces became the focus of the analysis that underpinned the US defense acquisition decisions, and this played perfectly into the strength of closed-loop combat simulations.[19] However, US Army analytic organizations realized that closed-loop combat simulations could not be relied upon as the single tool needed to do analysis. While the automated decision rules allowed for the development of stochastic models that could be run numerous times to ensure there was a representative set of battle outcomes, the automation of the human decision-making process was recognized to be too simplistic to rely on for a complete assessment of combat operations.

---

17 TRADOC Systems Analysis Activity. (undated). Battle Analyzer and Tactical Trainer for Local Engagements, White Sands Missile Range, New Mexico (development of the BATTLE simulation was initiated in August 1976).

18 Jeff Appleget and Fred Cameron," Analytic Wargaming on the Rise," Phalanx 48, No.1, (March 2015): 28–32

19 Jeff Applegte, Robert Burks, Fred Cameron, The Craft of Wargaming: A Detailed Planning Guide for Defense Planners and Analysts. (Annapolis, MD: Naval Institute Press, 2020): 17–18.

Both the Army's Center for Army Analysis (CAA) and the Training and Doctrine Command (TRADOC) Analysis Center (TRAC) developed analysis protocols that first used wargames to thoroughly examine different courses of action (COAs) or concepts of operations (CONOPS) before deciding on a single scheme of maneuver that was then instantiated in their closed-loop combat simulations.[20]

Wargaming has taken on a more prevalent role in DoD since the beginning of the US irregular warfare (IW) campaigns in Afghanistan (2001) and Iraq (2003), although there have also been major efforts to develop computer simulations to model IW operations such as counterinsurgency and stability operations. Initially, some modelers attempted to add IW complexity to existing kinetic-focused combat simulations. They added a third side to their simulation to represent the population. Others added civilians on the battlefield, so kinetic engagements between two uniformed, armed forces had the potential to cause "collateral damage" among the populous. They were missing the point. The whole dynamic of warfare changes when the center of gravity shifts from defeating an adversary's military to influencing a civilian population. In an IW campaign, kinetic engagements between armed combatants are infrequent, small, sporadic, and often counterproductive. The Counterinsurgency Field Manual FM 3-24 said it best: "You can't kill your way out of a counterinsurgency." Whole of government approaches and modeling civilian populations' attitudes and behaviors are required, and modeling these attributes presented significant challenges to defense modelers.[21] DoD stood up the Human Social Culture Behavior Program in 2009 to address some of these challenges, specifically "to develop, implement, and demonstrate forecasting and predictive models of human behavior for both analytic application and warfighter training in support of non-traditional warfare operations."[22] The United Kingdom's Defence Science and Technology Laboratory developed the Peace Support Operations Model (PSOM).[23] This is not a closed-loop simulation, but a simulation that requires periodic decision-making by human subject matter experts – a computer-hosted wargame. PSOM was used in 2010 and 2011 for training UN peacekeepers in Central Asia[24] and to assess

20 Appleget and Cameron, 30–31.

21 Appleget, J., C. Blais and M. Jaye, "Best practices for US Department of Defense model validation: lessons learned from irregular warfare models," *Journal of Defense Modeling and Simulation: Applications, Methodology, Technology*, 10(4) 395–410, 2013.

22 http://www.pacific-science.com/ad7/sites/default/files/HSCB%20Project%20 Summary_13Mar12_0.pdf, referenced 06/27/2019.

23 Body, H. and C. Marston "The Peace Support Operations Model: Origins, Development, Philosophy and Use," *Journal of Defense Modeling and Simulation: Applications, Methodology, Technology*, 8(2) 69 –77, 2010.

24 Nannini, Christopher J., Jeffrey A Appleget and Alejandro S Hernandez. "Game for Peace: progressive education in peace operations," Journal of Defense Modeling and Simulation: Applications, Methodology, *Technology*, 10(3) 283–296, 2013.

campaign options in Afghanistan.[25] A survey of simulation tools used to model conflict in Iraq and Afghanistan was completed by RAND in 2014.[26]

As the conflicts in Iraq and Afghanistan approached their first decade, the US DoD began to realize that both wargames and closed-loop combat simulations have important and distinctly different roles in the analytic process. The United States' involvement in Iraq and Afghanistan has highlighted that counterinsurgency and stability operations cannot be modeled well in existing closed-loop combat simulations. While agent-based simulations show promise for modeling human behavior in regions of conflict,[27] there are no closed-loop IW simulations that parallel the quantitative analytic capability of those used by DoD to assess major kinetic combat operations. In historical terms, modern-day wargames are much like the Prussians' Free Kriegspiel, while today's closed-loop combat simulations are more similar to Rigid Kriegspiel. Each tool has its purposes, and in most cases those purposes are not overlapping. Wargames should not be used for quantitative assessments and closed-loop combat simulations cannot replicate human commander's decision-making processes.

Wargames Today

As analytic wargames began to regain some traction in the 2010-time frame, they were attacked by combat simulation advocates. Analysts who teethed on closed-loop combat simulations derided wargames as "a simulation of one replication" or a "sample size of one," noting that you could not run a particular wargame multiple times, varying random variable values to generate quantitative output for statistical analysis. What they failed to understand was that a wargame's focus is on qualitative data, decisions produced by human players, while the computer-based closed-loop combat simulations are focused on quantifying the attributes of a force engaged in high-end kinetic combat.

---

25 Connable, Ben; Walter L. Perry, Abby Doll, Natasha Lander, Dan Madden, *"Modeling, Simulation, and Operations Analysis in Afghanistan and Iraq: Operational Vignettes, Lessons Learned, and a Survey of Selected Efforts" RAND Corporation*, National Defense Research Institute, Santa Monica,CA, 2014, p. 26.
26 Connable, Ben; Walter L. Perry, Abby Doll, Natasha Lander, Dan Madden, *"Modeling, Simulation, and Operations Analysis in Afghanistan and Iraq: Operational Vignettes, Lessons Learned, and a Survey of Selected Efforts" RAND Corporation*, National Defense Research Institute, Santa Monica,CA, 2014.
27 Appleget, Jeffrey, Robert Burks, and Michael Jaye, "A demonstration of ABM validation techniques by applying docking to the Epstein civil violence model," *Journal of Defense Modeling and Simulation: Applications, Methodology, Technology*, 11(4) 403–411, 2014.

Running a series of wargames to generate multiple replications or running wargames to compare and contrast different concepts or technologies is problematic. Running any simulation for multiple replications is typically done by holding most variables fixed and introducing randomness for certain, identified random variables (such as system-on-system probability of hit) so the statistics of the multiple replications can be calculated to examine the range of expected outcomes, given the introduced randomness. It is difficult, if not impossible, to produce multiple replications of a wargame because of the learning effect that the players experience, so the difference in replications is confounded by the players learning more about the operating environment and the opponent's method of prosecuting combat with every subsequent replication. If the first replication of the wargame produced clear winners and losers, would the loser use the exact same strategy for the second replication? Would the winner expect the loser to not learn and try the same course of action? You could attempt to eliminate the learning bias by having different players play each replication, but that would assume you could find a large supply of players with identical experiences. In practice, some organizations that utilize wargames do play them multiple times with the same players, but these cannot be considered multiple replications, at least not in the traditional sense. When the TRADOC Analysis Center ran the H-I-T-L simulation JANUS to develop a concept of operations (CONOPS), for later instantiation in the closed-loop simulation CASTFOREM, the blue side had around 30 "pucksters" plus a command and staff, and the red side had around 10 pucksters and their own command element. Pucksters typically maneuvered and fought elements of the force, such as a tank company in a brigade operation. In order to obtain a "record run" that could be instantiated in the simulation, they would need four to five runs to familiarize all the pucksters with the operational plans. One or both sides would usually petition to move the starting locations of their forces between runs, so these cannot be considered "replications" in the pure sense. Only the results of the final record run were used to inform the CONOPS instantiation in the simulation.[28]

At the Office Secretary of Defense (OSD) Office of Net Assessment, players play a wargame multiple times where the players' learning is one of the subjects of study. Understanding how a command and staff's thinking evolves when combating an adversary's new technology or concept in a wargame allows the evolution of doctrine without putting forces at risk.[29]

In DoD today, there are few in uniform who can design, develop, conduct, and analyze a wargame. What used to be an integral part of a professional military officer's education and experience is now an afterthought, at best. This has caused

---

28 Personal experience of Jeff Appleget, Commander of Troops, TRAC-WSMR, 2001-2004.
29 Communication with Phillip Pournelle, CDR, USN (retired), former office of Net Assessments analyst, 17 June 2020.

a fair amount of "BOGGSAT" wargaming to be conducted, a Bunch of Guys and Gals Sitting Around a Table. The term BOGGSAT is a pejorative term that implies that a group of people tasked to conduct a wargame produces results that meet the tasker with minimal rigor and resource expenditure. There are two reasons BOGGSATs occur. The first is that the command does not give the wargaming team the resourcing to do a proper wargame. The second is that no one on the wargaming team has any wargame design experience, so the team simply improvises as best it can. In many cases, both lack of resources and experience spur the occurrence of BOGGSATs. We often hear of BOGGSATs being used to conduct planning wargaming at some US Combatant Commands (CCMDs).

Some of our CCMDs have contracted out some of their wargaming requirements to make up for the lack of uniformed wargamers. This can present a challenge. Some contracting organizations have their own methods of doing a wargame; so if a command's wargaming requirements do not quite match the method of the contracted wargaming organization, the organization may only wargame the part of the required wargame that their methods can accommodate. Most wargaming requirements are unique, and a wargaming best practice is to design the wargame around the organization's wargaming requirement, instead of trimming the requirements of the organization to fit a predetermined wargaming method.

Wargames have multiple points of failure. Wargames fail when the wargaming team and the sponsor do not come to an agreement of the wargame's objective and key issues. This often occurs when the wargaming sponsor is a senior official whose subordinates are reluctant to force the official to clarify and refine the initial wargaming tasking. The best-designed wargame can be a failure if the wargaming team cannot secure the appropriate players. Wargames can also fail if not executed properly. Keeping the players immersed in the wargaming environment, ensuring the game stays on schedule, managing the game's adjudication and data collection, and solving the inevitable glitches that often occur require an experienced and adaptive wargaming team. Analytic wargames depend on accurate and detailed data collection, so a well-designed wargame with the best players can still be a failure if the data collection effort is flawed. Finally, a wargame may be well designed, flawlessly executed with clear and concise data collected, and the game's analysts may fail to conduct useful analysis.

In conclusion, there are many more wargames being conducted since 2015 in DoD than before, thanks to the reinvigoration spawned by the Office of the Secretary of Defense's stewardship. However, more does not necessarily mean better or useful. Wargames designed by teams with no wargaming experience or education will most likely encounter two or more of the points of failure enumerated above. If wargaming is to again become a part of the US DoD culture, wargaming education and wargaming experience must be directed and driven by DoD leadership.

Simulations Today

Introduction

Closed-loop simulations provide the means to assess the combat capabilities of a collection of entities (weapon systems and formations) given that the decision that those forces will engage in battle has been already been made. These simulations are not wargames, as there are no dynamic human decisions that impact the flow of events of the operations simulated in the computer model. While it is true that there are algorithms in closed-loop simulations that represent some decisions that humans make in combat, they are rudimentary, IF-THEN type of decisions.

Simulation Types

For the ease of simplicity, we will use ground combat simulations as the basis for our discussion. Ground combat simulations are used by both the US Army and the US Marine Corps and are arguably the most complex of the combat simulations used throughout DoD, both in the sheer numbers of combat platforms and in the complexity of the operating environment.

Aggregate Simulations

These simulations typically array forces linearly in a series of sectors (often referred to as "pistons") where, in each sector, algorithms will assess if an attack will occur, and if so who will be the attacker. In each sector, the simulation calculates a combat power score (also known as a firepower score) for each force to assess the combat power ratio that exists between the forces. The first simulations that used combat power scores calculated the combat power comparison assuming that each side had perfect information on all the forces, friendly and adversary, in that sector. In other words, not only did each side have perfect intelligence on its adversary's force composition, but each side also had perfect communications because it knew the status of each and every friendly unit in that sector. Also note that the combat power assessment assumed that the opposing commanders each had an identical assessment of the combat power value of each of the systems of all the forces. That is, there was no modeling of a commander's misperception that the adversary's force is more or less formidable than the specified combat power values. Also note that surprise could not be modeled with this construct. Each side was omniscient with respect to its adversary, so there was no way for a commander to maneuver an unseen force to a position of advantage to attack the enemy from an unexpected direction. Quite simply, if a force has a 3 : 1 or better advantage in combat power over their adversary, then that force would attack that adversary. Attrition of each side's combat power was then assessed based on the calculated combat power ratio, and each side's combat power was then decremented accordingly. Movement of both forces was then assessed based on the amount of combat power lost and the type of terrain the sector consists of. Movement may have been mitigated so that a unit's movement in one sector did not expose the flank of a friendly unit in an adjacent sector. As simulations became more sophisticated, combat power scores were modified. The Marine Corps' Tactical Warfare Simulation, Evaluation, and Analysis System (TWSEAS) and MAGTF (Marine Air Ground Task Force) Tactical Warfare Simulation (MTWS) took into account "perceived combat power," which limited each side's calculation to only its current knowledge of the opposing force. Some simulations calculated dynamic combat power values, updating values based on the remaining forces on the battlefield after each time step, so, for example, an air defense weapon might have no combat power once all opposing aircraft had been destroyed.

Entity Simulations

In an entity simulation where individual combatants (systems or personnel) are engaging, each entity is assigned to travel from a starting position, via a series of waypoints, to a destination that it will reach if it survives. If the entity detects an adversary's entity, and current rules of engagement (ROE) permit, it will fire at that entity. An algorithm will then assess the probability that the entity hit the adversary's entity (P(hit)), and if it did hit, the amount of damage the hit inflicted (P(kill/hit)) where "kills" are typically categorized as "catastrophic," "mobility," "firepower," or "mobility and firepower."

Simulations and Prediction

The late, great Air Force analyst, Clayton Thomas, described simulation-based analysis as an IF-THEN statement. Both the simulation and the data comprise the IF side. The simulation and the data are then used to produce the simulation's output, the THEN side. If the simulation represented reality and if the data were precise, then the result would be an accurate prediction.[30] In general, neither is true. In the following paragraphs, we will examine how well our simulations represent reality and how precise our data are.

Standard Assumptions

There are standard assumptions for most closed-loop computer simulations. Human factors, such as leadership, morale, combat fatigue, and training status are typically not explicitly represented in these simulations. If not represented explicitly, then the implicit assumptions are that both sides have exactly the same characteristics with respect to human factors. While there have been attempts to represent human factors

---

30 Chapter 13, "Verification Revisited" in Military Modeling for Decision Making, MORS 1997, edited by Wayne P. Hughes, Jr.

such as training status and morale, the value of such explicit representation needs to be justified. If the simulation is attempting to assess the value of a new weapon system and the study team is using a scientific-method approach, having outcomes differ due to human factors makes it more difficult to determine the true cause of the difference between two sets of replications. In most combat simulations, the human factors are either ignored or standardized and mirrored so as to make the basis of comparison as free from confounding factors as possible.

Data

The simulations described above usually need between three to six months to instantiate a new scenario, and will cost around one million US dollars to get the simulation ready to run (terrain and performance data developed, quality controlled, and input, and scheme of maneuver developed, instantiated, and tested). Data is always challenging. Performance data must be developed to account for every interaction that could happen between all systems that will be represented on the battlefield. Performance data development can be especially challenging when examining future scenarios with emerging technology. Even developing data to simulate today's forces comes with challenges. The US Army has perhaps one of the most robust processes to develop performance data, but even that process uses only about 10% of actual data. This data is collected from ranges such as Aberdeen Proving Grounds where, in a controlled environment, US Army weapon systems are fired at captured enemy systems to determine their vulnerability to US weapons. They also use captured enemy systems to fire at actual US Army systems to determine their vulnerabilities. Often several US ground combat vehicles are rolled off the production line with the expressed intent of testing their vulnerabilities to enemy systems. After test firing is conducted, engineers determine the damage caused and record that information, which becomes the basis for the performance data that is generated for ground combat simulations. The other 90% of the data is then "surrogated," that is, interpolated, extrapolated, or otherwise estimated from that existing test data. This data is often developed using engineering-level simulations. Ground combat weapon systems are relatively inexpensive and numerous so testing their vulnerabilities can be done given the availability of the appropriate enemy weapon systems and ammunition. The Navy and the Air Force are challenged to come up with test data that can be used to develop performance data against their platforms. Firing captured adversary anti-ship missiles at a multibillion-dollar Ford class aircraft carrier to see how many hits it can withstand before sinking just is not possible, so often the data used is more of an educated guess than a mathematical approximation. One of the biggest threats to today's naval vessels is the anti-ship missile (ASM), but there have been less than 300 recorded instances of ASM hits on vessels that could be used to develop data.[31]

---

31 Faucett, J.E. (2019) " Shore-based Anti-Ship Missile Capability and Employment Concept Exploration," Master's Thesis, Naval Postgraduate School,Monterey, CA, June 2019.

Simulating the Reality of Combat

Many, if not most, of today's computer-based combat simulations are extraordinarily complex (the Concepts Evaluation Model, a theater-level deterministic closed-loop combat simulation used by the US Army's Concepts Analysis Agency in the late twentieth century was over 250 000 lines of computer code). However, that complexity does not translate into a model that can accurately predict the outcome of combat. This complexity has given rise to two different schools of thought. Simulation skeptics refer to them as "black boxes," which means that the users of these simulations have little to no understanding of the simulation's processes that convert inputs into outputs. Some simulation advocates believe that with the complexity that these simulations have, all the processes of combat are modeled to a high degree of precision, thus the simulation's outputs must be believed without question or debate. Subscribers to either school miss the fundamental truth that the models in these simulations are abstractions and approximations of some specific aspects of combat that the simulation was originally designed to model. All simulations are comprised of one or more models, and all models are an abstraction of reality, with some processes modeled explicitly, some implicitly, and some processes not modeled at all because the simulation's designer did not intend for the simulation to address those excluded processes. Prospective simulation users need to do some research and come to at least a basic understanding of what a simulation under consideration for use in a particular study was originally designed to model, and what its strengths and shortcomings are before they select a simulation that will be useful for the purpose at hand. In a RAND paper examining non-monotonic, chaotic output from a very simple deterministic, Lanchester-based combat simulation (2 variables, 18 data elements, and 8 rules), the authors state "The typical model simulates combat between opposing forces at some level of abstraction. No combat model is seriously expected to be absolutely predictive of actual combat outcomes. It is common, however, to expect models to be relatively predictive. That is, if a capability is added to one side and the battle is refought, the difference in battle outcomes is expected to reflect the contribution of the added capability."[32] This concept of relative predictability underpins the usage of simulations to conduct Analyses of Alternatives, or AoAs, studies that are used to justify weapon system acquisitions in DoD.

The Capability and Capacity of Modern Computing to Represent Combat

Moore's law, the idea that the processing power of computers doubles every 18–24 months, has led to the school of thought that our simulations are accurate

---

32 Dewar, J.A., J.J. Gillogly, and M.L. Juncosa, *Non-Monotonicity, Chaos, and Combat Models*, RAND, Santa Monica, CA 1991, p. 1.

representations of kinetic combat. However, most of the combat simulations in use today have their roots in the 1960s or 1970s, and in most cases, the computer code has not been optimized to take advantage of this increase in processing power. But even more significant is the notion of artificial intelligence (AI) and the idea that we can accurately represent, to the minute detail, every aspect of combat. The advances in AI that should dissuade us of this notion begin with IBM's computer "Deep Blue" that was programmed to play chess. In 1997, Deep Blue was able to beat Garry Kasparov, at the time a reigning chess champion.[33] A more recent AI triumph was "AlphaGo." AlphaGo was programmed to play the ancient game of Go, and was able to beat world Go champion Ke Jie in 2017.[34] "As Deep Blue and AlphaGo have demonstrated, in games of finite size with well-specified rules, computers can use artificial intelligence (AI) techniques to top human performance."[35]

Let us now begin to examine ground combat and see if it fits the description of the type of games AI has successfully mastered, specifically finite size and well-specified rules.

Finite Size

For finite size, we need to consider the number of pieces, or entities, and the size of the game board, or terrain box.

Number of Pieces/Entities

Chess has six different types of pieces: pawn, knight, bishop, rook, king, and queen, for a total of 16 pieces per side, or 32 total pieces. The US Army's Armored Brigade Combat Team (ABCT) has over 4700 soldiers and over 1300 vehicles, including 90 tanks, 90 infantry fighting vehicles, 112 armored personnel carriers, 18 self-propelled howitzers, and 4 unmanned aerial vehicles.[36] This does not include other assets that would support the ABCT in combat, such as Army and Air Force attack aircraft, and Army logistics assets. When an ABCT is represented in a combat simulation, only the primary combatants are represented, which would be about 300 vehicles and 200 infantry, or 500 of the 6000 entities that would actually be on the battlefield. Assuming the ABCT's adversary was similarly equipped and a similar size, the total entities that would be represented in the simulation would be around 1000, as compared to the 32 chess pieces.

---

33 https://en.wikipedia.org/wiki/Deep_Blue_versus_Garry_Kasparov
34 https://techcrunch.com/2017/05/24/alphago-beats-planets-best-human-go-player-ke-jie/
35 "AlphaGo: Using Machine Learning to Master the Ancient Game of Go," 27 January 2016, Official Google Blog, googleblog.blogspot.com/.
36 The U.S. Military's Force Structure: A Primer, July 2016, p22-27, Congressional Budget Office, July 2016.

Terrain

The game of chess has 64 squares, alternating in color but identical otherwise. The chess board is two-dimensional, that is the 64 squares all occupy the same plane. The modern battlefield whether it be the mountainous terrain of Afghanistan, Korea, or Iran; the open and rolling terrain of the North German Plain; or the deserts of Iraq and Kuwait is a three-dimensional terrain: mountains, hills, plains, and valleys. That terrain is broken up by water: rivers, streams, lakes, swamps. There can be vegetation and man-made objects that sit on top of the terrain. The terrain and these objects, unlike the chess board, change over time. Rain and snow change the trafficability of the terrain for combat systems, and it can be different for different types of systems (wheeled versus tracked and dismounted infantry). The terrain can be deformed by digging trenches, building obstacles, and explosions. Vegetation can change over time – a deciduous forest has much better lines of sight after the trees have dropped their leaves. Man-made objects can be rubbled or destroyed. For the time being, we will ignore the airspace above the terrain, but fixed and rotary wing aircraft, manned and unmanned, are an integral part of modern ground combat, given the weather permits. And if we do represent aircraft, we cannot forget that rounds fired from artillery and mortars do share the same airspace as aircraft.

Rules

Let us now consider rules. For ease of understanding and simplicity, we will only cover the major rules for movement, attack, adjudication, and victory conditions.

Movement

The movement rules of chess are fairly straightforward: each of the six pieces has unique movement rules. Moves are alternated between the two players, that is one player moves, and that move is observed and then analyzed by the opponent before the opponent moves. There is really no time–distance factor in a chess move, only one piece can be moved in a turn, and it must conform to the move-ment rules for that given piece. In a move, the more mobile pieces (queen, bishop, rook) can move from one end of the board to the other, while the king is limited to one square per move. If we considered chess as a wargame, it would be an open wargame, all information is available to all players.

Ground combat systems' movement is dictated by the terrain and the objects that rest on the terrain. As in chess, the different weapon systems have different movement "rules." Tanks and other tracked vehicles are better in open and rolling terrain, and struggle in mountainous and forested terrain. Wheeled vehicles move well on roads and can move well in open and rolling terrain if it is firm, that is, not impacted by weather or torn up by tracked vehicles. Dismounted infantrymen can go almost any-where, but not very quickly. All movement may also be modified by a commander's orders. Units may be ordered to stop at a certain point or to proceed until a given objective has been accomplished. Movement in a ground combat environment can be simultaneous and almost constant. Unlike chess, there are no discrete, alternating moves – movement is fluid and nearly nonstop. The observation of one side's moves by the other depends on the sensors deployed to observe, and observations may be mitigated or obscured by deception as well as environmental effects.

Attack

In chess, an opponent's piece can be attacked if it occupies a space that the attacker can move one of its pieces to using the proscribed move for the attacking piece.

In ground combat, a system must first acquire an opponent's system before attacking. Unlike chess, where all pieces are seen by both sides and all possible attacks can be easily identified, combat systems have to find an enemy entity before attacking. Once that entity is acquired, it can then be attacked. The acquisition process can be from the attacking system, from a third party, or from a combination of systems on the battlefield. A tank or an infantryman typically acquires its own targets to attack, where indirect fire assets such as attack aircraft, mortars, and artillery usually attack targets that have been identified by another entity on the battlefield. Once it is decided, a system will engage a target, there may be different attack mechanisms that can be employed and must be specified. Systems often have more than one weapon. A tank has a main gun and a coaxial machine gun. An infantry fighting vehicle has a 25-mm chain gun and an anti-tank missile. An infantryman has a rifle, grenades, and a bayonet. A tank's main gun has different types of main gun ammunition for different purposes.

Adjudication

In chess, if the piece is attacked, the adjudication is simple: it is removed from the board, in essence destroyed. It does not matter what the two pieces are, a lowly pawn can attack and destroy an opponent's pawn or its queen. In ground combat, systems can be removed from the battlefield if destroyed, but systems can also be damaged (or wounded for human "systems"). A damaged system may still be able to move and/or fire, so the amount of damage and remaining functionality of the system must be determined. Damage can be cumulative, so a damaged system can be subsequently destroyed if attacked again, depending on the attacking system and the damage that it can deliver. Damaged systems can also be repaired and returned to the battlefield depending upon the amount of damage sustained and the availability of resources needed to repair the system.

Victory Conditions

A chess game ends when a player's king is in check and under attack by the opponent and cannot move to a space where it is safe from attack on the next move. In ground combat, victory conditions are highly dependent on the objectives of the combatants. Each side may have a "breakpoint" designated, that is, lose no more than x percent of your force executing this mission. Usually an attacker has a goal that either has to do with seizing terrain or destroying an enemy formation. A defender may have the goal of retaining the terrain it defends, inflicting a certain amount of damage to the attacker before yielding the terrain, or delaying the attacker for a length of time before yielding the terrain. Note that the objectives of attacker and defender may both be met (seize terrain and inflict a certain amount of damage before yielding terrain).

Summary

Today's closed-loop combat simulations are useful for assessing the goodness of adding capability to a formation through use of scientific method-like process for comparison. They cannot predict future outcomes of battle. They are tremendously more complex than chess or Go. The complex decision calculus a battlefield commander goes through that results in forces maneuvering and attacking is greatly simplified to IF-THEN types of decisions. In summary, closed-loop combat simulations are good at comparing different force options given the assumptions and simplifications are known and acceptable.

Campaign Analysis

In other literature, you will be able to find comparisons of wargames and computer-based combat simulations with lists of attributes and an assessment of which of the two tools is "better" vis-à-vis a particular attribute. This is a false dichotomy. Wargames and combat simulations are two different tools that are designed to produce two very different types of outputs. Wargames are used to investigate the human decision-making process and are not the primary tool to be used for making quantitative assessments or comparisons. Combat simulations are used to quantify the differences in forces, often using the scientific method process, but have little utility to investigate the human decision-making processes. Professional analytic agencies have recognized the utility of both tools and have often used a combination of tools to produce a more thorough and complete study of the phenomena of combat. This process can be described as "campaign analysis." The process of designing a campaign analysis is described by Kline, Hughes, and Otte:

> The campaign's objectives come from direction provided by political and military leadership at the national level. Derivation of a concept of operations to achieve those objectives, and metrics to measure their achievement, is done in collaboration with the commander and his staff. Assumptions are agreed to and provide a bound on the analytical study.[37]

---

37 Kline, J., Hughes, W., and Otte, D., 2010, "Campaign Analysis: An Introductory Review," *Wiley Encyclopedia of Operations Research and Management Science*, ed Cochran, *J. John Wiley & Sons, Inc* Vol 1 pp 558–566.

Once the goals of the campaign are thoroughly understood by all through a series of interactions between the stakeholders and the analysts, then the design process begins. A campaign analysis can be as simple as a wargame conducted to develop a CONOPS that is then instantiated in a closed-loop simulation for quantitative analysis, or it can be far more complex. The analysis done for the US Army's Future Combat Systems used over 50 different wargames and simulations.[38] As Kline et al. describe: "We concurrently select a model or series of models to represent the campaign environment. Broadly speaking, models bound the campaign in either a series of engagements (pulses of power) or a continuous operation where many small engagements create a larger effect (cumulative warfare)."[39] "Model categories range from closed-form probabilistic equations, computer simulations, optimization, and wargames to field experiments and operational rehearsals."[40] "For example, a wargame may help to develop concepts of operation and employment for the opposing sides. The wargame's interactions may be adjudicated by tactical simulations, equations, historic engagements or professional judgment. Once an employment concept or course of action is generated, it may be programmed in a larger campaign simulation to conduct analysis on many model variations."[41]

The realization that a campaign is a series of engagements or many small engagements that create a larger effect reinforces the realization that Lanchester made studying the battle of Trafalgar, and the conclusion that the CARMONETTE mathematical modelers came to about half a century after Lanchester: applying a closed-form computer simulation to combat should only be done for small, short engagements, because the human decision-making is such an important element of combat that to ignore it and presume that algorithms alone could accurately model the complexity of a campaign of modern combat's decision processes is pure folly.

Conclusion

Both wargames and closed-loop combat simulations will continue to have roles in analysis. Wargames will continue to help us understand trade-offs in CONOPS and courses of action development and will play a major role in the creation and development of new employment concepts and new tactics, techniques, and procedures

38 Future Combat System (FCS) Milestone B Analysis of Alternatives (AoA), TRAC-TR-03-018, ft Leavenworth, KS, 14 May 2003, P.Z.
39 Kline, J., Hughes, W., and Otte, D., 2010, "Campaign Analysis: An Introductory Review," Wiley Encyclopedia of Operations Research and Management Science, ed Cochran, J. John Wiley & Sons, Inc Vol 1 pp 558–566.
40 Kline, J., Hughes, W., and Otte, D., 2010, "Campaign Analysis: An Introductory Review," Wiley Encyclopedia of Operations Research and Management Science, ed Cochran, J. John Wiley & Sons, Inc Vol 1 pp 558–566.
41 Kline, J., Hughes, W., and Otte, D., 2010, "Campaign Analysis: An Introductory Review," Wiley Encyclopedia of Operations Research and Management Science, ed Cochran, J. John Wiley & Sons, Inc Vol 1 pp 558–566.

as new weapon systems and capabilities are integrated into our fighting forces. Closed-loop combat simulations will continue to provide the capability to assess the physics-based qualities of our forces and will allow us to tabulate, in the absence of any human input, the relative technological merits of new weapon systems and new formations.

Part II

Historical Context

2

A School for War – A Brief History of the Prussian *Kriegsspiel*

Jorit Wintjes

Julius-Maximilians-Universität, Würzburg, Germany

Introduction

In March 1914, the Bavarian general staff ran a major wargame making operational use of the whole Bavarian Army.[1] Given the overall political situation on the eve of the First World War as well as German war plans for a future conflict involving the major European powers of the day – France, Russia, Austria, and Britain – the scenario that was used for what may well have been one of the last big peacetime wargames of the Bavarian general staff is at first somewhat surprising: instead of being based on the assumption that Germany would have to face both France and Russia, it was in fact set both in the past and in an alternative reality.[2] Designed by major Rudolf Ritter von Xylander, aide-de-camp to the Bavarian Army's chief of staff,[3] it assumed a different outcome for the war of 1866 – Prussia, instead of trying to limit the impact of its victory on Austria and its allies, had in fact pushed for territorial concessions, gaining most of northern Bavaria up to the Danube; a Prussian general governor was installed in Nuremberg and forces conscripted in Franconia. The civilian population, the scenario specified, was rather unhappy about it all, with anti-Prussian sentiment rampant both

---

1 The surviving documents can be found in BayHstA GenStab 1102.
2 Anhang Lage Nr. 1, BayHstA GenStab 1102.
3 On the biography of Xylander see Kramer/Waldenfels 1966, 436–438.

Simulation and Wargaming, First Edition. Edited by Charles Turnitsa, Curtis Blais, and Andreas Tolk.
© 2022 John Wiley & Sons, Inc. Published 2022 by John Wiley & Sons, Inc.

in occupied Franconia and in Bavaria itself. Three years later, on 1 May 1869, Austria declared war to Prussia and the North German Confederation, leading to the mobilization of both the Bavarian and the Württemberg armies; both Bavaria and Württemberg, where a pro-Prussian party wielded considerable influence, initially remained neutral, declaring for Austria 10 days later, and then initially only in secrecy. While disagreements on the overall command structure of the allied forces kept the Württemberg army out of the war for a few days more, actual fighting began on the morning of 11 May 1869, with Prussian forces taking a key bridge near Ingolstadt. The wargame then simulated the following four days of conflict, coming to an end on the evening of 15 May 1869.[4]

While the scenario's historical and political premises thus were decidedly fictional – and in a rather creative way, one is inclined to add – other elements of it were not: the Bavarian Army was assumed to have the full mobilization strength of the contemporary Bavarian Army of 1914, as was the case with the armies of all other belligerents in the scenario. Although this made the scenario in a way hopelessly anachronistic – due to technological progress the Bavarian Army of 1914 was immeasurably more capable than that of 1869 and by using elements like spotter balloons and reconnaissance aircraft already pushing into the third dimension – it allowed the participating officers to exert command over forces they could directly relate to and which they might eventually be called up to lead into real battle. Using a scenario that was totally different from the actual Bavarian war planning – which by 1914 was focused on holding operations against the French Army in Lorraine[5] – forced the participants to think "outside the box," a fact stressed by the scenario designer after the war. Xylander, who after commanding two artillery regiments by 1918 had been in charge of 3rd Army's artillery, pointed out that not only had the need for operational security made it impossible to run full-blown wargames with the actual war plans, doing so would also have fostered overconfidence in and too much reliance on the plan.[6]

The March 1914 wargame is thus an excellent example for the true purpose of wargames – far from being an instrument with which to train officers for plans that were already prepared, their reason d'etre was in fact to prepare officers for making decisions even where no prepared plan existed and to enable them to outthink the enemy regardless of the overall situation. This general assessment is supported by Xylander's notes, which luckily survive.[7] Particularly in his general

---

4 By that time, Prussian forces had managed to push the Bavarians beyond the Danube; as Württemberg had, however, joined forces with the Bavarians, the Prussians were facing a developing counterattack on their left flank.
5 On the Bavarian Army's campaign in Lorraine and the Vosges see Deuringer 1922.
6 Xylander 1935, p. 8.
7 Folgen zur Anlage und Durchführung von Übungen, BayHstA GenStab 1102.

comments on the outcome of the wargame he stresses the negative impact of what he calls "schematism" on the decision-making of the participants, pointing out how important it was for officers to act quickly, decisively, and to think on their feet. Xylander also put considerable emphasis on proactive aggressive action, criticizing "pessimism" as the result of setbacks and warning rather drastically against giving the order to retreat, unless in dire circumstances.[8]

From a military history point of view, then, the wargame designed and run by Xylander is much more than a simple curiosity from the pre-war years. Instead, it provides key evidence for both the training of military decision-makers on the eve of the First World War0 and for the operational culture of the German – or in this case Bavarian – army. Given that wargames generally form an integral part of the training of military decision-makers, it is rather unsurprising that together with war plans and actual manoeuvres, they have in the past been explored for insights into the decision-making culture of armies.[9]

Despite the importance of wargaming for understanding military decision-making culture, however, the evolutionary history of the instrument has seen very little if any attention.[10] The history of wargaming in general is still largely unexplored, which is all the more surprising as this history is quite an unusual one – originally a Prussian invention dating back to 1824, wargames were officially introduced to other armies of the industrialized world only after the Prussian victory of 1870/71. Once spread beyond Prussia's border, however, they almost immediately caught on, and by the end of the following decade wargames were in widespread use and have – in some way or another – ever been since. While a detailed history of wargaming would therefore be a welcome or even necessary addition to the existing body of military history literature, it is not possible to provide even only a brief overview over all of that history within the confines of the present chapter. Instead, it will concentrate on the initial history of wargaming in Prussia and Germany from 1824 to the beginning of the First World War, a history which is so far poorly explored and still includes quite a few blank spots. It is nevertheless hoped that the present overview can serve as a first introduction to a field that is well worth studying and where still much remains to be found.

Before turning to the actual history of the Prussian *Kriegsspiel*, however, two brief preliminary remarks are necessary – one on the terminology of what was to

---

8 "Um Gotteswillen keine Rückzugsbefehle [. . .] wenn nicht unabwendbar" (Folgen zur Anlage und Durchführung von Übungen, BayHstA GenStab 1102).
9 See eg the analysis by Robert Citino of inter-war wargames run by the Reichswehr (Citino 2005).
10 For brief overviews see Wintjes 2017, Peterson 2016, 4–15, Wintjes 2016, Wintjes 2015, Van Crefeld 2013, 145–153, Peterson 2012, 221–240, Vego 2012, Berger 2000, Hohrath 2000, Pias 2000, 180–183, Perla 1990, 35–45, Knoll 1981, Young and Lawford 1967, 3–4, McHugh 1966, 27–58.

become the very first professional conflict simulation officially introduced into a European army, the other one on the nature of the surviving sources.

As for the terminology, it is the word *Kriegsspiel* itself that is quite possibly one of the reasons why it has not generated much interest among many historians –while its second element, *spiel* (game), carries a certain notion of lack of seriousness, the combination of *spiel* (game) and *Krieg* (war), certainly one of man's more serious activities, is for many instinctively problematic – war, after all, is not a game, and even if one of the greatest theoreticians of war, Clausewitz, compared war to a game of cards,[11] this was to highlight a specific aspect of war rather than ascribe a playful, for want of a better word, character to the experience of war in general. In this context it is well worth noting that the perceived difficulties of the phrase "*Kriegsspiel*" or "wargame" are no inventions of twenty-first century political correctness; rather, it was already the early nineteenth century that was acutely aware of the incongruity of the two terms, even if there may have been some nuances to modern perceptions of the term. Thus already before the invention of the Prussian *Kriegsspiel*, simulative games covering different aspects of warfare – which will be briefly covered in the following section – existed, and one of the inventors expressively stated that he did not think his invention was a *Spiel*, i.e. something designed purely or even mainly for recreational purposes;[12] as late as the 1870s, attempts in Prussia were made to replace the term with a different one, though without success.[13]

As for the sources of *Kriegsspiel* history, they fundamentally fall into two different categories, those on the games themselves and those on the use of the games. The former includes a wide variety of material, which can be broadly categorized in rulesets, secondary literature, maps, and the actual gaming material including tokens, rules, dice, etc. As throughout the nineteenth century nearly all of the relevant written material was not classified but openly published and available on the book market, rulesets as well as secondary literature – articles, reviews, and the like – generally survive, if in some cases only in very small numbers. Beyond published material, further rulesets will have existed in the form of "house rules" or modification of published rules, much as they do today in wargaming clubs; this unpublished material is much harder to find if it has been preserved at all. A case in point here is a set of rules developed in the Magdeburg garrison in the late 1840s, which was eventually circulated as a private print, of which apparently not a single

---

11 Clausewitz *Vom Kriege* 1.21.
12 Venturini 1803, p. 2.
13 In 1879 Ernst von Reichenau suggested to use the term *Planmanöver* ("manoeuvre on a map") instead (Reichenau 1879, pp. 5–6).

copy survived.[14] Maps, of which by the mid-1870s quite a number existed, were also freely available on the market, but given the costs involved in producing them they were only printed in small numbers.[15] As a consequence, there is at present only a small number of maps extant, including an 1866 map of the Königgrätz battlefield and its surroundings and an 1872 atlas of the Metz region.[16] Even rarer than the maps is the actual gaming material or the *Apparat*, as it was called in German. While ads in contemporary journals yield valuable information on what was inside a box of gaming materials, no German *Apparat* seems to have survived, though British examples from the 1880s exist both in public and private collections.[17]

As for sources on the implementation of the games, these include any specific instructions by the umpires, all material relevant to the scenario, orders and reports written during the actual game, sketch maps and final write-ups of the game itself, and any discussion afterward. Much of that material would have been regarded as scrap paper immediately afterwards, with only scenario notes and final write-ups likely to find their way into an archive. As moreover much of the Prussian army's archival material was destroyed during WW2, surviving sources are few and far between. An important exception is the Bavarian Army, where not only scenario notes survive in significant numbers, but in some cases – including the one mentioned at the beginning of this chapter – also much of the material associated with the actual running of the game, including orders and reports sometimes even including marginalia by the umpires.

Kriegsspiel Prehistory

The concept of depicting warfare on a tactical, operational, or strategic level is hardly a new one, and even older are attempts at finding analogies to warfare in certain types of strategy board games. Games like chess or draughts have long been seen as precursors to modern wargames,[18] and while the latter has a pedigree

---

14 Anonymus 1870, p. 100.

15 Surviving inventories of contemporary regimental and staff libraries offer some insights on the number of maps available; thus the Saxonian general staff had in the late 1870s' *Kriegsspiel* maps of the Paris theater in the 1871 war (in 1/5000 scale), a "traditional" 1/8000 map of the Königgrätz battlefield and its surroundings, another 1/8000 map of the Gitschin battlefield in Bohemia, a map covering the eastern approaches to Metz (in 1/6250 scale) and another 1/6250 map of the Metz region. (General Staff 1878, p. 127).

16 Both are currently in the possession of the British Library; for the Königgrätz map see Cartographic Items Maps 27210.(14.), for the Metz Atlas see Cartographic Items Maps 33.e.8.

17 See e.g., National Army Museum 1967-09-51.

18 On the Greek game *petteia* and the similar Roman *ludus latrunculorum* see Lamer 1927.

going back to similar games in Greek and Roman antiquity, the former can claim a particularly close association to warfare – or so many early modern observers thought. Thus, Frederick the Great is famously to have said that nothing can depict the movements and formations of forces on the battlefield better than the games of chess.[19] While this arguably says more about early modern warfare and its perception by contemporary military decision-makers than about the actual character of chess, it is worthy of note that the line of development that would eventually lead to the wargames of the nineteenth century rested mostly on chess, but not exclusively.

Other types of games associated with warfare can be found as early as the sixteenth century, and it is a card game published between 1559 and 1562 by the Hessian military engineer Reinhard Graf zu Solms that for the first time tried to depict units in different formations; moreover, Solms's game was also apparently the first game invented by a military professional, and he explicitly designed it with the education of young noblemen in mind.[20] The game was therefore not primarily designed with the purpose of entertaining the players in mind, but rather as an educational instrument – that Solms also considered it to be useful command and control tool, with which a general could communicate his intentions to his officers before a battle only serves to underscore this point[21]: for all its weaknesses – the game lacked any representation of the ground as well as any combat mechanisms – Solms's invention was at least with regard to its intention the first professional wargame, matching at least to some extent the ambition of its title – *Kriegsspiel*. While the Solms's *Kriegsspiel* appears to have gained little attention, the idea of using card games to depict military action lived on, and throughout the late seventeenth and eighteenth centuries card-based wargames enjoyed some popularity, with those printed by the French engraver Gilles Jodelet de La Boissière probably being the most popular one, going through several reprints.[22]

By the end of the eighteenth century, experiments with board wargames derived in some way or another from chess had increased significantly. Whereas in the latter half of the seventeenth century the *Newerfundenes grosses Königsspiel* invented by the Ulm merchant Christoph Weickmann and published in 1664 remained a curiosity,[23] around a century later several inventors wrestled with the

---

19 Wahl 1798, p. 15.
20 Solms 1562; on the biography of Reinhard Graf zu Solms see Wintjes 2015, pp. 23–25 with further literature; for a description of the game game see Wintjes 2015, pp. 25–31.
21 Solms 1562, 1v-2r.
22 Still surviving in several examples is the second edition (Boissière 1698) of the jeu de la guerre, which is said to have been ordered by Louis XIV as a means of instruction for his son; Boissière also invented a fortress game (jeu des fortifications), which saw both reprints and translations (see e.g. Stapf 1680).
23 Weickmann 1664; for a description of the Weickmann game see Hilgers 2008, 32–38.

challenges of depicting as many aspects of contemporary warfare as possible by means of a board wargame. Indeed, one could be forgiven for assuming that by the end of the eighteenth century, designing wargames was somewhat of a fashion and popular not only among military men but in fact with individuals of surprisingly varied backgrounds. Of particular importance are the activities of the Brunswick mathematician Johann Christian Hellwig, who served at the Brunswick court as a master of pages and was also involved in teaching officers of the Brunswick army.[24] Hellwig, who initially called his game *Taktisches Spiel* before switching to *Kriegsspiel* for its last edition, published the first version of his wargame in 1780, with a supplement to the rules appearing two years later.[25] By 1803, he had considerably revised his game and published what was essentially a new and substantially altered version. Hellwig had taken chess as a starting point for his development and wrestled mostly with the problem of introducing topography as a meaningful element to his game. In order to do so he devised a customizable board composed of differently colored square tiles, which represented different types of terrain; a prototypical gaming board of 1617 squares was included in the second edition of his game.[26] However, while the board represented a significant development from the original chess board of 64 squares, many of the movement and combat resolution mechanics still resembled chess rules to a great extent. While the game distinguished between different types of artillery, cavalry, and infantry, the basic principle of conflict resolution, so to speak, was still the simple taking of pieces. Also, little thought was given to the question of scale – evidently the game board was meant to depict an operational theater more than a battlefield, with the figures representing large bodies of soldiers, but Hellwig did not relate his gaming pieces to any particular unit types or sizes.

Johann Christian Hellwig was not the only Brunswick wargame designer active during the last decades of the eighteenth century. At about the same time, Johann Georg Julius Venturini, an erstwhile engineer officer with the Brunswick army and noted military theoretician, developed a *Kriegsspiel,* which initially was published in 1797 and then, in a significantly altered and updated version, republished in 1803 shortly after his death.[27] This latter edition includes several important innovations including a discussion of ground scale – the game uses a map composed of squares that are supposed to represent an area of 2000 ×2000 paces[28] – and specific instructions as to what the gaming pieces actually

---

24 On Johann Christian Hellwig and Kriegsspiel see Nohr/Böhme 2009.
25 Hellwig 1780, Hellwig 1782, Hellwig 1803.
26 A copy survives in the Brunswick university library (Sign. 2300-6214).
27 Venturini 1797, Venturini 1803.
28 Venturini 1803, 3.

represent – infantry and cavalry brigades as well as artillery batteries.[29] The rules make allowance for the splitting up of brigades and the creation of composite forces of infantry, cavalry, and artillery elements, and they deal on a surprisingly detailed level with the logistics of warfare, right down to including wagon trains and field bakeries. Indeed, it is the logistical problem that a large-scale campaign of the late eighteenth and early nineteenth century represents that is in the focus of the Venturini *Kriegsspiel*, which is best characterized as an operational-to-strategic-level wargame. Players get to lead a campaign into enemy territory, which may last for several years, and have to take great care not to leave their forces unsupported during winter – which can result in the total loss of units.[30] While Venturini also included quite sophisticated movement and combat rules distinguishing between ranged and close combat, using numerical superiority as the key deciding factor, it is rather obvious that the game was mainly aimed at teaching the overarching importance of logistics in war. And it was indeed for that purpose – teaching – that Venturini intended it. He saw his own game as an educational tool to instruct the players in the operational art of war, or *Feldherrenwissenschaft*, as he put it.[31] In doing so, Venturini took a path quite different from that of almost all other wargame designers at the time, who put the entertainment value, so to speak, of their games first and any educational use only second. Venturini was different, so different in fact that he explicitly noted that his invention, although it was called *Kriegsspiel*, should not be called a *Spiel* at all.[32]

While Venturini was probably well ahead of his time in terms of conceptualizing what a wargame actually should be, his game found little reception. This may in part have been due to his early death, though the competition of Hellwig's game certainly will have played a part as well. Hellwig eventually bequeathed his *Kriegsspiel* to the son of one of his close friends and colleagues, Jakob Eleazar de Mauvillon, an officer in the Brunswick army, professor of military science and noted German liberalist.[33] Jakob Eleazar's son Friedrich Wilhelm de Mauvillon served in the Brunswick army as well and apparently used the game for the instruction of officer cadets by recreating historical battles with it.[34] In 1822, Mauvillon republished the last edition of the Hellwig *Kriegsspiel* with a new introduction[35]; while wargaming activity in Brunswick may have continued for some time afterward, Mauvillon, who had entered Prussian service in 1813 and finally

---

29 Venturini 1803, 6–7.
30 Venturini 1803, 17.
31 Venturini 1803, 1–2.
32 Venturini 1803, 2.
33 On Jakob Eleazar de Mauvillon see Hoffmann 1981.
34 Mauvillon 1822, 295–296.
35 Mauvillon 1822.

retired in 1822, apparently found no successor, and his death in 1851 marks the final end of the Brunswick era of wargaming development.[36]

Already in the eighteenth century, however, there were wargaming enthusiasts elsewhere and with rather different backgrounds. In Bohemia, Johann Ferdinand Opiz, a bank official, had developed a *Kriegsspiel,* which was first published by his son in 1806.[37] In the publication, Opiz claimed the game to have originated in the 1740s, which would make it the earliest of the late eighteenth-century board wargames.[38] Whether there is any truth to that claim or not, Opiz did introduce an important new element to the *Kriegsspiel* – the use of dice, which were used to determine casualties produced by infantry and artillery fire;[39] in order to keep track of these casualties, Opiz also introduced the use of tables, which the players had to constantly update.[40] The introduction of dice was not only a significant departure from chess-based mechanisms, it also for the first time made it possible to introduce the factor of chance to a wargame. Opiz was very explicit about this – in war, he noted, not only courage, reason, instinct, and the intellect of the general mattered, but also luck and chance.[41] While this may simply echo Frederick the Great who is supposed to have considered *fortune* to be a key quality in any officer, looking back at Opiz from the late nineteenth century, when Clausewitz's ideas on friction in war not only had found wide acceptance, but when wargaming was actually considered to be an extremely suitable instrument for confronting officers with the insecurity caused by friction, makes him sound almost visionary. The Bohemian bank official can certainly claim to his credit the first conscious introduction of an element to the wargame that could have a crucial impact on the outcome but lay outside the direct control of the players. In other respects, the Opiz *Kriegsspiel* was similar to the Hellwig and Venturini ones; it was also played on a board made up of differently colored squares each representing a particular type of terrain, and in terms of scale it was close to the second edition of the Venturini game.[42] In the case of the Opiz game, the pieces represented regiments or battalions of cavalry and infantry, which were further distinguished by type and strength – a unit of line infantry was supposed to have 1,200 men on strength, one of skirmishers only 800 men.[43] While Opiz did not specify the ground scale, it is obvious that he had a tactical setup in mind – not only does the "ideal army" that he suggests for two players include a total of 14,400 horse, 41,600 foot and

---

36 On Friedrich Wilhelm de Mauvillon see Poten 1884, 714–715.
37 On Johann Ferdinand Opiz and his son Giacomo see Wintjes 2015, 21–22.
38 Opiz 1806, 41–42.
39 Opiz 1806, 74–81.
40 Opiz 1806, 76.
41 Opiz 1806, 43.
42 Opiz 1806, 44–46.
43 Opiz 1806, 58–60.

30 guns,[44] he also stresses the tactical nature of the game several times.[45] Clearly, his *Kriegsspiel* depicted combat on a far smaller scale than the Venturini wargame.

Johann Ferdinand Opiz and his *Kriegsspiel* are nowadays almost totally forgotten, and already during the early years of the twentieth century it was held in little regard. Yet despite this obscurity, at the time Opiz was at least as influential as Hellwig and Venturini, giving inspiration to the last of the great wargame inventors of the late eighteenth and early nineteenth centuries, Georg Leopold von Reisswitz. In 1812 Reisswitz, a Prussian government official,[46] published a "tactical game" (*Taktisches Spiel*), which managed to gain some official attention. After he presented his game to the two Prussian princes Friedrich Wilhelm and Wilhelm, it was apparently played quite intensively for some time among a circle of close friends of the young princes, though nothing is heard of the game after 1814.[47] The 1812 publication of the Reisswitz game is an important piece of evidence for the history of wargaming as Reisswitz included an extensive introductory note shedding some light on his biography and the developmental process of his game.[48] It offers some interesting insights into how he actually went about designing the game – he mentions asking serving officers, talking to other wargames designers working on similar games and using the already published games of Hellwig, Venturini, and Opiz as a source of inspiration.[49] More importantly, however, it also allows a fascinating glimpse into an early nineteenth-century wargaming community in which individuals from all walks of life were engaged in wargames design, ranging from active officers to one of the founders of the Breslau Bible Society.[50] The Reisswitz game itself was quite conventional in that it made use of a gaming board made up from square tiles, though these were now not merely color-coded but also of different heights and shapes, thus allowing for the creation of a three-dimensional terrain.[51] Given the tactical nature of his game, it is hardly surprising that the map that could be built up from the various terrain tiles was a fairly small one. As the first wargame inventor, Reisswitz actually noted the scale of his board, which came at a slightly eclectic 1:2373.[52] Also innovative was his way of treating casualties – like Opiz, Reisswitz accounted for units suffering casualties over time instead of simply being taken away. However,

---

44 Opiz 1806, 59.
45 See eg Opiz 1806, 42.
46 On the biography of Georg Leopold von Reisswitz see Wintjes 2016, 58–60.
47 Wintjes 2016, 63.
48 Reisswitz 1812, v–xvi.
49 Reisswitz 1812, xiv.
50 Wintjes 2016, 59.
51 Reisswitz saw this three-dimensional terrain pieces as his main contribution to the development of wargaming (Reisswitz 1812, ix–x).
52 Reisswitz 1812, 11.

while Opiz relied on tables that the players had to update, Reisswitz introduced the concept of exchanging tokens once a unit has suffered a certain number of casualties, thus allowing a fairly detailed visual depiction of the course of an engagement.[53] Despite these innovative features, the Reisswitz *Kriegsspiel* was a step back behind the Opiz game in one key respect: when it came to determine the outcome of actual combat, Reisswitz consciously chose not to employ dice but to rely on mechanisms similar to those of Venturini and Hellwig. While Reisswitz claimed it to be self-evident that a game could not depict the "irrational" factors that may or may not lead to success in combat,[54] in eliminating any element of chance, Reisswitz basically reduced his game to a militarized version of chess, though admittedly a complex and detailed one played on a sandbox-type of board.

Looking back from Georg Leopold von Reisswitz wargame to the early card games of the sixteenth and seventeenth centuries, some patterns are at least vaguely discernible. Throughout the 250 years that separate Reinhard Graf zu Solms and Georg Leopold von Reisswitz, wargame inventors struggled with four crucial questions, which even today anyone designing a wargame has to face: the representation of units, the design of the map, the mechanism of movement, and the resolution of combat. While card games represented an early attempt at solving the first problem, they did not offer solutions for the others, and by the eighteenth century, wargame designers mostly looked to chess for inspiration. As a result, late eighteenth- and early nineteenth-century wargames almost invariably used maps composed of square tiles that were directly connected to the movement mechanism – which often simply specified the number of squares a gaming piece was allowed to move. Combat resolution in many cases also was more or less directly based on chess, simply resulting in removing – or "taking" – a gaming piece. The gaming pieces themselves, which in Hellwig's original game still looked quite similar to chess pieces, eventually developed into small rectangular blocks not too dissimilar from the symbols used by printers and engravers for depicting units on for example a battle map. In all, then, by the early nineteenth century, wargames were still on a fundamental level highly modified games of chess; as such, their capability of depicting even only certain aspects of military action in a way that was actually useful to military decision-makers was quite limited, a notion shared already by contemporary observers. When Georg Leopold von Reisswitz published his *Kriegsspiel* in 1812, there was therefore little to suggest that wargaming could ever be something else than a fringe activity pursued mostly by interested civilians and officers in the spare time. Yet this was all about to change within less than a decade.

---

53 Reisswitz 1812, 49.
54 Reisswitz 1812, 48.

A School for War – the Prussian *Kriegsspiel*

In some ways, the spring of 1824 was a turning point in military history, as it marked the very beginning of professional wargaming activity throughout the Prussian Army, which would eventually lead to wargaming being adopted by armies throughout the world. The nowadays nearly omnipresent techniques of simulating operations, testing plans in command exercises, or sharpening decision-making skills by using competitive wargames took their beginning from a series of events taking place in Berlin in spring 1824, events which had a certain comic element to them. The key figure was a young artillery lieutenant of the guards division by the name of Georg Heinrich von Reisswitz, son of the afore-mentioned Georg Leopold von Reisswitz.[55] Young Reisswitz was apparently a man of rather peculiar tastes, at least for an officer of his age – instead of spending his spare time in the accepted manner with alcohol, women, and gambling or any combination thereof, he locked himself up in his room for hours on end playing the violin; his horsemanship, always considered a mark of a good officer as well as a true gentleman was, to quote one of his friends, displaying more vigor than grace, which is probably a slightly ambiguous compliment.[56] Most of his spare time however was devoted to refining an invention of his, a new *Kriegsspiel,* which was partly based on that of his father, though differing from it in several key aspects. Reisswitz soon managed to gather a bunch of kindred spirits, young offic-ers who shared both his fascination with the *Kriegsspiel* and the lack of any mean-ingful combat experience typical for many young officers in the Prussian Army of the 1820s and 1830s.

One would presume that it took not too long before the divisional commander got notice of one of his younger officers displaying what many contemporaries probably considered to be an irritatingly ungentlemanly behavior – shunning alcohol, gambling as well as women, they instead occupied a number of rooms in the officers' mess equipped with maps on which they pushed strange little blocks of wood or lead in order to, so that they could, well, "play war." In order to find out what this was all about, Reisswitz's divisional commander, who happened to be crown prince Wilhelm, ordered a presentational game.[57] Unknown to Reisswitz and his merry band of like-minded fellow officers, who were probably happy to oblige, crown prince Wilhelm also invited the chief of the Prussian, von Müffling, to the presentation, together with other generals.[58] Müffling's enthusiasm for the

---

55 Still the most important source for the life of Georg Heinrich von Reisswitz is Dannhauer 1874.

56 Dannhauer 1874, 528.

57 Dannhauer 1874, 529; see also Wintjes 2016, 64–65.

58 Dannhauer 1874, 529.

whole thing was apparently quite limited – after all, everybody knew what a *Kriegsspiel* was, and while it might pose an interesting intellectual challenge, it nevertheless had very little to do with what officers normally had to do for a living.

On came the day of the presentation, and when the young *Kriegsspieler* entered the room, they were not only greeted by the crown prince but also by Müffling, a few other generals – and a decidedly frosty atmosphere:[59] clearly, Müffling and his fellow officers expected yet another *Kriegsspiel*, so many of which had been invented in the preceding decades, all of them in one way or another related to chess and primarily meant to entertain. Nevertheless, Reisswitz and his comrades set up the *Kriegsspiel* and a game with Müffling as a participant – and an hour into the game he famously exclaimed "This is no game, it is a school of war!"[60] The chief of the Prussian general staff was genuinely enthused about the *Kriegsspiel*, quickly setting things into motion so that after a few months, a royal edict ordered every single regiment of the Prussian Army to acquire a set of rules, a suitable map, and gaming materials – the *Kriegsspiel* was officially introduced into the Prussian Army. The instant success of the *Kriegsspiel* must have surprised Reisswitz and his comrades, who now had to spend much of their time frantically trying to turn the rules into something that could actually be published; the result was a testimony both to their diligence and their modesty as it named only Reisswitz as the author, although he had spent much of the summer of 1824 in Russia, thus unable to contribute much to the actual publication.[61]

The *Kriegsspiel* was an enormous success – but why did it succeed where all other previous *Kriegsspiele* had failed? Müffling clearly considered the *Kriegsspiel* to be quite different from all earlier wargames, and so it indeed was, in three important ways. First of all, whereas playing earlier *Kriegsspiele* sometimes meant wading through hundreds of paragraphs, Reisswitz's *Kriegsspiel* had no rules – or, to be more precise, it did have rules, but as the game was fundamentally driven by umpires, the so-called *Vertraute*,[62] only these umpires had to know the actual rule mechanics. All the participants had to know was the application of tactics and how to give orders; in other words, participants in the Reisswitz *Kriegspiel* needed to be proficient in nothing but what they had to know as officers in any case. As a result, while those serving as umpires had to have a sound knowledge of the rules – which were quite involved, though not nearly as complicated as some of the earlier *Kriegsspiele* –[63] it took only a small number of officers in any garrison to actually start playing the *Kriegsspiel*.

---

59 Dannhauer 1874, 529.
60 Dannhauer 1874, 529.
61 Dannhauer 1874, 530; see also Wintjes 2016, 65-66.
62 Reisswitz 1824, 3–4.
63 In the original 1824 Reisswitz game the rules extended over 59 pages in 78 paragraphs.

Having the umpires solely responsible for the actual working of the game not only made access to the game easy for participants, it also had a second important implication. Only the umpires moved units on the map, only they decided over the outcome of combat, which they then communicated to the participants, and all communication between umpires and participants had to be done in writing, with speaking strongly discouraged. Participants had to piece together a picture of the situation on the ground by processing the information given to them by the umpires, who were in total control over the flow of information. With such a setup, the Reisswitz *Kriegsspiel* succeeded where all earlier *Kriegsspiele* had failed (if they had tried to depict it at all) – the employment of umpires made it possible to depict a certain degree of friction and fog of war in a way impossible in any other simulation where the participants directly interacted with each other.

The third important difference between the *Kriegsspiel* and earlier games was the map it was played on – Reisswitz had at some point in the developmental history of his *Kriegsspiel* decided to abolish the square-gridded map system his father was so proud of that he called it the core of his invention,[64] and to use a topographical map instead. The consequences were simple yet had a considerable impact: in order go play the *Kriegsspiel*, one had to be able to read such a map! While from a twenty-first century point of view this is generally seen as part of a key set of capabilities an officer should have, it is worth pointing out that in the early 1820s, topographical maps based on a scientific survey of the ground were very much a new technology.[65] Across Europe, armies introduced topographical bureaus only from the 1820s onward, and particularly in the 1820s and 1830s it could not be expected from a young officer to be able to read a topographic map without prior training – the often-repeated joke about an officer with a map being the most dangerous person on the battlefield certainly rings true for a period where many officers simply did not know how to make proper use of maps. The *Kriegsspiel*, then, presented itself as an ideal instrument for training officers in the use of topographical maps, and that alone would have probably caused some enthusiasm among the generals. In Müffling, however, Reisswitz met not only a general aware of the need for using topographical maps and finding some means of teaching his officers their use, topography was in fact very close to Müfffling's heart – after all, he had been responsible for the very first topographical survey of a Prussian province which resulted in the publication of the first Prussian atlas of topographical maps published from 1815 onward. Müffling furthermore was the driving force behind the establishment of the Prussian army's topographical

---

64 Reisswitz 1812, ix–x.
65 For an overview over the development of cartography in the late eighteenthand early nineteenth centuries, see Imhof 1965, 9–16.

bureau,[66] and one may safely assume that whatever skepticism he brought with him into the room where the first *Kriegsspiel* was played vanished quickly once he saw the tropographic map on which the game was played.

While Müffling may have been drawn toward the Reisswitz *Kriegsspiel* mostly by the use of topographic maps, another key figure in the history of the *Kriegsspiel*, Helmuth von Moltke, who took an avid interest in the *Kriegsspiel* from early on,[67] realized the great potential that lay in its general setup: having umpires decide over the outcome of an action, and moreover making them responsible for the communication between participants, not only allowed them to fully concentrate on the decisions they had to make without sparing too much thought on complex rules, the employment of umpires channeling the information available to the participants also made it possible to confront them with at least some of the uncertainty they were eventually to face on the battlefield – the an battlefield where, to quote Clausewitz, everything was simple and the most simple thing was difficult.[68] The *Kriegsspiel* could be used to teach officers how to process information, how to track the movement of friendly and enemy units, how to get an at least remotely accurate picture of the overall situation, and perhaps most importantly, how to write orders in a way that the recipient actually understood both the order and his superior's intention. In short, the *Kriegsspiel* was the best instrument at hand with which to prepare military decision-makers for making decisions on the battlefield.

Despite the enthusiasm displayed by Müffling, and although after the introduction in 1824 the game quickly caught on particularly among younger officers, with wargaming clubs emerging in different garrisons, the *Kriegsspiel* soon also faced considerable criticism. In order to understand both this development and how this criticism was eventually overcome, it is necessary to turn briefly to the state of the Prussian Army of the 1820s and 1830s. While Prussia could look back on a proud military tradition, had more recently successfully participated in the Napoleonic wars and undergone a series of important reforms aiming at raising in particular the quality of army officers, by the mid-1820s the Prussian Army was very much a second-rate force. Financial constraints prevented the army from undertaking large-scale maneuvers, and even proficiency in the use of weapons suffered from the lack of funds; one of the army's most noted tacticians of the

---

66 Niemeyer 1997, 226.
67 It is interesting to note that the association of Moltke with the *Kriegsspiel* was such that less than two decades after its adoption by the Prussian Army he was credited as one if its co-inventors (Anonymus 1839, 490).
68 Clausewitz *Vom Kriege* 1.7.

time, Carl von Decker,[69] famously complained that in eight years as the commander of an artillery battery, he never got to fire a single shot, thus lacking any practical experience with the weapon system he was to employ against an enemy in war.[70] While the lack of opportunities to go on exercises out into the field was probably not much of an issue in the years immediately following the Napoleonic wars, as most officers could still look back on actual combat experience, throughout the 1820s and 1830s the number of younger officers with such experience was rapidly falling so that by the end of the 1830s even battalion or regimental commanders did not necessarily have seen active service in the wars against Napoleon. As the *Kriegsspiel* did at least offer an opportunity to simulate going on maneuver, one would assume that it was popular particularly among these younger officers lacking any other experience. And indeed, most of the individuals involved in developing the *Kriegsspiel* throughout the first two decades after its official introduction in 1824 were young officers, early in their career and often with an artillery background. While from a modern perspective this appears to be quite natural, given that one generally assumes younger people to adopt modern technology faster and more readily than older generations, in real, regimental life it apparently caused more than only one irritation – und understandably so.

One of the *Kriegsspiel*'s strengths was without doubt that it separated the participants from the actual mechanism. Basically, only the umpires needed to know the rules, while all the participants needed to know was how to give orders, process intelligence and information, and piece together a picture of the overall situation – in other words, all the participants needed to know was how to properly do their job as officers. The umpires on the other hand were responsible for running the game, which required an intimate knowledge of the rules, but which also put them in a position where they, willingly or not, had to judge the actions of the participants. With younger officers more likely to invest time and energy into getting a good understanding of the rules, *Kriegsspiel* games could produce situations in which a young lieutenant serving as umpire had to notify one of his superior officers of the adverse consequences of a suboptimal decision the latter had made. Depending on the temperament of the participant, this was a socially awkward situation, even without assuming that the superior officer in question was one of those with combat experience. It is thus hardly surprising that one criticism leveled against the *Kriegsspiel* was that it incited younger officers to forget their station – behind which loomed the much graver accusation of fostering indiscipline among younger officers.[71] As the years went on, however, and the composition of the

---

69 On the biography of Decker see Meerheimb 1877 1895, 93.
70 Müller 1873, 88.
71 Troschke 1869, 277; Dannhauer 1874, 531.

officers' corps gradually changed, the critics fell silent, and by the end of the 1840s the *Kriegsspiel* appears to have been widely accepted in the Prussian Army.

Apart from the generational change within the Prussian Army this was also to no small extent due to the dedication of a fairly small group of officers who from early on promoted the Kriegsspiel; eventually, one of them would rise to become Prussia's most eminent general – the already mentioned Helmuth von Moltke. He had started to serve with the Berlin garrison in 1828 and soon became acquainted with the circle of wargamers that had originally formed around Georg Heinrich von Reisswitz. By the time Moltke took up the *Kriegsspiel*, however, Reisswitz himself was already dead. While the success of his invention had gained him considerable prominence, it understandably had been the cause for jealousy among his fellow officers. As Reisswitz apparently was not an easy character in the first place, he eventually seems to have made more enemies than even royal patronage could overcome. When in 1826, he was slated for promotion to captain, he did not obtain a captaincy in the guards but was sent to Glogau in Silesia instead. Reisswitz took this extremely badly, interpreting it as being sent into a provincial exile. A year later he was dead, having shot himself during a stay with his father in the winter of 1827.[72] His death had little impact on the subsequent history of the *Kriegsspiel*, as two of his comrades still serving in Berlin, Carl von Decker and Johann von Witzleben, were not only active in playing the game – the Berlin *Kriegsspiel* club dates back to the late 1820s and was likely founded by the two – they also energetically worked on a supplement of the rules, which was finally published early in 1828.[73] This supplement not only represented the result of an extensive Kriegsspiel activity of the Berlin garrison, its authors also sought information from various sources, apparently involving a fairly wide range of officers in its creation.[74] The result addressed some imbalances found in the game; for example, artillery, when firing cannister, was deemed to be far too powerful and its effectiveness therefore reduced.[75] While it is tempting to suspect Reisswitz's overenthusiasm for his own arm of service – after all, he was an artilleryman – behind this, one should note that Reisswitz explicitly referred to a brief publication of official firing tests by Gerhard von Scharnhorst, from which he had taken the data used in establishing his tables. The imbalance found by Decker and Witzleben was therefore probably simply due to the mismatch between a weapon system's performance on a shooting range and in actual battle.[76] The 1828 supplement also added several new mechanisms to the original *Kriegsspiel*, one of

---

72 Dannhauer 1874, 531–532; see also Wintjes 2016, 66–67.
73 Anonymus 1828.
74 Anonymus 1828, 68.
75 Anonymus 1828, 69.
76 Reisswitz 1824, 9–10; for the results of the official tests see Scharhorst 1813.

which has been a staple of both professional and enthusiast wargamers ever since – the *Chancewurf* or chance throw. Realizing that participants every now and then could come up with the idea of doing something not covered by the rules, Decker and Witzleben stipulated that the umpires could decide by a simple throw of a six-sided die whether to allow the action or not.[77] The 1828 supplement was not a set of rules in themselves; together with the original rules, which were present in every regimental library, it formed the first *Kriegsspiel* used throughout the Prussian Army; when thinking about the wargaming activity in the Prussian Army throughout the late 1820s, 1830s, and 1840s, it is important to bear in mind that not the Reisswitz rulebook alone, but rather the 1824 rules in conjunction with the 1828 supplement formed the basis for this activity. Together they formed what could be called the *Ur-Kriegsspiel*, and while the introduction of wargaming into the Prussian Army will forever be connected with Georg Leopold von Reisswitz, Carl von Decker and Johann von Witzleben were also of considerable importance in forming the original *Kriegsspiel*.

The Prussian *Kriegsspiel* 1824/28 – 1862

Three developments had a significant impact on the first decades of the *Kriegsspiel*'s existence. The initial reluctancy by some older officers to embark on an activity that could encourage younger officers not only to make decisions well above their pay grade but also to actually judge the actions of higher ranking officers has already been mentioned. While there is extremely limited information available on the degree of wargaming activity in Prussian garrisons throughout the 1830s and early 1840s, it is clear that by the end of the latter decade the *Kriegsspiel* had overcome almost all resistance within the army. This does not necessarily mean that each and every garrison was a center of lively wargaming activity; instead, there will have been significant differences between, say, Magdeburg, where one of the leading figures in the local wargaming club in the 1840s was Helmuth von Moltke,[78] and smaller provincial garrisons where officers showed less inclination to spend their spare time studying maps and writing orders.

The second key development with a direct impact on the history of the Kriegsspiel was technological progress, which was finally making itself felt in the army and which would eventually transform the nature of combat on

---

77 Anonymus 1828, 75.
78 Dannhauer 1874, 530–531; Moltke served in Magdeburg as the chief of staff of Army Corps VI between 1840 and 1848.

the battlefield almost beyond recognition.[79] It has already been mentioned that the original *Kriegsspiel* of Reisswitz, Decker, and WItzleben was basically designed to depict a Napoleonic battle, with infantry fire losing any effectiveness whatsoever beyond 100 yards. Although the Prussian Army prided itself of the new service musket it had introduced in 1809, which was supposed to be superior to foreign muskets in terms of accuracy and effective range, it suffered from the same defects that hampered all smoothbore flintlock muskets of the period – flintlocks had a disturbing tendency to fail or misfire, dramatically increasing in adverse weather, while the ballistic capabilities of musket balls, which were poor to begin with, did not exactly improve from the mismatch between ball diameter and barrel bore that was often the result of hastily produced ammunition. By the early 1830s, however, two key inventions had transformed the musket from an unreliable short-range weapon useful particularly for shrouding the battlefield in a dense fog into a dependable mid-range weapon with the capability of striking an enemy at ranges well beyond 200 yards. Percussion locks, which entered the Prussian Army in the shape of the 1831 musket, dramatically reduced the rate of misfires to the point that the individual soldier could actually trust his weapon to fire in nearly every weather and climate. When a few years later many Prussian muskets were converted to the Minié system, the combination of percussion locks and rifled barrels allowed effective infantry fire at ranges of 400 yards or more. By the 1840s, technological progress had gained considerable momentum, and the Prussian Army decided to adopt a revolutionary new technology – breech-loading infantry rifles. The Dreyse needle gun, originally introduced in 1841 but only arriving with the army in large numbers in 1848, not only offered the increased capabilities of percussion lock rifled weapons but also allowed for rapid reloading, thereby increasing the volume of fire a body of infantrymen could create. In addition, breech-loading also made it much easier to reload in cover, thus pointing to significant changes in infantry tactics away from the large solid blocks of soldiers typical for much of the Napoleonic period.[80]

With all these changes underway, it is rather obvious that the *Kriegsspiel* faced the danger of becoming as obsolescent as the old flintlock smoothbore muskets Prussian infantrymen had used when it was introduced in 1824. However, Prussian wargamers were acutely aware of this danger, and as a consequence the original 1824/28 rules saw considerable modification in the late 1840s. In the Magdeburg garrison, Gustav Weigelt, another young artillery officer, put together a set of house rules that were then privately published in 1848 and circulated in small numbers

---

79 For the relationship between technological progress and the developmental history of the *Kriegsspiel* see Wintjes 2017.
80 Wintjes 2017, 12–21 with further literature.

among the members of the Magdeburg garrison.[81] Given that these rules were developed at a time when Helmuth von Moltke was serving in Magdeburg and was probably one of the leading members of the local wargaming club, the Weigelt rules would make an interesting companion piece to the other set of rules published in the latter half of the 1840s that originated with the Berlin wargaming club. Unfortunately, yet given the private nature of their publication probably not surprisingly, they do not survive; indeed, apparently already in the early 1860s it was difficult to access a copy, something noted by a reviewer of the 1862 *Kriegsspiel* rules, which will be discussed below.[82] The rules put together by the Berlin wargaming club, however, do survive, and they offer some interesting insights into how the *Kriegsspiel* had fared in the decades between its introduction into the army and 1846, when the Berlin rules were published for the first time.[83] Apparently, one criticism surfacing every so often was the overall complexity of the game; the original rules together including the 1828 supplement ran to 95 pages, and some rules such as those for determining how a fire spread in a forest village leave the observer wonder whether umpires actually stuck to the rules, just guessed at the progress of the fire or simply declared a squall of rain to stop the fire. As a consequence, the anonymous editors of the 1846 rules tried to simplify the rules considerably, arriving at a more manageable book of 51 pages. They also introduced an important innovation that removed the need for producing the special *Kriegsspiel* dice Reisswitz originally had invented. Instead, they used a "normal" six-sided die, providing tables in which combat effects were related to die throw results.[84] Easily the most important change to the original rules however was the acknowledgement of the increased efficiency of infantry and artillery fire – the 1846 rules allowed for the first time to accurately depict what combat was supposed to look like on a mid-nineteenth century battlefield, thereby providing a key update to the older rules.[85] While little is known about the actual use of the 1846 rules, the fact that they were reprinted in 1855 suggests a considerable demand, and one may safely assume that in the late 1840s most if not all of the wargaming clubs in Prussian garrisons transitioned from the *Ur-Kriegsspiel* to the 1846 *Kriegsspiel*. In all, the Weigelt and Berlin rules stand for an important characteristic of the overall history of the *Kriegsspiel* – the willingness of those involved in regularly running wargames to keep the rules up to date, even in an age when rapid technological progress sometimes made the inventions of yesterday obsolescent within a decade or less. At the

---

81 Tschischwitz 1862, 1.
82 Anonymus 1870, 100.
83 Anonymus 1846.
84 Anonymus 1846, 5
85 Anonymus 1846, 1.

same time, the considerable effort put into producing accurate rules, which led to no fewer than nine sets of rules being published in the two decades between 1860 and 1880, also is testimony to the importance ascribed to the *Kriegsspiel*; for many if not all Prussian officers, the *Kriegsspiel* had an important place in their profession, and while producing actual rules seems to have been an activity mostly younger officers engaged in, the work of Julius Verdy du Vernois, who in 1876 published an important study on the *Kriegsspiel* while serving on the general staff before becoming minister of war in 1879, shows the appreciation for the *Kriegsspiel* among the leadership of the Prussian Army.[86]

While the adaptation of the rapid technological progress typical for the mid-nineteenth century was one key characteristic of the *Kriegsspiel*'s history that is, given the importance ascribed to it by the Prussian army, almost self-evident, another main characteristic is not as easily understood or explained. Whereas the *Kriegsspiel*'s popularity was ever increasing in the Prussian Army so that by the early 1860s most officers will have had some experience in playing or running *Kriegsspiele*, and while this activity not only resulted in the publication of rules but also in the establishment of a market for maps, gaming aids, and sets of tokens, which made access to the latest version of the *Kriegsspiel* extremely easy – one just had to go to a high street bookseller and order everything that was necessary for running a *Kriegsspiel* – other European armies showed very little if any interest. It is quite possibly the most peculiar phenomenon in the history of professional wargaming that for almost half a century it was something done exclusively by the Prussian Army. While this may be partly explained by the second-rate nature of the Prussian Army for much of the 1830s, 1840s, and 1850s, it is nonetheless somewhat puzzling, all the more so as other European armies were clearly aware of the Prussian activities. For example, in Austria the *Kriegsspiel* was known, and some officers even advocated its introduction into the Austrian Army, but without success. In Sweden the original 1824 game appears to have raised some interest to the point of a translation being published in 1830, yet the *Kriegsspiel* did not find official endorsement.[87] A Bavarian officer suggested as early as 1830 a game for use in the Bavarian Army, producing rules of his own[88]; however, Wilhelm von Aretin's *Strategonon* was conceptually a step backward from the Reisswitz–Decker–Witzleben *Kriegsspiel*, sharing more similarities with the Venturini and Opiz games than with the former. That there was more to the history of the *Kriegsspiel* than the official introduction into armies – or rather the lack thereof – is shown by a rather intriguing piece of evidence: in 1828, possibly even before the Decker–Witzleben supplement appeared, a small booklet was published in

---

86 Verdy du Vernois 1876.
87 Anonymus 1830.
88 Aretin 1830.

Munich containing some additional instructions for umpires.[89] Other bits and pieces of evidence suggest that by the mid-1850s at latest, some officers outside Prussia had taken up wargaming and were using the readily available *Kriegsspiel* rules.[90] Of course, in their case wargaming was not an activity that was officially encouraged in the case of officers serving in regular garrisons or even mandatory for those attending a military academy – it was simply a hobby. Only the wars of 1866 and 1870/71 would finally change this, as will be discussed below.

The Golden Age – 1862 to c. 1875

In 1862, *Kriegsspiel* history entered what can arguably be called a golden age. In the years that followed the publication of a new set of rules in 1862, there were not only eight further sets of rules produced in Prussia itself, the *Kriegsspiel* also finally succeeded beyond Prussia's borders, and by 1875 many first world armies were either already actively engaged in wargaming or on the verge of introducing it. It is due to these years that 150 years later wargaming is widely seen as an indispensable tool in military establishments all across the world – which raises two important questions: why did the level of *Kriegsspiel* activities increase so dramatically after 1862, and what was the reason for the *Kriegsspiel* finally being adopted almost in a hurry by armies all across the globe, armies which had been decidedly uninterested during earlier years?

While the latter question will be addressed in more detail below, the answer to the former is directly related to the technological progress, which had been a defining characteristic of *Kriegsspiel* history already in the decades following its introduction into the Prussian Army. By the early 1860s, the pace of that progress had further increased, with the army transitioning from smoothbore muzzle-loading artillery to breechloaders in the mid-1860s and a new rifle yet again increasing the fighting power of the infantry from 1871 onward. In addition, battlefield communication, which until well into the 1860s had relied on the centuries-old method of carrying orders by mounted messengers, was revolutionized by the introduction of the telegraph.[91] For the *Kriegsspiel* these developments meant that the challenge of obsolescence was more pressing than ever, requiring constant updating of rules.

It fell again to a young lieutenant to bring up the first set of rules. Wilhelm von Tschischwitz, who, unusually for *Kriegsspiel* inventors who mostly had an

89 Anonymus 1828b.
90 For example, Austrian army reformer Friedrich von Beck-Rzikowski had apparently learned about the *Kriegsspiel* in the early 1850s and was an early Austrian proponent of wargaming.
91 Wintjes 2017, 15.

artillery background, served with a Silesian infantry regiment, produced a new set of rules in 1862.[92] His original intention was twofold: first, interest in the *Kriegsspiel* was constantly rising, and even the 1855 reprint of the 1846 Berlin rules seems to have been hard to come by a decade later. More important, however, was the need to update the rules, as the 1846 Berlin *Kriegsspiel* had accommodated neither the needle gun nor rifled breech-loading artillery, which by the early 1860s was entering service with the Prussian artillery.[93] The *Kriegsspiel* was once again in danger of becoming obsolescent, and Tschischwitz's main aim was updating the rules so as to represent the realities of a mid-1860s' battlefield. While working on the rules he also decided to prune them even further than the Berlin authors had done, eventually resulting in a publication of a mere 21 pages. Comparing the original *"Ur-Kriegsspiel"* of 1824/28, the 1848 Berlin *Kriegsspiel* and Tschischwitz's rules published in 1862 clearly shows how Tschischwitz quite energetically tried to simplify the rules in order to improve overall gameplay; that he decided not to get rid of the rather involved rules on burning down perishable pieces of the topography[94] – which, as various tests at this author's institution have shown, result in ever more frantic bouts of dice-throwing – is one of the peculiarities of his rules. In all however, the Tschischwitz rules seem to have been an outstanding success, and as the wargaming activity in the Prussian Army saw a marked increase in the years before the war of 1866, that activity must have been based mostly on the Tschischwitz rules. The war itself culminated after a brief campaign in Germany in which the unwillingness of Prussian commanders to stick to the overall plan was more than matched by the ineptitude of some of their opponents, and an equally brief campaign in Bohemia during which the Prussians managed to outmaneuver the Austrians, in the Battle of Königgrätz.[95] In a strange turn of events, the battle brought together a number of key individuals of the *Kriegsspiel*'s history – commanding the Prussian Army was of course one of the *Kriegsspiel*'s greatest supporters, Helmuth von Moltke; in the midst of some of the heaviest fighting was Gustav Weigelt, commanding the 7th division's artillery; and in the 2nd army young lieutenant Wilhelm von Tschischwitz led the 10th company of IR 23 into the battle. It is, then, a rather fitting coincidence that the only "traditional" *Kriegsspiel* map currently known to have survived is a map of the Königgrätz battlefield and its surroundings.

---

92 On Tschischwitz and his rules see Wintjes 2019.
93 See Wintjes 2017, 18–20 with further literature.
94 Tschischwitz 1862, 20.
95 Wawro 1996, 222–228.

As a result of his experiences, Tschischwitz produced a second edition of his rules in 1867,[96] with changes including a new layout for the tables determining the casualties caused by infantry and artillery fire and revised data for rifled artillery, the introduction of which the Prussian Army had hastened after the end of the war, having entered the campaign in Bohemia still with a large number of older smoothbore muzzleloaders on strength – and suffering at the hands of the better equipped Austrian artillery. The 1867 Tschischwitz rules again saw wide circulation, and as the decade drew to an end, the wargaming activities in the Prussian Army further intensified; it is probably fair to say that in the army that eventually went to France in 1870, nearly all officers had some experience with the *Kriegsspiel*, and many of them will have made that experience with the Tschischwitz rules. The popularity of the rules is attested by a third edition that was published in summer 1870, presumably directly before the outbreak of the war,[97] and a final fourth edition published in 1874, which was based on the experiences of the 1870/71 war and included several changes.[98] On the basis of his four editions alone, Wilhelm von Tschischwitz can already claim to be a key figure in the history of the *Kriegsspiel*, as no other *Kriegsspiel* author had comparable success with his rules; given that the Tschischwitz rules formed the basis for the first translations into English and French, which will be briefly discussed below, and thus stood at the beginning of the spreading of wargaming beyond Prussia's borders, Tschischwitz stands out as one of the most important individuals in the history of professional wargaming, second only, if at all, to Georg Heinrich von Reisswitz.

The great popularity of the *Kriegsspiel* in the Prussian Army, which reached a high water mark in 1874 on the 50th anniversary of its introduction, not only brought Wilhelm von Tschischwitz to prominence, it resulted also in the emergence of a competing set of rules authored by Thilo Wolf von Trotha.[99] Trotha, who published the initial version of his rules in 1870, followed by a corrected reprint in 1872 and a revised edition in 1874,[100] was an infantry officer just like Tschischwitz, but there the similarities seem to have ended. Trotha, who was 22 years older than Tschischwitz and had retired from active service in 1868, not only employed significantly different mechanisms but had also significantly different ideas about infantry combat, the latter being in some cases the reason for the former. Whereas Tschischwitz seems to have seen a distinct trend towards the

96 Tschischwitz 1867.
97 Tschischwitz 1870.
98 Tschischwitz 1874.
99 On Thilo Wolf von Trotha see Anonymus 1876.
100 The second edition of the Trotha wargame differed from the first only in a very small number of corrections.

firefight becoming paramount in infantry combat, Trotha argued for the firefight being merely the precursor to the decisive part of infantry action – close combat using cold steel.[101] While looking back from the twenty-first century it would be easy to praise Tschischwitz's position as visionary and conversely ridicule Trotha's as backward, one has to bear in mind that the discussion about the relative importance of the firefight and close combat was in the 1870s still very much undecided, and even as late as 1906 did the later French marshal Foch advocate the bayonet charge straight into the enemy as the only way of gaining victory on the battlefield.[102] Although Trotha's rules do not seem to have seen as widespread use within the army as had the Tschischwitz *Kriegsspiel*, they nevertheless occupy an important place in the overall history of professional wargaming as they, together with the Tschischwitz rules, served as a source for foreign wargame designers.

Apart from causing Tschischwitz, who had seen active service in France as one of IR 23s battalion commanders during the encirclement and siege of Paris,[103] and Trotha, who had been recalled from retirement to command a logistics base in Hessia,[104] to update their rules, the victory of 1870/71 as well as the successful campaigns of 1866 had a profound influence on the development of the *Kriegsspiel* in that they changed the character of the officers' corps considerably. When the *Kriegsspiel* was originally introduced to the Prussian Army in 1824, the army was steadily losing officers with combat experience, and by the 1840s, very few younger officers will have had any experience of commanding troops in the field, let alone in battle. The *Kriegsspiel* provided the opportunity to gain at least some token experience in moving forces around in space and gave some idea of the friction that could exist on the battlefield. By 1871, however, things had actually been reversed: two major conflicts separated by only four years will have enabled the vast majority of the Prussian officers to gain combat experience of some sort, causing them to view the *Kriegsspiel* in a very different way from their predecessors from the 1840s and 1850s. Participants could now relate the action on the map to their own experiences, and umpires will at times have been sorely attempted to simply decide over the outcome of actions based on their personal experience – and not merely in the case of nerve-racking fires raging on the map. The ability of participants and – much more important – umpires to contribute their own experiences to the *Kriegsspiel* set the stage for a fundamental discussion

---

101 Trotha 1870, 5.
102 "Les lauriers de la victoire flottent à la pointe des baionnettes ennemies. C'est là qu'il faut aller les prendre, les conquérir par une lutte corps à corps, si on les veut." (Foch 1903, 320–321).
103 Wintjes 2019, 17–18.
104 Anonymus 1876, 247.

of its character and overall setup, a discussion that would flare up in the mid-1870s and result in the next important developmental step in the history of professional wargaming.

The Changing *Kriegsspiel* – c. 1875 to 1914

Despite the attempts first by the Berlin wargamers and then by Tschischwitz, the *Kriegsspiel* had always retained a fairly complex set of core rules and mechanisms. And while the participants did not need to have any deeper knowledge of the rules, such knowledge was indispensable for the umpires, without which a game could not be run. In the changing army of the post-1871 world, this had two very different but equally important consequences. On the one hand did the widespread, recent combat experience many officers had make them naturally inclined to compare every decision by the umpire with that experience and to ascribe adverse turns of event to a perceived "unrealism" of the rules – instead to their own mismanagement of affairs; in other words, the umpires, and with them the *Kriegsspiel* as a whole, slowly started to lose authority. On the other hand were officers serving as umpires equally inclined to judge the outcome of combat as stipulated by the – complicated – rules against their own experience, and when they came to the conclusion that the results produced by the game ran contrary to their own experiences, they called into question the necessity of a complex set of rules – which not only made the running of wargames difficult, but in the end would serve only to produce "unrealistic" results. Clearly the *Kriegsspiel* had arrived at a crucial point in its history, and it is a testimony to the overall importance of the instruments to the Prussian Army that there were no calls for actually abandoning the practice. Instead, the mid-1870s saw energetic attempts at reforming the *Kriegsspiel,* which had lasting effects both on it scope and on how it was actually run.

In 1873, Jacob Meckel, who as a young lieutenant with IR 68 had also taken part in the Battle of Königgrätz before serving with IR 82 in France in 1870, published a study on the *Kriegsspiel,* which raised some of the issues mentioned above.[105] After the war, Meckel had been posted as an instructor to the Hanover *Kriegsschule* before being transferred to the Berlin military academy, where he soon acquired a reputation as a gifted tactician. Consequently, his ideas on the *Kriegsspiel* soon found supporters, with the later Prussian minister of war Julius Verdy du Vernois being the most vocal one, publishing his own study on the

---

105 Meckel 1873; on the biography of Jacob Meckel see Kerst 1970.

Kriegsspiel in 1876.[106] In his study, Meckel suggested two key innovations to be made to the existing rules: One, instead of relying on a core of complex rules for resolving combat, umpires should henceforth simply decide over the outcome based on their own experience. The advantages in terms of gameplay were obvious: no longer was there any need of consulting tables – and of knowing how to consult them in the first place! – indeed, basically all the problems wargame designers had struggled with ever since the late eighteenth century with regard to the resolution of combat were solved in an instant; invariably, the *Kriegsspiel* could be run faster and much more smoothly. Verdy had called this way of running the game the *abgekürztes Verfahren*; it later became widely known as *Freies Kriegsspiel*, as it was freed from the constrains of the "traditional" rules.[107] The other key innovation was to move the *Kriegsspiel* beyond the depiction of combat on a tactical level. While there is some evidence for operational-level wargames being played in the Prussian Army as early as the late 1840s,[108] the *Kriegsspiel* itself was clearly designed with decisions on a tactical level in mind, illustrated for example by the existence of tokens designed to represent individual persons. While Meckel still considered the *Kriegsspiel* to be eminently useful for teaching battlefield tactics, he advocated the development of similar, yet slightly different sets of *Apparate* for wargames on a larger scale; basically, he suggested three *Kriegsspiele* to be introduced: a tactical game called the *Regiments-* or *Detachement-Kriegsspiel*, which was roughly to the scale of the old *Kriegsspiel* (though Meckel preferred 1/6250 to the old wargaming scale of 1/8000[109]); an operational game, the *Großes Kriegsspiel*, making the depiction of campaigns like the Prussian foray into Bohemia in 1866 possible; and a strategic wargame – unsurprisingly called *Strategisches Kriegsspiel* – for even larger scenarios. A year after his initial publication, Meckel published specifics on his idea for these new *Apparate* in 1874, before near the end of that same year finishing work on his own rules, which appeared early in 1875.[110]

Meckel's ideas on the *Kriegsspiel* proved to be instantly popular, for several reasons. For once, they found quasi-official endorsement when 1876 Julius Verdy du Vernois published his study on the *Kriegsspiel*, claiming the future to belong to the "free" *Kriegsspiel*; it is fairly safe to assume that as a result, the wargaming activity at the Berlin military academy as well as at many *Kriegsschulen* switched from the "traditional" to the new, "free" *Kriegsspiel*. At the same time, the "traditional" *Kriegsspiel* lost one of its most ardent supporters when Thilo Wolf von Trotha died

---

106 Verdy 1876.
107 Verdy 1876, viii.
108 Beutner 1889, 308–309.
109 Meckel 1874, 4–6.
110 Meckel 1875.

suddenly and unexpectedly in the same year.[111] Then, the abolishment of a complex set of rules and mechanism made scaling the *Kriegsspiel* up from the tactical level much easier – one can only imagine how complex "traditional" rules for a wargame on the scale of the Bavarian scenario mentioned in the introduction would have needed to be; instead, the umpires relied only on their own experience and some tables for the speed of marching columns. And finally, in an army awash with officers who had seen combat in one or even two major wars, the idea of the umpire basing his decision on his own experience rather than on a complex set of rules was guaranteed to catch on quickly. Whether however the new, "free" *Kriegsspiel* entirely replaced the old, traditional one, as was claimed both at the time by ardent supporters of the "free" *Kriegsspiel* and by modern interpreters, is rather questionable. While the rules published by Meckel were in official use up to the outbreak of the First World War, having been republished with slight revisions in 1903 by Ludwig von Eynatten,[112] other *Kriegsspiel* rules emerged even after 1876, which relied on "traditional" mechanisms, most notably the Regiments-*Kriegsspiel* published by an otherwise unknown lieutenant Naumann, which was already seen by contemporaries as a direct reaction to the "free" *Kriegsspiel*.[113] A further set of rules for the *Regiments-Kriegsspiel* would appear as late as 1901, published by the Hessian cavalry officer Carl von Zimmermann;[114] while acknowledging the practice of the "free" Kriegsspiel, Zimmermann advocated the continued use of dice for two reasons – they allowed to retain an element of luck or chance and offered the possibility of avoiding situations that were potentially socially awkward.[115]

These two reasons, which explain to some degree the reluctancy to let the "traditional" *Kriegsspiel* go, deserve closer attention. As for luck and chance, Meckel's reform had a significant impact on the character of the game. By giving up the complex, dice-driven rules, Meckel had also abolished the element of chance that lay in them. From his experience, that of a teacher at a military academy, this was probably of only small significance. Yet in fact it led to the *Kriegsspiel* changing somewhat from an open simulation to one entirely controlled by the umpires – which may have been useful in certain educational contexts, but which could also serve to stifle independent thinking as it gave the umpires an extremely powerful tool at hand with which to impress their view on how things should be done on the participants; at the same time the degree of uncertainty that lay within the

---

111 Anonymus 1876, 247.
112 Meckel/Eynatten 1903.
113 Anonymus 1878, 154–155; apparently, Naumann's Kriegsspiel gathered sufficient attention to merit the publication of a revised edition, which appeared in 1881.
114 Zimmermann 1901.
115 Zimmermann 1901, 38.

"traditional" rules was replaced with the firm conviction of the umpires', which, while presumably based on their own combat experience, made them to a certain degree calculable – which makes one wonder at which point participants tried to out-think the umpires instead of their opponents. A present-day example this author came across when running a "traditional" *Kriegsspiel* based on the 1867 Tschischwitz rules at the *Führungsakademie der Bundeswehr* may illustrate the problem:[116] In a divisional-level scenario, a team of participants decided to employ a battalion of light infantry in an extremely unorthodox way; any respectable Prussian officer would probably have instantly decided against this course of action, and going by the "traditional" rules the chances of success were tenuous at best. Yet due to a succession of exceptional dice throws the battalion ended up successfully ambushing a large number of enemy cavalry, resulting in a spirited discussion on the relative merits of doing things "by the book" or not. In a "free" *Kriegsspiel*, such a situation would never have arisen, and as a consequence the ensuing discussion would never have taken place. Whatever the undeniable practical merits of the "free" *Kriegsspiel*, it was – and is – ultimately much more suited to confirming its participants in what they already believe to know than to making them question their own beliefs and actions.

Another reason why wargaming clubs in particular may have stuck longer with the "traditional" rules is the social issue raised by Zimmermann, which was caused by making the umpire the sole judge of anything that happened on the map: While this was not a problem in the educational context of for example a military academy where senior officers would teach younger officers, one wonders how wargamers elsewhere handled the problem of a potentially junior umpire calling the actions of one of his superiors into question. The *Kriegsspiel*, which was originally meant to be as much of an open, intellectual exercise as anything else, was in the danger of becoming an instrument for the most senior officer in the room to force his opinion on everybody else. As information on the activities of wargaming clubs in the German army in the decades after the war of 1870/71 is currently almost completely absent, it is unclear how this problem was solved.

The last quarter of the nineteenth century saw, apart from the establishment of the "free" *Kriegsspiel*, two other developments of considerable importance for the history of professional wargaming. One – the spreading of the *Kriegsspiel* beyond Prussian borders – will be discussed in the following section, the other deserves brief mentioning here. It was noted above that Meckel introduced the idea of applying the *Kriegsspiel* to different scales, using it for tactical, operational as well as strategic purposes. Over the next two decades, this would not remain the only diversification, so to speak, of the *Kriegsspiel*. Rather, it moved both into a

---

116 See also Wintjes/Pielström 2019.

different dimension, with the first naval wargames appearing in England in the 1870s,[117] and into a slightly different genre – the *Festungskriegsspiel*, or fortress wargame, appeared which basically tried to depict the rather static activity of a siege. A first set of rules was published in 1872 by Paul Louis von Neumann, a Prussian artillery officer and the son of Rudolf Sylvius von Neumann, one of Prussia's leading artillerymen of the mid-nineteenth century.[118] Neumann, who in 1872 served as an instructor at the *Artillerie-Schießschule*, saw his *Kriegsspiel* primarily as an instrument for teaching younger officers the intricacies of attacking and defending a fortified place.[119] As a consequence, he did not provide a complex set of rules, relying instead on the umpires in much the same way as the proponents of the "free" *Kriegsspiel*. Given its subject the "Apparat" of the *Festungskriegsspiel* differed considerably from that of the "standard" *Kriegsspiele*; played on a map to the scale of 1/2500, the gaming materials included tokens for various types of trenches, gun mountings, ammunition facilities, and even craters caused by the explosion of mines. While little information survives on how widespread the use of Neumann's rules actually was, the general concept of a *Festungskriegsspiel* certainly caught on, with the Bavarian general staff running such wargames right until the outbreak of First World War.[120]

Kriegsspiel Beyond Borders – 1871 to 1914

The wars of 1866 and 1870/71 not only had a profound impact on the history of the *Kriegsspiel* in that they fundamentally changed the character of its main audience from a group largely lacking combat experience to one largely lacking members without combat experience, they were also instrumental in wargaming finally gaining wide acceptance in other armies. Already in 1866 most contemporary observers had been surprised by the rapid and total success of Prussian forces moving swiftly through Bohemia and beating the Austrian Northern Army decisively at the Battle of Königgrätz. Prussia, which up to that point had fairly little to its credit in terms of military successes, had managed to defeat what many had considered to be one of the major European powers, even if that power had faced more than one serious crisis in the two decades preceding the war. It should therefore not come as a surprise that, when barely four years later Prussian forces

---

117 For one of the earliest naval wargames see Colomb 1879; the history of early naval wargaming is by and large still unexplored.
118 Neumann 1872.
119 Neumann 1872, 1–2.
120 For the last *Festungskriegsspiel* about which information survives see BayHStA Generalstab 1315.

together with their allies – among them elements of the Bavarian Army, which in 1866 had fought the Prussians – marched into France, many contemporary observers expected the French Army to prevail easily, as it was seen as the most professional and most experienced army in continental Europe. In addition to that its infantry was equipped with the Chassepot rifle, the latest in small arms providing infantrymen with hitting power against targets at twice the range of the Prussian needle rifle, and – unknown to most foreign observers – in the mitrailleuse, an early machine gun, the French Army had a key trick up its sleeve with which to break infantry assaults for good. Instead, with the French war effort severely hampered by inefficiency and outright chaos during the initial stages of the war, France was rapidly defeated, the newly formed German Empire replacing it in its position as continental Europe's foremost military power.

Many contemporary observers must have been profoundly irritated. And already before the actual end of hostilities, attempts at understanding the reasons for the resounding Prussian success soon identified several key aspects of the Prussian way of waging war that had contributed significantly to the outcome of the war – quite apart from such trivialities as Prussian breech-loading field artillery, being a generation ahead of its French counterpart and consequently able to pummel French positions with impunity. From a twenty-first century point of view, many observations appear to be glaringly obvious – for example the realization that staff officers who had already learned in peacetime to work together as a team could easily outperform a staff put together only at the commencement of hostilities; or the general usefulness of a general staff as a standing organization both in times of war and in times of peace. As a consequence, staff work – and the preparation of officers for it – moved into the focus of military reformers in armies across Europe. Below the level of army or army corps HQs, observers were soon convinced that Prussian officers had also an edge in general preparation, which is where the *Kriegsspiel* came into the equation. Wargaming was seen as an eminently useful instrument to make officers face the uncertainties of battle and to teach them not only map-reading skills and an understanding of the difficulties of moving troops in a confined space, but also how to process information and to write orders properly.

The spreading of the practice of wargaming throughout armies in Europe and beyond is every bit as interesting as the history of the *Kriegsspiel* itself; however, it is at present even more under-researched than the latter. For the purpose of the present chapter, only the introduction of wargaming to the British army will therefore be covered in some detail, whereas only very brief mention will be made of how other armies took up the *Kriegsspiel*. While the small professional British army differed in many respects from large, conscription-based continental armies like the Prussian or French one, it was – together with Italy and Belgium – among the earliest adopters of the *Kriegsspiel*; compared to other armies introducing wargaming in the 1870s, the British army had an interesting direct connection to the developmental history of the Prussian *Kriegsspiel*: During the war of 1870/71 an experienced artillery officer by the name of Rudolf von Roerdansz served as the military attaché in London, apparently keeping the military circles in the British capital well-informed about what happened on the ground in France. Roerdansz,[121] in one of the peculiar turns of events so typical for the history of professional wargaming in the nineteenth century, was the son-in-law of another eminent artilleryman in the Prussian Army, Gustav Weigelt – who around 20 years earlier as a young officer of the Magdeburg garrison had produced the first of the new *Kriegsspiel* rules. While the degree to which Weigelt still took an inter-est in the development of the *Kriegsspiel* is difficult to gauge, his son-in-law was clearly well-versed in it, giving a lecture at the Royal United Services Institute late in 1871 describing both the rules themselves and the use of wargaming in the Prussian Army. Roerdansz's lecture was not only well received, it also gathered considerable media attention, with several newspapers covering it in detail.[122] Apparently, it took little further convincing, as by the end of the year the decision must have been made to introduce wargaming to the British army. A few months later in March 1872 Prince Arthur, the Duke of Connaught and Strathearn, who had learned about the *Kriegsspiel* in Germany and taken great interest in it, gave another highly publicized talk to officers of the Dover garrison.[123] Prince Arthur not only described the *Kriegsspiel* but ran a presentational game which, while probably not the very first wargame run in Britain, is at present the earliest about which any details are known; in accordance with the then-emerging scare of a possible continental invasion, the scenario assumed the landing of an enemy force in the vicinity of Hythe, pushing toward Dover, and a British force based on Dover trying to outmaneuver the enemy between Folkestone and Dover.[124] By the time Prince Arthur ran his game in the officers' mess at Dover Castle, the *Kriegsspiel* had already been officially introduced to the British army for about four weeks, yet the first regular wargaming activity seems to have been taken up in Aldershot only after Prince Arthur's talk.

In introducing the *Kriegsspiel*, the British faced the same fundamental problem everybody else outside Prussia had to deal with – the creation of rules suitable for the use in the respective armies. In the case of the British army – as well as several others – the translation of the current Prussian set of rules was seen as the most

---

121 On the biography of Roerdansz see Kleist 1895, 732.
122 *Pall Mall Gazette*, 23 December 1871, 10; see also *The Broad Arrow*, 23 December, 826 and 30 December, 841-842; reports of the Roerdansz lecture were also reprinted overseas, see e.g. *The Adelaide Observer*, 16 March 1872, 3.
123 For detailed accounts see e.g.*Glasgow Herald*, 14 March 1872, 5 or *The Broad Arrow* 16 March 1872, 161.
124 *The Broad Arrow* 16 March 1872, 161.

efficient solution. Apparently already late in 1871 the British military attaché in Berlin was ordered to procure a set of rules, materials, and maps so that the process of translation could commence. As noted above, however, in the early 1870s an intensive discussion of the *Kriegsspiel* rules had begun in Prussia, and while the concept of a "free" *Kriegsspiel* lay still in the future, there were two competing sets of rules on the market by 1871 – the Tschischwitz rules originally published in 1862 and republished in 1867 and 1870, and the Trotha rules published for the first time in 1870. The army decided to go with the Tschischwitz rules, and Evelyn Baring, the later Lord Cromer, at the time a young Royal Artillery captain serving with the topographical and statistical department of the War Office, was tasked with producing a translation. In doing so, he was apparently supported by Rudolf von Roerdansz, and in mid-February 1872 the *Rules for the Conduct of the War-Game* were published.[125] The first publication includes a note by the then-adjutant general to the forces, Richard Airey, noting the order given by the secretary of war as well as the commander in chief of the British army for every officer to "avail himself to the utmost of this useful means of instruction."[126] Given that the British army would in the following years put particular stress on the employment of wargaming for teaching officers how to write orders, one cannot deny a certain historical irony to the fact that with Richard Airey the introduction of the *Kriegsspiel* was overseen by the very man who had written and signed one of the British army's most infamous orders of the nineteenth century, which nearly 18 years earlier had set one of the finest bodies of British light cavalry on a direct course both to glory and the grave.[127]

The further history of professional wargaming in the British army is beyond the scope of this paper; in the decades following the 1870s, the history of wargaming in the British army would eventually run through several different phases characterized both by a waning of official interest as well as an increase of wargaming activities in the volunteer movement from the latter half of the 1880s onward, which continued right to the beginning of the First World War. While the British Army thus was an early adopter of wargaming – if something like that can be said for an activity practiced by the Prussians for half a century – others soon followed suit. The Italian and Belgian armies have already been mentioned; the introduction of wargaming to the latter saw the first translation of the Tschischwitz rules into the French language; these rules then served as the basis for the 1874 introduction of wargaming into the French Army.[128] It should be noted that,

---

125 Baring 1872.
126 Baring 1872, 2.
127 The literature on the famous Charge of the Light Brigade is vast; still eminently useful as an introduction is Woodham-Smith 1953.
128 Petre 1872.

just as in the British case, the translation was not a literal one; rather, the rules originally designed for the Prussian Army were adapted to the army in question, resulting in rules that did not betray their "Prussian pedigree" but differed in key respects from each other just as the army organizations using them did. Already a year before the introduction of the *Kriegsspiel* to the French Army did the Russian Army take up wargaming with a first publication of a set of rules in 1873.[129] In Austria, it did not take much longer to introduce the *Kriegsspiel*: After a key publication in 1874 inspired both by the "traditional" rules and the new "free" *Kriegsspiel* advocated by Verdy du Vernois and Meckel, its use was officially endorsed by the army in 1876, and while its introduction was met with some of the same criticism the original *Kriegsspiel* had faced in the Prussian army in the 1830s and 1840s, younger officers seem to have taken to it with considerable enthusiasm.[130] Just as in the wake of the successes of 1866 and 1870/71 Prussian Army fashion influenced not only the larger European armies but arrived even at arguably rather smallish establishments such as the Papal guards, who from 1872 onward began to wear *Pickelhaube* helmets, wargaming was introduced in smaller armies as well; thus in 1881 the Spanish army published its own set of rules,[131] again based on a translation of Prussian rules, the author combining elements from the Tschischwitz and Trotha *Kriegsspiele*. Across the Atlantic, the US army took up wargaming in the early 1880s as well, with William Roscoe Livermore producing a first set of rules in 1883, again combining elements of the Tschischwitz and Trotha rules.[132] Whereas the armies mentioned above all adapted the Prussian rules to some extent to the organization and equipment of their own armies, the Swiss apparently did not see such a need and had by the mid-1870s used the original Prussian *Kriegsspiel*[133] rules for some years.

In all, during the decades following the Prussian victory of 1870/71, the concept of wargaming proved to be the greatest export success of the Prussian Army – a success that went beyond the mere spreading of ideas and concepts; a contemporary handbook pointedly noted in 1875 that the Voß'sche Buchhandlung, the main seller of *Kriegsspiel* rules and materials, was doing excellent business, selling examples to places as far away as Japan.[134] By 1914, then, all major armies across the globe had adopted wargaming in some way or another.

---

129 Löbell 1875, 727.
130 Bilimek 1883, 1–2.
131 Ramos 1881.
132 Löbell 1875, 727.
133 Löbell 1875, 727.
134 Löbell 1875, 728.

Conclusion

By the time the Bavarian general staff was wargaming its what-if defense against a Prussian invasion on the eve of First World War, wargaming was as indispensable a tool for military training and operational planning as it is at the beginning of the twenty-first century. The history of professional wargaming, or to be more specific the history of the Prussian *Kriegsspiel* was one of an outstanding success, given how quickly it was eventually adopted by armies all across the globe. What makes this success story all the more interesting is that for about half a century, it was actually nothing of the sort – throughout the decades following the 1820s the *Kriegsspiel*, even after it had overcome opposition in the Prussian Army and found wide acceptance there, was essentially an activity undertaken by a second-rate power with little to show for other than a glorious past; if anything, the *Kriegsspiel* was an expression of a certain Prussian quirkiness more than anything else. That professional wargaming had taken hold in the Prussian Army itself was the result of circumstances in which chance played not a small role: Not only took it an officer of almost Holmesian peculiarity to invent it, but it found in Müffling possibly the one officer in the Prussian Army who was most inclined to fall for something that involved topographical maps; it then managed to fascinate a young officer who had seen service in the Danish Army and was himself a rather interesting character – being once characterized as a man who could be silent in five different languages – and would eventually turn out to become Prussia's greatest general with a formative influence on the Prussian Army throughout the latter half of the nineteenth century. Moreover, the *Kriegsspiel* appeared at a time when not only the Prussian officers' corps was slowly changing its composition, with the number of officers with combat experience receding, but when technical arms pressed for a greater professionalization of "soldiering" in general and "officering" in particular; it is hardly surprising that it was mostly artillery officers who were engaged in the development of the *Kriegsspiel* as they increasingly sought to stress the scientific aspect of their trade. Finally, in a century in which armies all across Europe and beyond were driven by technological progress, often only able to react to a seemingly never-ending stream of new inventions that made obsolescent what only yesterday had been state-of-the-art technology, Prussian *Kriegsspiel* designers put great effort into keeping rules and tables up to date. It is a clear testimony to the importance the Prussians assigned to the *Kriegsspiel* that an instrument originally designed for simulating Napoleonic warfare was constantly adapted to the ever-changing realities on the battlefield.

Of course, the ultimate proof of the pudding, as they say, is only in the eating. All the developments within the Prussian Army sketched out above would have counted for nothing when it came to other armies taking over professional wargaming, had the Austrians thoroughly thrashed the Prussians in 1866 or the

French managed to march to Berlin in 1870. It was due to the Prussian successes in these two wars more than to anything else that professional wargaming owed its worldwide success. Foreign observers, astounded by the unexpected, swift success of Prussian arms, quickly came to the conclusion that the way the Prussians went about in war had much that was worth transferring to their own armies. And among other things like organized staff work, foreign observers identified the use of wargaming as one element responsible for the Prussian successes. While it is for various reasons – the lack of suitable sources being perhaps the most important one – impossible to estimate how much the *Kriegsspiel* actually contributed to the victories of 1866 and 1870/71, there can be no doubt about how contemporary observers thought of its importance at the time, and from the early 1870s onward, wargaming was seen as an activity of great importance in nearly all armies.

Given the way this brief overview has begun with an operational-level wargame of the Bavarian general staff, it is perhaps fitting to end it again with the Bavarian Army. During the brief period between its return from the battlefields of First World War to its garrisons and its final disbanding in the latter half of 1919 took up its peacetime activities, despite the turmoil Bavaria – and Germany – found itself in[135]; accordingly, the file on Bavarian Army staff wargames closes with a brief sketches of what is likely the last wargame ever run by the Bavarian Army, covering a Czech invasion of Silesia[136]; the scenario description showed clear signs of its time, including in the notes on the combat effectiveness of various units remarks on their political reliability.[137]

Works Cited

Anonymus 1878: Anonymus, Das Regiments-Kriegsspiel, *Militair-Wochenblatt* 63, 1878, 154–156.mes: From Gladiators

Anonymus 1876 Anonymus, Thilo v. Throtha [obituary], *Militair-Wochenblatt* 61, 1876, 247–248.

Anonymus 1870 Anonymus, Anleitung zum Gebrauch des Kriegsspiel-Apparats zur Darstellung von Gefechtsbildern, mit Berücksichtigung der jetzt gebräuchlichen Waffen, von T. v. Trotha. [review], *Militair-Wochenblatt* 55, 1870, 100–101.

Anonymus 1846 Anonymus, *Anleitung zur Darstellung militairischer Manöver mit dem Apparat des Kriegs-Spiels*, Berlin 1846.

Anonymus 1839 Anonymus, Preußen, *Allgemeine Militär-Zeitung* 62, 489–492.

---

135 For an overview see Zorn 1986, 145–148, 168–185.
136 BayHStA Generalstab 1479.
137 For example, ID 5 personnel was characterized as durch Berliner ziemlich verseucht ("contaminated by Berlin personnel") (BayHStA Generalstab 1479).

Anonymus 1828 Anonymus [Carl von Decker and Klamor August Ferdinand von Witzleben], Supplement zu den bisherigen Kriegsspiel-Regeln, *Zeitschrift für Kunst, Wissenschaft und Geschichte des Krieges* 13, 1828, 68–105.

Aretin 1830 Wilhelm Freiherr von Aretin, *Strategonon. Versuch, die Kriegführung durch ein Spiel anschaulich darzustellen*, Ansbach 1830.

Baring 1872 Evelyn Baring, *Rules for the Conduct of the War-Game*, London 1870.

BayHstA GenStab Bayerisches Haupt- und Staatsarchiv, Akten des kgl. bayerischen Generalstabs.

Berger 2000 Bernhard Berger, Gespielte Vorbereitung auf den Ersten Weltkrieg. Die operativen Kriegsspiele Österreich-Ungarns, *Österreichische Militärische Zeitschrift* 38, 595–604.

Beutner 1889 Friedrich Wilhelm Beutner, *Die Königlich Preußische Garde-Artillerie, insbesondere Geschichte des 1. Garde-Feld-Artillerie-Regiments und des 2. Garde-Feld-Artillerie-Regiments. Erster Band*, Berlin 1889.

Bilimek 1883 Hugo Ritter von Bilimek-Waissolm, *Die Leitung des Kriegsspieles und die Grenzen seiner Mittel*, Wien 1883.

Boissière 1698 Gilles de La Boissière, *Le Jeu de la guerre ou tout ce qui s'observe dans les Marches et campements des armées, dans les batailles, combats, sièges et autres actions militaires, et exactement représenté avec les définitions et les explications de chaque chose en particulier*, Paris 1698.

Citino 2005 Robert M. Citino, *The German Way of War*, Lawrence/KS 2005.

Colomb 1879 Philip Howard Colomb, *The Duel: a Naval War Game*, Portsmouth 1879.

Dannhauer 1874 Ernst Heinrich Dannhauer, Das Reißwitzsche Kriegsspiel von seinem Beginn bis zum Tode des Erfinders, *Militair-Wochenblatt* 59, 1874, 527–532.

Deuringer 1922 Karl Deuringer, *Die Schlacht in Lothringen und in den Vogesen 1914: Die Feuertaufe der Bayerischen Armee (2 Bde.)*, München 1929.

Foch 1903 Ferdinand Foch, *Des principes de la guerre : conférences faites à l'École supérieure de guerre*, Paris 1903.

General Staff 1878 Königlich Sächsischer Generalstab (ed.), *Katalog der Bibliothek und Karten-Sammlung des Königl. Sächs. Generalstabs*, Dresden 1878.

Hellwig 1803 Johann Christian Ludwig Hellwig, *Das Kriegsspiel: Ein Versuch, die Wahrheit verschiedener Regeln der Kriegskunst in einem unterhaltenden Spiele anschaulich zu machen*, Braunschweig 1803.

Hellwig 1782 Johann Christian Ludwig Hellwig, *Versuch eines aufs Schachspiel gebaueten taktischen Spiels. Praktischer Teil*. Leipzig 1782.

Hellwig 1780 Johann Christian Ludwig Hellwig, *Versuch eines aufs Schachspiel gebaueten taktischen Spiels*. Leipzig 1780.

Hilgers 2008 Philipp Hilgers, *Kriegsspiele. Eine Geschichte der Ausnahmezustände und Unberechenbarkeiten*, München 2008.

Hoffmann 1981 Jochen Hoffmann, *Jakob Mauvillon. Ein Offizier und Schriftsteller im Zeitalter der bürgerlichen Emanzipationsbewegung*, Berlin 1981.

Hohrath 2000 Daniel Hohrath, Prolegomena zu einer Geschichte des Kriegsspiels, in Angela Giebmeyer/Helga Schnabel-Schüle (ed.) *„Das Wichtigste ist der Mensch". Festschrift für Klaus Gerteis zum 60. Geburtstag*, Mainz 2000, 139–152.

Imhof 1965 Eduard Imhof, *Kartographische Geländedarstellung*, Berlin 1965.

Kerst 1970 Georg Kerst, *Jacob Meckel. Sein Leben, sein Wirken in Deutschland und Japan*, Göttingen 1970.

Knoll 1981 Werner Knoll, Die Entwicklung des Kriegsspiels in Deutschland bis 1945, *Militärgeschichte* 20, 1981, 179–189.

Lamer 1927 Hans Lamer, Lusoria tabula, *Realencyclopädie der Classischen Altertumswissenschaften* 13.2, 1927, 1900–2029.

Löbell 1875 Heinrich von Löbell, *Jahresberichte über die Veränderungen und Fortschritte im Militairwesen 1874*. Berlin 1875.

Mauvillon 1822 Friedrich W. Mauvillon, *Ueber die Versuche, die Kriegsführung durch Spiele anschaulich darzustellen, und deren Anwendung zum Unterricht in Militairschulen*, Militairische Blätter 3, 1822, 289–344.

McHugh 1966 Francis McHugh, *The United States Naval War College Fundamentals of War Gaming (3rd ed.)*, Newport/RI 1966.

Meckel 1873 Jacob Meckel, *Studien über das Kriegsspiel*, Berlin 1873.

Meckel 1874 Jacob Meckel, *Der verbesserte Kriegsspiel-Apparat*, Berlin 1874.

Meckel 1875 Jacob Meckel, *Anleitung zum Kriegsspiele. Erster Teil: Direktiven für das Kriegsspiel*, Berlin 1875.

Meckel/Eynatten 1903 Fritz von Eynatten, *Anleitung zum Kriegsspiel von Meckel, neu bearbeitete Auflage*, Berlin 1903.

Meerheimb 1877 Richard von Meerheimb, Decker, Karl von, *Allgemeine Deutsche Biographie* 5, 1877, 8–10.

Müller 1873 Hermann von Müller, *Die Entwickelung der Feld-Artillerie in Bezug auf Material, Organisation und Taktik, von 1815 bis 1870*, Berlin 1873.

Neumann 1872 G. Neumann, *Directiven für das Festungs-Kriegs-Spiel*, Berlin 1872.

Niemeyer 1997 Joachim Niemeyer, Müffling, Karl, *Neue Deutsche Biographie* 18, 1997, 266–267.

Nohr/Böhme 2009 Rolf F.Nohr/Stefan Böhme, *Johann C. L. Hellwig und das Braunschweiger Kriegsspiel*, Braunschweig 2009.

Opiz 1806 Giacomo E. Opiz, *Das Opiz'sche Kriegsspiel, ein Beitrag zur Bildung künftiger und zur Unterhaltung selbst der erfahrensten Taktiker. Ausführlich beschrieben von dem Erfinder Johann Ferdinand Opiz*, Halle 1806.

Perla 1990 Peter Perla, *The Art of Wargaming*, Annapolis 1990.

Peterson 2016 Jon Peterson, A Game Out of All Proportions: How a Hobby Miniaturized War, in: Pat Harrigan/Matthew G. Kirschenbaum (ed.), *Zones of Control. Perspectives on Wargaming*, Cambridge 2016, 3–31.

Peterson 2012 Jon Peterson, *Playing at the World. A History of Simulating Wars, People and Fantastic Adventures from Chess to Role-Playing Games*, San Diego 2012.

Petre 1872 Augustin Petre, *"Kriegsspiel", jeu de la guerre, guide des opérations tactiques exécutées sur la carte*, Bruxelles 1872.

Pias 2000 Claus Pias, *Computer Spiel Welten*, PhD dissertation, Weimar 2000.

Poten 1884 Bernhard von Poten, Mauvillon, Friedrich Wilhelm von, *Allgemeine Deutsche Biographie* 20, 1884, 714–715.

Ramos 1881 D. Máximo Ramos, *El Juego de la Guerra*, Madrid 1881.

Reichenau 1879 Ernst August von Reichenau, *Über Handhabung und Erweiterung des Kriegsspiels*, Berlin 1879.

Reiswitz 1812 Georg Leopold von Reiswitz, *Taktisches Kriegs-Spiel; oder, Anleitung zu einer mechanischen Vorrichtung um taktische Manoeuvres sinnlich darzustellen*, Berlin 1812.

Reisswitz 1824 Georg Heinrich von Reisswitz, *Anleitung zur Darstellung militairischer Manöver mit den Mitteln des Kriegsspieles*, Berlin 1824.

Scharnhorst 1813 Gerhard von Scharnhorst, *Über die Wirkung des Feuergewehrs*, Berlin 1813.

Solms 1562 Reinhard Graf zu Solms, *Kriegsbeschreibung voll. I-IX*, Lich 1559-1562.

Stapf 1680 Johann Ulrich Stapf, *Das Festung Baues Spiel*, Augsburg 1680.

Troschke 1869 Theodor von Troschke, Zum Kriegsspiel, *Militair-Wochenblatt* 54, 1869, 276–277 and 292–295.

Trotha 1870 Thilo Wolf von Trotha, *Anleitung zum Gebrauch des Kriegsspiel-Apparates zur Darstellung von Gefechtsbildern*, Berlin 1870.

Tschischwitz 1862 Wilhelm von Tschischwitz, *Anleitung zum Kriegsspiel*, Neisse 1862.

Tschischwitz 1867 Wilhelm von Tschischwitz, *Anleitung zum Kriegsspiel. Zweite, verbesserte Auflage*, Neisse 1867.

Tschischwitz 1870 Wilhelm von Tschischwitz, *Anleitung zum Kriegsspiel. Dritte verbesserte Auflage*, Neisse 1870.

Tschischwitz 1874 Wilhelm von Tschischwitz, *Anleitung zum Kriegsspiel. Vierte verbesserte Auflage*, Neisse 1874.

Van Crefeld 2013 Martin van Crefeld, *Wargames: From Gladiators to Gigabytes*, Cambridge 2013.

Vego 2012 Milan Vego, German War Gaming, *Naval War College Review* 65, 2012, 106–147

Venturini 1797 Johann Georg Julius Venturini, *Beschreibung und Regeln eines neuen Krieges- Spiels, zum Nutzen und Vergnügen, besonders aber zum Gebrauch in Militair-Schulen*, Schleswig 1797.

Venturini 1803 Johann Georg Julius Venturini (+), *Vervollkomnete Darstellung des von ihm erfundenen Kriegsspiels*, Braunschweig 1803.

Verdy du Vernois 1876 Julius Adrian von Verdy du Vernois, *Beitrag zum Kriegsspiel*, Berlin 1876.

Wahl 1798 Samuel Friedrich Günther Wahl, *Der Geist und die Geschichte des Schach-Spiels bei den Indern, Persern, Arabern, Türken, Sinesen und übrigen Morgenländern, Deutschen und andern Europäern*, Halle 1798.

Wawro 1996 Geoffrey Wawro, *The Austro-Prussian War: Austria's War with Prussia and Italy*, Cambridge 1996.

Weickmann 1664 Christoph Weickmann, *Newerfundenes grosses Königsspiel*, Ulm 1664.

Wintjes/Pielström 2019 Jorit Wintjes/Steffen Pielström, "Preußisches Kriegsspiel". Ein Projekt an der Julius-Maximilians-Universität Würzburg, *Militärgeschichtliche Zeitschrift* 78, 2019, 86–98.

Wintjes 2019 Jorit Wintjes, *Das Preußische Kriegsspiel*, Opladen 2019.

Wintjes 2017 Jorit Wintjes, When a Spiel is not a Game. The Prussian Kriegsspiel from 1824 to 1871, *Vulcan* 5, 2017, 5–28.

Wintjes 2016 Jorit Wintjes, "Not an Ordinary Game, But a School of War" Notes on the Early History of the Prusso-German Kriegsspiel, *Vulcan* 4, 2016, 52–75.

Wintjes 2015 Jorit Wintjes, Europe's Earliest Kriegsspiel? Book Seven of Reinhard Graf zu Solms' Kriegsregierung and the 'Prehistory' of Professional War Gaming, *British Journal for Military History* 2, 2015, 15–33.

Woodham-Smith 1953 Cecil Woodham-Smith, *The Reason Why*, London 1953.

Xylander 1935 Rudolf Ritter und Edler von Xylander, *Deutsche Führung in Lothringen 1914. Wahrheit und Kriegsgeschichte*, Berlin 1935.

Young and Lawford 1967 Peter Young /J.P. Lawford, *Charge! Or, How to Play War Games*, London 1967.

Zimmermann 1901 Carl von Zimmermann, *Winke und Rathschläge für die Leitung des Regiments-Kriegsspiels*, Berlin 1901.

Zorn 1986 Wolfgang Zorn, *Bayerns Geschichte im 20. Jahrhundert*, München 1986.

3

Using Combat Models for Wargaming

Joseph M. Saur[1]

Taurus TeleSYS, Yorktown, Virgnia, USA

"All models are wrong but some are useful."[2]

'In 2015, the Military Operations Research Society (MORS) held a three-day conference in Tysons Corner to examine the relationship between combat models and wargaming. Was wargaming not sufficiently "rigorous"? Could wargames be made more repeatable? Could the results be quantified somehow? To all of these charges, Dr. Bill Lademan, Director of Wargaming, Marine Corps Warfighting Laboratory (MCWL) had one answer: "Wargaming is not broken; don't try to fix it!"

---

1 Mr. Saur served for 20 years as a Surface Line Officer in the U.S. Navy, and is a distinguished graduate of the Naval War College. Since retirement, he has earned a Master's Degree in Computer Science, and has worked as a programmer, modeler, tester, researcher, and program manager. Academically, he has served as an adjunct professor:
  ECPI: Undergraduate: simulation, gaming, computer science, project management
  Regent University: Graduate: cybersecurity; Undergraduate: computer science, project management
  Georgia Tech Professional Education Short Course: *Fundamentals of Combat Modeling*
  Marine Corps University Command & General Staff Elective: *Fundamentals of Wargame Design*
2 George Box. A later version gives a somewhat different picture: *"Remember that all models are wrong; the practical question is how wrong do they have to be to not be useful."* From *Empirical Model-Building and Response Surfaces"*, George Box and Norman Draper, 1987.

Simulation and Wargaming, First Edition. Edited by Charles Turnitsa, Curtis Blais, and Andreas Tolk.
© 2022 John Wiley & Sons, Inc. Published 2022 by John Wiley & Sons, Inc.

In truth, comparing combat models to wargames is somewhat akin to comparing hammers to screwdrivers: They both have their legitimate uses; they are not easily interchangeable, and the choice of one or the other depends on the specific circumstances being addressed. Combat models can be very useful if one needs to understand the potential impact of a proposed change to the status quo: the impact of a proposed new weapons system, the wisdom of adopting new tactics, or the danger of a potential kinetic conflict. Wargames, on the other hand, are most useful when attempting to understand the factors and issues surrounding a potential conflict, when ensuring that our plans cover all (or at least most!) possible enemy moves, and when training leaders to understand their tactical and strategic options in a given scenario.

To better understand these two options, we will look at three things: the nature of combat models, the nature of wargames, and the distinctly separate uses (and usefulness) of each. But first, a bit of history.

Some would assert that combat models came in with Robert McNamara's time as Secretary of Defense. And certainly, the emphasis his Whiz Kids put on the implementation of systems dynamics models did have a significant impact, but they were certainly not the first.[3]

In the 1879 Gilbert and Sullivan operetta, *The Pirates of Penzance*, the character Stanley sings *I Am A Modern Major General*, and lists all of the "information vegetable, animal, and mineral" that he is an expert in, including mathematics, history, art, and architecture. Unfortunately, he then maintains that his knowledge of *tactics* and *strategy* is rather limited. It would appear that, in that period, the search was on for the *science of war*; the formula or formulae that would lead one to victory if rigorously applied. And indeed, a number of individuals did identify a formula for attrition that could be applied.

In 1902, LT J. V. Chase, USN, created a calculus-based formula that was submitted to the US Naval War College in a paper headed, *Sea Fights: A Mathematical Investigation of the Effect of Superiority of Force in*; this strongly resembled the basic Lanchester formula.[4] In 1916, Bradley Fiske independently published a similar set of results in his book, *The Navy as a Fighting Machine* in 1916. And in 1915, M. Osipov published a series of articles in the Tsarist Navy journal *Voenniy Sbornik* (Military Collection) entitled *The Influence of the Numerical Strength Of Opposed Forces on Their Casualties*.[5] In this article, Osipov presents two formulas: one that resembles Lanchester's basic formula, and one that resembles his square formula.

---

3 I strongly suspect that an early caveman would count the number of spearmen he had and the number his opponent could muster and use a very basic formula to decide whether fighting today was a good idea. . .!

4 Bradley A. Fiske, *The Navy as a Fighting Machine*, U.S. Naval War College, 1916.

5 M. Osipov, "The Influence of the Numerical Strength Of Opposed Forces on Their Casualties", translated by Robert I. Helmbold and Allen S. Rehm, published in "*Warfare Modeling*", ed. Bracken, Kress, Rosenthal, MORS, 1995

And that brings us to Fred Lanchester. In 1916, Lanchester, a British electrical engineer, published, *Aircraft in Warfare: The Dawn of the Fourth Arm*, in which he discussed the strategic and tactical uses of aircraft in combat. This included long-range reconnaissance, artillery spotting, aerial combat, and both tactical and strategic bombing. As part of his analysis, he included two chapters on the mathematics of attrition. His basic law describes the results of combat between unequal forces when all combat is hand-to-hand; the example used is warfare between forces armed only with swords and shields. If Red has 100 men, and Blue 150, the extra 50 have no one to actually fight, and so must wait until one of their comrades is eliminated before joining in. By contrast, the square formula applies in situations where the troops are armed with missile weapons (e.g. rifles, bows, artillery) so that the extra 50 can add their fire to that of the first 100, and the weaker side is eliminated more rapidly. Given the then ongoing events (World War I), Lanchester's book was widely read, and so the formula we know now bears his name.

Both models assume a fight to the finish (i.e. neither side pulls out at some pre-set percentage of losses), a somewhat unlikely condition, but that is not a major issue. The real problem is the number of historical battles that tend to defy the law! Three examples:

- At the Battle of Rorke's Drift, 150 British soldiers successfully held off 3-4000 Zulu warriors.
- At the Battle of Watling Street, Gaius Suetonius Paulinus's 10 000 Romans defeated some 230,000 Britons under Boudica.
- At the Battle of Leyte Gulf, a small US force (6 escort carriers, 3 destroyers, and 4 destroyer escorts) convinced a much larger Japanese force (4 battleships, 6 heavy cruisers, 2 light cruisers, and 11 destroyers) to break off the battle by virtue of their aggressive and determined attacks.

The real issue with Lanchester, however, is the assumption that one can quantify the effectiveness of both sides with equal correctness, and that brings us to our first topic.

The Nature of Combat Models

Let us start with models and simulations. (Most of us tend to confuse the two, so we will try to sort these out.) The old Defense Modeling and Simulation Organization (DMSO) defined a model as, "A physical, mathematical, or otherwise logical representation of a system, entity, phenomenon, or process." A model, then, tends to a static description of the relationship of the various parts, whether those parts be physical, mathematical, or steps in a process. Once the model is run over time (to observe its dynamic behavior), we have a simulation: "A method for implementing a model over time." So, then a computer-based flight simulator is an algorithmic model that describes the behavior of the aircraft under various conditions, and, when activated, provides a simulation of flight that reacts to the player's commands.

Physics-based models tend to be relatively straightforward. For example, if we wanted to predict the range of a howitzer under normal conditions, we would use a mathematical formula that would take the weight of a standard projectile, the explosive power of the standard powder charge, the elevation angle of the gun, and the force of gravity, and would use all of these factors to predict the average range. (Since, other than the force of gravity, each of these factors is assumed to be a "standard" value, there is some distribution of the actual values, and this leads to a certain amount of normal dispersion.) These types of models, where there is only one correct answer, are known as *deterministic models*[6]. Lanchester's is such a model.

Once we have our answer, we can validate the results: We will take a howitzer out to the range and fire a dozen or so rounds downrange at a specified elevation. We can then measure the actual ranges and compare these to our predictions. This is relatively easy. The hard part comes when we add humans into the mix.

Human behavior is not as predictable as the flight of a projectile: People can think, and they can perversely do exactly what they know you do not want them to do! (Ask any parent. . .!) And human behavior under stressful conditions – such as when one is being shot at – can be even harder to predict. After the battle of Gettysburg, for example, ordnance sergeants found rifles on the battlefield that had been loaded multiple times, but never fired. "How could that be?," you ask. Well, the loading process had two distinct parts: inserting the powder and bullet into the barrel, but then placing a percussion cap where the external hammer would hit it. Whether the soldier was new and poorly trained, and thus did not place a new cap each time because he did not realize he needed to, or the action was deliberate because he could not bring himself to shoot at another human being, or was simply too stressed to perform all the steps in the correct order, we will never know, but the point is that it is estimated that only one casualty was inflicted for every 250 or so bullets consumed (and presumably fired!) Bottom line: a combat model, because it depends heavily on human behavior, is never as neat, clean, and mathematically precise as a physics-based model, so we must rely on statistical averages and distributions rather than precise values.

---

6 Deterministic model: *A model in which the results are determined through known relationships among the states and events and in which a given input will always produce the same output.* DMSO Glossary of Terms; https://apps.dtic.mil/dtic/tr/fulltext/u2/a349800.pdf

But back to combat modeling. For this section, then, we will use football as our analogy for combat, and *Madden Football* as our combat model; we will start by examining the parts needed to create the latter.

For those not familiar with it, *Madden Football* is a videogame that allows one to "play" his or her team against one controlled by the computer (to be more technically correct, against one controlled by software written by a team of programmers). The player can choose the play and/or formation to be used (i.e. to act as both the offensive and defensive coordinators), and during execution, to determine whom to throw to, when to throw, whom to tackle, and so on, by acting as either the quarterback on offense, or defensive lead when defending. So, underneath the hood, what is necessary to make this work?

First of all, we need to define what we need to make up a team. We need a quarterback, a fullback, some halfbacks, receivers, an offensive line, and so on for a total of about 50 different positions (including backups). These are our "pieces," and are akin to the types of pieces we would need in a military model: infantry units, artillery units, etc. For each (football) player type, now we need statistics: What characteristics define a quarterback?

- His accuracy
- His completion rates
- His ability to run

Fortunately, we have help here from the Fantasy Football enthusiasts who religiously collect and publish on the Web all the numbers we might need.

So, now we have the first leg of our game: the pieces.

Now, how do we use these statistics? Well, we need a *Combat Results Table*, or CRT. This can be simply a table, or a set of mathematical and/or statistical algorithms for determining the outcome when quarterback A passes to receiver B who is covered by defender C. Are the algorithms deterministic? Or is it based on a comparison of statistics with dice rolls thrown in to decide whether the action was successful or not?[7] As the results are not determined by a deterministic formula, these types of models are called *stochastic*[8] As one might imagine, these algorithmic approaches can be both complicated, and difficult to balance: The results must be in line with what we see on Sunday afternoon to be acceptable to the game player; they must not appear to be "slanted" in favor of the offense, the

---

7 For those interested in understanding how computers "roll dice," information on the programmatic approaches used to create pseudo-random numbers can be found on the Web: https://en.wikipedia.org/wiki/Random_number_generation

8 Stochastic Model: *A model in which the results are determined by using one or more random variables to represent uncertainty about a process or in which a given input will produce an output according to some statistical distribution.* Sometimes known as Markov models, or Monte Carlo models. DOD M&S Glossary, https://apps.dtic.mil/dtic/tr/fulltext/u2/a349800.pdf

defense, or a specific team. Some steps that would need to be considered (for actually building a CRT) are not being addressed here, but those include the modeling of the different factors will be considered to determine the values and entries into the CRT – the modeling question. This involves understanding the many different parameters (of our pieces, and their actions in the game) that affect the results and modeling a reasonable relationship among them.

Once we have our CRT, we have the second leg of our game.

Finally, we need to define our team behaviors: On first down, what percentage of the time do they pass? Run? When on defense, how often do they blitz? Does their behavior change when they are behind? Ahead? Again, the Fantasy Football statisticians are very helpful. And again, we need a set of algorithmic approaches to combining the statistics for the opposing teams: If Team A looks like they are planning to pass, how does Team B statistically respond? And how do the 11 players on each team interact to, again, create a believable outcome consistently?

So now we have the third leg of our game.

But now, we need to recognize something: We have, in fact, created a box, and nothing that we have not defined to a fare-thee-well will ever occur in our games. If we do not include actions that did not show up during the regular season (but which did get used during the actual game), we run the risk of skewing the results. For example, in World War II, no one ever considered the possibility of *kamikaze* attacks, as that was not something that had been seen before.[9] Uncovering such unforeseen activities is one of the problems that wargaming can be helpful with, although it is difficult to predict them before playing the wargame (perhaps multiple times).

To better understand the issue, let us look at Superbowl LII, which was fought, I mean played, on 4 February 2018 between the New England Patriots and the Philadelphia Eagles. To begin, we will take one instance of Madden Football, and instantiate it with all of the players, player stats, and team stats for the Patriots, as they existed that morning. For the statistics, we could use the numbers for the previous season, or even include stats for the previous years, but in a weighted manner:

$$(yr15 + (2 * yr16) + (4 * yr17)) / 7 = wgtd\_avg$$

Now, let us take a second instance of Madden Football, and again create the Patriots as they existed that morning, and then use a third bit of code, external to the two instances, to play one computer against the other. Given that the two teams, Pats A and Pats B, are statistically identical, we should expect, in a run of 1000 games, that each team would be successful 500 times. (If they are not, we need to examine our electronic dice!) In addition to the raw numbers, however, is the distribution: If everything is working correctly, we should have a normal Bell

---

9 https://en.wikipedia.org/wiki/Kamikaze

curve, with the majority of games won by one point, and trailing off (on both sides) as the number of points increases.

This is similar to the results seen elsewhere. In the 1972 book *Venture Simulation in War, Business and Politics*, Alfred H. Hausrath describes the work of George Gamow, and his hex-based combat model: Tin Soldier. His work starts with a 10 × 10 hexagonal matrix as shown here, with armies of x's and o's (he refers to them as tanks) fighting each other. Tanks move in random order, and in random directions. When tanks of opposing sides find themselves in adjacent hexes, a digital coin gets flipped, and one or the other goes away. Each game continues until one side or the other runs out of tanks.

Assuming we start with equal forces (10 Blue x's and 10 Red o's), our graph looks much like the one above, but as we begin to play with the size of the forces, the graph continues resembles a Bell curve, but the distribution begins to change. When the forces are, for example, 5 Blue vs. 10 Red, the results look like this, where in 100 games the outnumbered Blue forces only win 7 of the engagements, while Red wins the other 93.

But we seem to have gotten off the subject of football, so let us return to the stadium. We have shown that two completely equal teams will end up splitting the games between them, with the majority of the games decided by only a very few points. Now it is time to bring in the Philadelphia Eagles: We will leave Pats A on their copy of Madden Football, but we will replace Pats B with the Eagles, and again run 1000 games. I suspect that there are a few individuals who make a reasonable living doing just this out in Las Vegas. As I write this, there is an online story[10] describing the use of Madden Football to predict a Rams win in 2019; I am looking forward to seeing whether it works out.[11]

For the sake of argument, let us assume that the results are along the line of Pats 600, and Eagles 400. What would this tell us? Well, we now know that *the odds of the Pats winning are roughly 6:4*. Unfortunately, it does not tell us how the game actually went (the Eagles won, 41-33), but it does tell us that, if we could somehow get the actual Patriots team, and the actual Eagles team to play 1000 games, starting the exact same starting conditions, the Patriots would win roughly 600 games. But, you cry, that is impossible, and yes, it is: On the 2nd game, they would have learned from the 1st, and would play differently, and they would continue to learn until on the 1000th game, they would likely play completely differently.

Of course there are other issues. First of all, our behavior data (stats) are an average of the last season or so, but on 4 February, some players beat their normal

---

10 http://www.msn.com/en-us/sports/nfl/madden-nfls-official-simulation-predicts-rams-win-vs-patriots-in-super-bowl-liii/ar-BBSRqFs?ocid=ientp
11 In fact, the Patriots did win; I would suggest that the issue was the apparent use of only a single game run to make the prediction, rather than using multiple runs to establish the odds.

stats, and some did not. Some were injured, and were pulled out of the game. And the teams did not follow their normal behaviors: Neither team punted, but in all but one case, both teams were able to convert on fourth down. Both teams used not-seen-before trick plays, including both teams attempting to pass *to* the quarterback! (Worked for the Eagles, but not for the Pats.) So, in a sense, our beginning assumptions about the validity of the stats we used seem to be a bit "squishy."

What may not be immediately apparent is the relative nature of those odds. We saw that the odds for the Pats vs. Eagles were 6:4. What if we substituted the Jets (whose record that year was 5 wins and 11 losses) for the Eagles? We might find that the odds for that matchup would be more like 8:2. And if we matched the Jets against the Cleveland Browns (who won 0 of 16 games), the odds might be Jets 9.5, Browns .5. The point is that the odds of any matchup can only be determined if we have relatively good statistics for both of the two opponents, and the stats have to have been collected over a period of time, and in contests at roughly the same level. What does that last caveat mean? Well, let us take a quick look at college football: On 22 September 2018, unranked Old Dominion (a 28.5 point underdog) defeated 13th ranked Virginia Tech. How did that happen? Well, if we try to create a predictive model similar to the one above, we find a number of issues:

- Instead of three to four years of statistics for each player, we might have only three to four *games* worth of data, and could be based on very uneven performances. Football at this level is much more of an emotion-driven game – they are playing because they want to; not because it is their job.
- We might find the same with the coaches: Neither (if I recall correctly) had many years with the school, and so their overall record was again iffy.
- Unlike the pros, the coaches had little or no control over changes to the team roster: players graduate, transfer, and/or dropout.

We have similar issues when attempting to estimate the effectiveness of potential opponents. We can, sometimes, and from a distance, observe their exercise maneuvers, but we are not always aware how "free" their imaginary opponent is to win. Historically, there have been instances where the plan originated with the Kaiser, or the King, or the Chairman, and this made defeating their plan somewhat dangerous to one's future.

Let us turn to the First Gulf War (Desert Storm): In late 1990, prior to the war, the DoD combat modeling community was asked for a prediction of the number of American casualties. They had two problems to contend with:

- The first had to do with the experience level of the modelers. In 1989, after the collapse of the Soviet Union, the Warsaw Pact, and the removal of the Berlin Wall, Washington called for a "Peace Dividend." What that meant on the ground was a budget reduction for many contracts, including those supporting the DoD and Federally Funded Research and Development Centers (FFRDC) combat modeling shops. As a result, many of the more senior analysts were no longer available to assist in estimating Iraqi effectiveness.

- Secondly, there were no actual statistics. The only combat that we had seen was during the Iran–Iraq War, and given the level of professionalism displayed by the Iranians, the data that could be gathered would only give an indication of the Iraqi ability against another Arab army.

Now, I confess that I was not there, so this next would have to be considered speculation, but some thoughts:

- We did have lots of data on the Warsaw Pact nations dating back to World War II. We knew how they were equipped, how they trained, and what doctrine they followed. We were not certain how well they had retained their WWII abilities and/or professionalism as we had not seen them in combat since. (In fact, we were quite surprised by some of the mistakes we saw the Russians make in their 1979 invasion of Afghanistan!)
- We knew the Iraqis had Soviet equipment, training, and doctrine, so could possibly try to equate them with one of the Warsaw Pact armies and use those statistics.

Bottom line was that the predictions were completely off by many orders of magnitude: While the actual number was 612 (145 killed; 457 wounded), the predictions appear to have ranged from 20 000 to 30 000![12]

A similar issue involved the use of a computer model to assist with medical planning. Based on the high number of casualties predicted, 108 medical units and 23 000 medical personnel were deployed[13]. Unfortunately, the medical planning model used to predict the *type* of casualties was based on the wound distribution seen during the Second World War, so that the prediction was that the most likely issue would be a sucking chest wound. Given that everyone in Desert Storm was wearing body armor (which was not true during WWII), the likelihood of that kind of wound was actually quite low. As a result of the prediction, however, large quantities of consumable supplies (including antibiotics) were shipped to the desert, only to be abandoned later.

One last point: Up until now, we have focused on *data-related issues*; let us take a moment to consider *implementation issues*. We have to keep in mind that most of our model builders are *not* military veterans, but are instead modelers,

---

12 Mystics and Statistics Blog by the Dupuy Institute, http://www.dupuyinstitute.org/blog/2016/05/17/assessing-the-1990-1991-gulf-war-forecasts/
13 GAO Report to the Chairman, Subcommittee on Military Personnel and Compensation, *Committee on Armed Services*, House of Representatives: Operation Desert Storm; Full Army Medical Capability Not Achieved, August 1992.

mathematicians, and/or computer scientists, and they sometimes might simplify the implementation in ways that make perfect sense from an algorithmic perspective, but give strange military results. An example: When LtGen Paul Van Riper was President of the Marine Corps University (MCU; 1989–1990), the school obtained a newly revised computer-based combat model to use to support some of the classes. One of the new features was the addition of the Army's Kiowa helicopter, which had a radome perched above the rotor. The idea was simple: hide behind a hill or some trees with only the radome showing; when a target was spotted, pop up, fire a missile, and drop down under cover. Well, the MCU staff decided to experiment, and created a simple scenario that pitted one Kiowa against one Marine Cobra attack helo; they then ran the scenario 100 times. At the end of the runs, they were surprised (shocked?) to see that the Kiowa had won each and every engagement! 100 to 1? Does this make sense? They began to dig into the model, and finally discovered that, *in the implementation of the tactic described above*, the hill (or trees) would pop up with the Kiowa! It could shoot through its cover, but would be undetectable to the Cobra, which could not get a shot off as it could never acquire a target. . .! Another issue that has two aspects:

- The implementation is "behind the screen," and is therefore invisible to the user. Unless the user either stumbles across a situation similar to the one described above, he or she may take the results at face value, which leads to the second issue:
- Learning the wrong lesson! Had the staff not conducted this experiment, they might have taught an entire generation of students that the Kiowa was all but invincible against other helicopters. . .! [14]

So, at this point we have to ask: if combat models are so bad, should we continue to use them, and if the answer is "Yes!", how should we use them?

Well, let us go back to our 2:1 odds Tin Soldier run: As we saw, Blue (5 tanks) lost 93 games to Red (10 tanks). Hmm; during the Cold War, was not NATO outnumbered 2:1 in terms of tanks? Well, yes it was. Did we want to bet on winning with those odds? No, we were not. So, what to do?

If we take the Blue tanks in the 2:1 odds model, and give them the ability to destroy Red tanks, not only in the 6 adjacent hexes, but in the 12 hexes beyond, we have essentially created the conditions surrounding the Battle of Karbala Gap.

On the night of 2–3 April 2003, the 10th Armored Brigade from the Medina Division and the 22nd Armored Brigade from the Nebuchadnezzar Division, supported by artillery, launched night attacks against the American bridgehead at Musayib. The attack was savagely repulsed using tank fire and massed artillery

---

14 Personal conversation with LtGen Van Riper; January 2015 at MCU.

rockets, destroying or disabling every Iraqi tank in the assault. The next morning, Coalition aircraft and helicopters rained death on the Republican Guard units, destroying many more vehicles as well as communications infrastructure. The Republican Guard units broke under the massed firepower and lost any sense of command and cohesion. By the end of the day, the tanks of the US 3rd Infantry Division had overrun Lt. Gen. Hamdani's headquarters and Hamdani and his staff fled. American forces lost no men killed in this action while Iraqi losses are estimated at 230–300 killed. The plot tells us the same thing: Blue wins 94 games out of 100; Red, only 6.

What does this tell us? Well, if we (NATO) can shoot further than the Warsaw Pact, we have a chance! So for the period from about 1945 until 1989 (i.e. the Cold War), NATO modelers assisted the planners by helping to identify how proposed changes in equipment, doctrine, deployment, and other variables might help to improve the odds. In other words, to move the line separating Blue wins from Red wins from the left side of the graph (Red wins most of the time) to the right side (Blue wins most of the time). Bottom line: How to defend budget requests for more of something: A-10's, Davy Crocketts, air superiority, Electronic Warfare, and so forth and so on. Given that the Cold War never went kinetic, one could argue that it appears to have worked! (One can only assume that there were Soviet combat modelers who were also looking at the potential outcome of a NATO–Warsaw conflict, and were advising caution on their side. . .)

To illustrate using our football analogy, we could create the two Madden instances we used before; one represents your current team, the other, some hypothetical or real adversary. After 1000 games, look at the odds. If you do not care for the outcome, try changing your team by using different quarterbacks, different receivers, and so on; or your play calling by using more passes and no punts, and run another 1000 games, and see what difference this made. Again, I suspect that something like this might not be unheard of in the offices of some NFL team owners!

At this point, you may be wondering whether we could apply statistical modeling to predict human behavior. People have thought about it. In 2004, I participated in a two-year DARPA experiment entitled "Integrated Battle Command." The concept was simple: create a family of models that would completely represent a human society. At the time, the DoD shorthand for the features of a society was PMESII: political, military, economic, social, infrastructure, and information. Set the models to interact with each other so that the results of the military model, for example, would be passed on to the other models, and the second- and third-order effects would become manifest. For example, if a military action resulted in the destruction of a bridge, the infrastructure model would break the transportation network at the correct point, and the economic impact (i.e. lack of food or fuel) would create social and political instability. As I said, we worked at it for about two years, and reached a number of conclusions that can be summarized here:

- Yes, it is possible to create such a family of interconnected models. However,
- No, it is not feasible to create a *validatable* economic model of a Third World country that has just been overrun by a First World power, is in the middle of an externally-funded insurgency, and is suffering from sectarian violence (in other words, Iraq in 2004!). The word "*validatable*," in this sense, implies that one can find similar situations in history, enter the appropriate data from the past, and have the model reproduce the historical outcomes. What examples would one use, and where would one find the data needed to instantiate the models?

This was not the first time the basic idea was considered. In 1951, Isaac Asimov published the first of five novels centered about the concept of *psychohistory*. In the first book, Hari Seldon, both a mathematician and a psychologist, creates a complex and interconnected set of models that purported to predict future human trends and events. The Foundation, a group of individuals who were the keepers of the model, would attempt to keep the model current and to feed in new data. Eventually, they are seen as meddlers who need to be stopped. Hari's models have two basic assumptions:

- That the population whose behavior was being modeled needed to be sufficiently large; in the novels, it was a Galactic Empire with millions of inhabited planets.
- That the population should remain in ignorance of the results of the application of psychohistorical analyses because if people are aware of the analysis, they will change their behavior.

Has the fact that psychohistory is a fictional science stopped various researchers from attempting to use statistical analysis to predict things? No; and we have a few examples. As you read these, ask yourself the following questions:

- The Pieces: What entities would we need to cavort about the digital landscape to make this model work? As we look to the future, how would we deal with leaders passing, and new leaders emerging? What about political and/or economic -isms? What properties would we assign, and where would we find the statistical data? (Most of the models that purport to predict future issues appear to assume an army of graduate students feeding data into the model. . .)
- The CRT: How would we define interactions between the entities? What historical evidence do we have that might be useful in assessing the impact, for example, of a charismatic leader on a population? Would level of education be an issue? How would climate change factor in? And how would we deal with wars, both large and small?

- Team behaviors: How would we quantify national behaviors? How might they change over time, and can that be predicted, or must it be part of the data input effort? How about small splinter groups that end up making an impact far out of proportion to their size?

So, as we consider these questions, let us look at what has been offered:

Europe's Plan to Simulate the Entire Planet

"The 'Living Earth Simulator' will mine economic, environmental and health data to create a model of the entire planet in real time. When it comes to global crises, we're not short of complex systems that look close to the edge: the climate, the food supply, energy security, the banking system and so on. Add to this the threat of war in many parts of the world and the possibility of global pandemics and it's a wonder that anybody gets out of bed in the morning. Science has certainly played an important role in understanding aspects of these systems, but could it do more?

Today, Dirk Helbing at the Swiss Federal Institute of Technology in Zurich outlines an ambitious project to go further, much further. Helbing's idea is to create a kind of Manhattan project to study, understand and tackle these techno-socio-economic-environmental issues. His plan is to gather data about the planet in unheard of detail, use it to simulate the behavior of entire economies and then to predict and prevent crises from emerging. Think of it as a kind of Google Earth for society. We've all played with Google's 3D map of the Earth that uses real data to reveal not only the town where you live and work but your home and back garden too."[15]

China Exclusive: China's "Magic Cube" Computer Unlocks the Future

"BEIJING, Oct. 3 (Xinhua) – Using a supercomputer like a magic cube as tall as a two-story building, Chinese scientists want to calculate the future of the earth. They hope to calculate almost everything in natural earth systems from the growing of a cloud to the changes of climate hundreds or thousands of years in the future with the buzzing, blue 'magic cube,' in the Zhongguancun Software Park in northern Beijing.

---

15 *technology review*, 30 Apr 2010, https://www.technologyreview.com/s/418756/europes-plan-to-simulate-the-entire-planet/

Several research institutes under the Chinese Academy of Sciences (CAS), including the Institute of Atmospheric Physics, the Institute of Computing Technology, the Computer Information Center and Sugon Information Industry Company, have jointly unveiled the special supercomputer named the prototype of Earth System Numerical Simulator and the software 'CAS Earth System Model 1.0' running on the device. . .

Ding Yihui, a member of the Chinese Academy of Engineering and an expert of China Meteorological Administration, said the prototype and the software are a breakthrough for China's development of the earth systems simulator, and will provide a solid basis for the integrated study of weather and climate.

Zhang Minghua, a researcher with the Institute of Atmospheric Physics, said the 'CAS Earth System Model 1.0' includes the complete modules representing the climate and biological systems, all scientifically interconnected.

'It will play an important role in reducing greenhouse gases and improving the climate,' Ding said.

Cao Zhennan, assistant to the CEO of the Sugon Information Industry Company, said earth system simulation needs a high-performance computer. The 'magic cube,' with a total investment of 90 million yuan (about 14 million U.S. dollars), has a peak computing power of at least 1 petaflop, making it one of China's 10 most powerful computers. Its total storage capacity is over 5PB.

Using the prototype of the Earth System Numerical Simulator, it takes about one day to calculate changes over six years in the atmospheric cycle, the water cycle, the rock and soil cycle, the biological cycle and other natural cycles, said Zhou Guangqing, director of the information center of the Institute of Atmospheric Physics.

According to Zhu Jiang, head of the Institute of Atmospheric Physics, the final simulator is expected to have a computing capacity 10 times higher than the current prototype."[16]

A Model to Predict War

"*Bruce Bueno de Mesquita teaches international politics at New York University and consults for the Pentagon and the private sector. He applies his predictive computer models to national security issues, public policy debates, mergers and acquisitions, legal proceedings and questions regarding regulation, corporate fraud and more.*

Your computer model has been used by the Department of Defense to help predict political outcomes across the globe. How does it work? The model starts by assuming that everyone cares about two dimensions on any policy issue: getting an outcome

---

16 English.news.cn 2015-10-03 11:39:45, https://chinarecentdevelopments.wordpress. com/2016/03/03/supercomputer/

as close to what they want as possible and getting credit for being essential in putting a deal together – or preventing a deal. The model estimates the way in which individual decision-makers tradeoff between credit and policy outcomes. Some are prepared to go down in a blaze of glory seeking the outcome they want, knowing they will lose. Others have their finger in the wind, trying to figure out what position is likely to win and then attach themselves to that position in the hope of getting credit for promoting the final agreement.

How has the political science world taken to the introduction of mathematics and computers? The response has been divided. Those more oriented toward quantitative modeling are generally supportive, sometimes enthusiastic and sometimes well-informed critics. Those whose approach tends toward area studies or historical case study analysis tend to be dismissive of technical approaches to studying politics.

Your model is based on game theory. What is this exactly, and how does it pertain to your model? Game theory is a mathematical structure for examining how people, for instance, interact strategically, each taking into account the expected responses of others and recognizing that others are taking into account that their responses are being taken into account, etc. Game theory just involves people who are assumed to be pursuing their interests and who do what they believe – perhaps mistakenly – will give them the best available outcome, given the constraints under which they must make choices. These constraints can involve limitations to their resources, their beliefs about the intentions of others, and many other sources of uncertainty and of risk."[17]

Do any of these appear to have been proven to be either accurate or correct? Quite frankly, not that I am aware of, although I will admit that, had they shown merit, that fact might not have been widely publicized. . . When one considers the number of individual researchers needed to provide ongoing support to such a model (modifying the input data as things change over time), the cost of such an effort would be prodigious. It would appear, however, that the lack of perceived success, does not seem to stop researchers from asking for (and I would assume, receiving) funding for such projects. . .

And if you are wondering, yes, there have also been DoD efforts in this regard.

Afghanistan Stability/COIN Dynamics – Security[18]

This was an attempt, in 2010, to apply the then recently released Army Field Manual 3-24: Counterinsurgency (COIN) to the conflict in Afghanistan. This slide

---

17 Computerworld Business Intelligence, 29 June 2009, https://www.computerworld.com/article/2551150/bruce-bueno-de-mesquita.html
18 https://www.theguardian.com/news/datablog/2010/apr/29/mcchrystal-afghanistan-powerpoint-slide

was an attempt to depict, using a systems-dynamics approach, the interconnections of the many variables that would be needed to model the then-current conditions. In something as large and as complex as this, however, there are a number of obvious issues:

- How close to "right" is the model? How can we validate it? Can we isolate subsections, and validate each separately? If so, then can we begin to integrate subsections and validate the interactions? How long will that take?
- Where do we get the data? Do we have data for "Relative Message Impact Govt vs Ins"? How do we differentiate between "Population Supporting Gov't and SF," "Population Sympathizing w/Gov't," "Neutral 'on the Fence'," "Population Sympathizing w/Insurgents," and "Population Actively Supporting Insurgency"[19]?
- Each of the connecting arrowed lines would require some sort of measuring device if one were to attempt to construct a "current status" dashboard, and the number of meters needed would be potentially overwhelming.

So let us rehash: combat models appear to have a number of demonstrated roles (with some caveats):

- They allow us to compare the relative strength of different teams, whether those teams be armies, navies, or sports organizations. The caveat is that the data and/ or statistics needed to accurately compare the teams may be difficult to quantify, may be wrong or misleading, or may be lacking entirely.
- They allow us to examine the impact of changes to one side or the other, or to the underlying environment in which the team interaction occurs. Again, the caveat here is the potential lack of accurate data. The scientist or engineer alleges that his or her new widget will perform a particular function x% faster/ better/cheaper than the previous system, but. . . this assertion has yet to be demonstrated in the hands of an actual player under actual field conditions (in the case of combat models, a soldier, sailor, airman, or Marine).
- They allow us to look at the totality of a system, but only up to a point. The DoD effort to model COIN operations in Afghanistan created a visually impressive design, but as the saying goes, while it compiles perfectly well in PowerPoint, it would be impossible to actually create and/or run the model in the real world.

So, what have we learned so far? Well, combat models appear to assist in quantifying the potential impact of a proposed change to the status quo through the introduction, for example, of a new weapons system, or in the tactical use of either current or future systems. They can help us to understand the potential odds in a possible conflict. In each case, however, it is the accuracy and availability of actual

---

19 The variables shown in dark green in the middle right of the model.

data that are often the weak links in our analysis, and no amount of "sharpening our pencils" will change that reality.

So now, let us take a look at the nature of wargames, how they differ, and how they might help.

The Nature of Wargames

Our assertion is that wargames, as affairs involving live decision-makers, are better able to explore the nuances of a problem in ways that combat models cannot. To understand how, we need to examine how wargames differ from combat models, where they intersect and interact, and how they complement each other.

Wargames, first and foremost, are games, and these are defined by DMSO as, "*A physical or mental competition in which participants, called players, seek to achieve some objective within a given set of rules.*" Let us look at this more closely:

- A game is a conflict.[20] Each player has specific goals in mind, which may or may not be symmetrical. In historical games, for example, the goal of one player is to replicate the historical outcome, while the other player attempts to rewrite history, and win!
- Each game has built-in obstacles.[21] When playing Monopoly, I might like to capture all the railroads on the first circuit, but the odds are against my rolling a 5, then a 10, then a 10, then a 10.
- There are rules.[22] Even my three-year-old granddaughter can tell when her older siblings are cheating at Tag, and she has never read the rules. But she knows they exist.
- There is closure.[23] The game will end, and how it ends has been defined.
- There is contrivance.[24] The built-in inefficiencies (dice rolls, card draws, etc.) are acceptable because, in the back of our minds, we can say, "It's only a game!"
- There is, however, an aspect of emotional half-belief in the imaginary world in which we are engaged in a quest to "save mankind!"[25] Despite the half-belief, thought, there is a recognition that this is "only a game," and if Plan A fails today, can always try Plan B tomorrow.

---

20 Talk by Sivasailam Thiagarajan; North American Simulation And Gaming Association (NSASGA) 2007 *Conference*, Atlanta, GA.
21 Ibid.
22 Ibid.
23 Ibid.
24 Ibid.
25 Talk by Bernie DeKoven, NASAGA 2007.

- Finally, we distinguish between "true games" and "real games." The former are based on physical configurations (e.g. Tetris) or statistics (e.g. poker); the latter may require (or may deliver) a knowledge of the real-world occurrence that aids the player in understanding what is going on (e.g. *"Axis and Allies: Eastern Front"*.[26]

So, getting to the bottom of it: combat models run on computers or on paper using predetermined data, and are limited to the moves, the statistics, and the behaviors that have been programmed into them. By contrast, wargames are played by people, and that difference is key.

Just as with our football model above, for a wargame, we need to define all the player types, whether at the battalion or battery level, the corps or division level, the task force or fleet level, or army or country level. What is the size of the unit? How much space does it occupy?[27] What statistics and/or effectiveness numbers can be measured? Are those statistics available? (While the Fantasy Football players can help us with Madden, even Jane's cannot always help us here!)

Once we have decided on the players and their statistical elements, how do they interact? Again, some sort of algorithm and/or CRT is necessary. This can be as simple as removing one figure from each side until one side or the other has a 2:1 advantage. Of course, if the odds are even at the start, every figure goes away! And if the starting odds are 100:99, at the end, there are only three figures remaining: one Red prisoner, and two Blue guards! Not exactly what we might want.

In most games, however, adjudication of the results is actually a bit more complex. It involves analysis of the strengths and weaknesses of the different unit types involved, their weapons capabilities, their training, tactics and doctrine, and an analysis of any historical evidence we might have concerning their actual behavior on the field in conflicts against specific opponents (what we really did not have with the Iraqis.) For recreational games, for example, the tables for musket firing might look something like this:

- Short range: 0–3"; long range: 3–6". Die roll = full casualties at short range; halved at long range. Roll one die for every x men (see below).
  - Grenadiers: one per every 6 men.
  - Line: one per every 8 men.
  - Engineers: one per every 10 men.
  - Militia: one per every 12 men.

Now this example is obviously for a musket-era tabletop game, but the principle is the same for more complex wargames that might be played at a War College. For each unit type and/or weapons system involved in the game, what is the potential

---

26 Talk by Otto Schmidt, Fall-In Wargame Conference, 2008.
27 A serious consideration as this often determines the scale of the map and/or hex grid.

impact on every unit of the opposing force? How will we decide if the attacker "makes his point"? How will we decide, for example, if a ship is hit, what capabilities are incapacitated? And so on – the tables can obviously get quite detailed, complex, and long. But is the analysis of a single kinetic fight the best use of wargames? Probably not; that might be better examined using a model such as described above.

So, then, how can we use a wargame? Well, there are a certain number of obvious cases:

- Examining our strategic plans: In the 1980s, I was involved in a series of wargames at the Naval War College where the Commander of the US Second Fleet (COMSECONDFLT) was "dusting off" the general war plans that he was assigned to support. The plans had all been devised some years earlier, and were based on the size, composition, and organization of the Atlantic Fleet as it then existed. Well, over time, all that had changed, and these games involved a number of months of replanning based on current forces, with the actual game intended as an opportunity for the various organizational participants to become familiar with their role(s) under the plan.
- Testing our operational plans: In a more restricted sense, the standard Military Decision-Making Process (MDMP) includes a step entitled "Course of Action (COA) Analysis Game," where the newly formed plan is subjected to a tabletop test. The intent is to exercise an operational or exercise plan at the level of the planning organization. Generally, the Intelligence Officer ("the Two") provides an estimate of the opponent's (Red's) most likely COA. The Operations Officer ("the Three") then plans to counter the Red COA, but in doing so, may ignore other possible Red opportunities or approaches. One student of mine, who had orders to be his brigade's Chief of Staff (CoS) indicated that it was his intention to bring in what he termed "disruptive junior officers" early in the planning process. He would then challenge them to think of other ways in which Red could act, other ways that they could embarrass or defeat the plan under development. With those alternative approaches in hand, his intent was to insist that they be considered in the planning effort, and that the final plan include subordinate plans and contingencies to deal with these possibilities.
- Training: Simple games can be used to familiarize junior officers (JO), for example, with concepts, systems, and missions that their ship does not have. As Ops Officer of USS Sylvania, we would run games where the JOs could be put into roles they would not normally occupy, and given hypothetical scenarios to help them expand their thinking. One scenario involved a JO acting as the Tactical Action Officer (TAO)[28] of a missile-equipped frigate (FFG) transiting the Straits

---

28 The TAO is an officer on watch who has the authority to engage a threat with weapons in the absence of the Commanding Officer.

of Hormuz. He is informed that there is an Iranian destroyer coming over the horizon at 25 000 yards, and is illuminating his ship with fire control radar in search mode; what would he like to do? This is an example of a "what-if" scenario, and proved effective in teaching the concepts, concerns, and considerations involved.

- System development: This game is more for communication between serving officers (operators) and the research and development (R&D) scientists, engineers, and program managers who work in the service laboratories and acquisition organizations. Usually set approximately 10–20 years in the future, the operators are given a scenario, and game forces that are usually a mix of current systems, and others that may not currently exist, but are either under consideration or in development. During gameplay, the operators would interact with the headquarters' (R&D, acquisition) representatives. The intent is for the latter group to understand the warfighters' needs: "Are the new/proposed systems working out as had been foreseen?" "What tweaks might improve them?" "What else would make sense in this scenario?" I have participated in games of this sort for both the Navy and Air Force, and have seen instances where the results of the game did have an impact on future weapons systems development.[29]

A better example of the use of a wargame might be the examination of the factors and issues regarding the South China Sea. The US position maintains that these are international waters, and insist on the right to both freely transit, and conduct military training maneuvers and exercises during those transits. The Chinese government, on the other hand, claims that these are Chinese territorial waters based on the "nine-dash line" assertion made by the Kuomintang Government of the Republic of China in 1947. This claim was denied in 2016 by an arbitration tribunal constituted under Annex VII of the United Nations Convention on the Law of the Sea (UNCLOS).[30]

So, what would we wargame?

Let us assume that the question being asked by the US Commander in Chief, Pacific (CINCPAC), has to do with what his options are regarding future Freedom of Navigation (FON) transits in light of increasingly hostile rhetoric and

---

29 In this regard, the Army has taken a slightly different route, and has been mixing gaming with modeling in their Early Synthetic Prototyping (ESP) program. When a new system is proposed, it can be modeled digitally, and inserted into an Army video game environment. A number of soldiers are then asked to use the proposed hardware in an urban tactical environment, and report on its impact and efficacy.

30 Map from "Nine-dash Line: A South China Sea matrix game" at https://paxsims.wordpress.com/2016/10/01/nine-dash-line-a-south-china-sea-matrix-game/

aggressive behavior by the Chinese Coast Guard. Well, let us think of this as a wargame, and put together some ideas:

The Players – Who Might Be Involved?

- US players:
  - CINCPAC/CINCPACFLT: told to continue to conduct FON exercises, and to maneuver/train exercise in those parts of the sea outside of all internationally recognized territorial waters (in essence, outside the 12-mile limit of any of the surrounding countries).
  - The President (POTUS), the Secretary of State: concerned about the impact on trade if the Chinese escalate tensions or attempt to regulate or impede the free flow of goods, more concerned with blunting Chinese expansionist claims and actions.
  - US Businesses (Apple, Walmart, others): almost totally dependent on China for raw materials and retail goods to sell and are concerned that any action might imperil their access to Chinese sources.
- Chinese players:
  - North/South Fleet Commanders: told to continue to harass US warships, and to inform them repeatedly that they do not have Chinese permission to be there, or to maneuver/train/exercise. Want to do more but are restrained by the Ministry of Trade and Foreign Minister, who are concerned that an accident (the P-3 incident), or overly aggressive actions by overly enthusiastic warriors could get out of hand.
  - Chinese government, Foreign Minister, Ministry of Trade: firmly committed to achieving recognition of Chinese claims but concerned about the potential impact on trade, more concerned with perceived US efforts to control events in Southeast Asia (hegemony). They want to goad the US, but not to start a shooting war. As mentioned, concerned about accidents and "rogue" warriors.
  - Taiwanese government representative: in this case, firmly behind the PRC (as it was originally their claim!)
  - Businesses: they sell to Apple, Walmart and other; totally dependent upon their ability to ship goods.
- Littoral players:
  - Japan: concerned with the possible impact on the transit of imported oil, other raw materials on which Japan depends.

- Vietnam, Cambodia, Singapore, Indonesia, Brunei: concerned with the possible impact on trade, but also fearful of Chinese expansionism. Vietnam, at least, has already fought them twice in recent years.
- The Philippines: they brought the original claim to UNCLOS; they believe that some of the islands that have been militarized by China in recent years fall within their territorial waters.
- Other players:
  - The UN, UNCLOS: concerned that if China gets away with such an obviously illegal land grab, others may try something similar.
  - Public/media opinion: see Chinese aggression as expansionist and are concerned that they will attempt to regulate and/or throttle seagoing trade, thereby increasing the cost of goods at the retail market.

The CRT – How Do We Adjudicate Political, Economic, Information and Other Non-Kinetic Actions? How DO WE ADJUDICATE KINETIC INTERACTIONS (Which, in This Case, We Hope Do Not Occur!)?

This obviously gets tricky, as there are no standard statistics available to go by. Most of the methods I have seen used rely heavily on professional judgment and consensus. Keep in mind that the format of this game can be quite simple (the oft-derided BOGSAT, or "Bunch of Guys/Gals Sitting Around a Table"). Or it can be quite elaborate (a full-fledged war college exercise with multiple separate command posts, organized White (Adjudication and/or Control) Team, or something in between. Either way, it does differ from what we often think of as a naval wargame: no ships! No charts or maps (although we might have a representation of the own shown above, but with no tactical play. This is a political/military/economic game, where interactions are not subject to measurement, where there are no historical metrics to go by, and where all adjudication, as pointed out earlier, is by consensus.

At the same time, these games are not without impact. In the book *Wargaming for Leaders*, the authors describe a similar game played in 1987, shortly after the announcement of the Strategic Defense Initiative (SDI; usually referred to as "Star Wars"). In it, the participants wanted to assess the Soviet reaction should the system prove effective to some degree. In the baseline case, where SDI either did not exist, or was completely ineffective, the Russian team was able to hit everything on their target list, and still have half their nuclear arsenal in reserve for second strikes if needed. However, when SDI was 15% effective (i.e. only 85% of the first strike would get through), the Russians were forced to add additional assets to high-priority targets (US leadership, US missile sites, etc.), so that they allocated their entire inventory when only 2/3 of the way through their target list. The participants were stunned![31]

The results of this game, played at the request of DoD, demonstrate the value of such wargames. It also points out the truth of Thomas Schelling's *Impossibility Theorem*: "One thing a person cannot do, no matter how rigorous his analysis or heroic his imagination, is to draw up a list of things that would not occur to him."[32] Wargames, in this sense, demonstrate the value of the adage, "Two heads are better than one!"

Now, part of the current *"wargames or combat models?"* controversy can be illustrated by an analysis of this wargame:

- No repeatability: Once the game has been played, the participants could not play the game again *with no knowledge of the results of the previous game(s)*. Unlike the computer that runs a combat model, human players have memories, and so wargames tend to be a one-shot deal. This can be an issue if one is attempting to assess the odds of victory (as we did earlier), but does help us to understand more of the options our prospective opponent might have.

- Qualitative vs. quantitative analysis: When the computer combat model runs, it is often difficult (if not impossible) to know why something has happened: What algorithm was invoked? What was the dice roll? Essentially, how did we get from this set of conditions to that set of outcomes? Everything is hidden behind the computer screen. The modelers are getting better at teasing out the dynamic logic, and computers are wonderful at producing statistics, but we still cannot ask the machine, "Why/how did that happen?" By contrast, in a wargame, we have full access to the players.
  - Rapporteurs (essentially, note takers) can be stationed in each command post for a traditional game and record the nature of any discussion: What issues/items were considered? How were they considered? What were those that were ignored or passed over? Why? This helps to give the after-game analysis team the information needed to identify key issues, and to provide the game sponsor with a better understanding of the end results.
  - Surveys and questionnaires can be given to the participants either at the end of the game for their immediate reaction, or at some point later on for their thoughts after some reflection, or both. Again, we want to understand why what happened did so in order that we can advise the sponsor more effectively.

---

31 Mark Harmon, Mark Frost, Robert Kurz; *"Wargaming for Leaders"*, McGraw-Hill, 2008
32 For a description of a game similar to this, see the PAXSIMS entry entitled, "Nine-dash Line: A South China Sea matrix game" at https://paxsims.wordpress.com/2016/10/01/nine-dash-line-a-south-china-sea-matrix-game/

The bottom line, then, is that wargames will never provide the kinds of quantitative charts (odds, timelines, casualty numbers, etc.) that a combat model can give, but given the kinds of issues (e.g. the South China Sea) that are typically addressed in wargames, is that really a problem?

- Subjective adjudication and control: There may be as many as three separate teams involved here, or these different functions might be exercise by a single individual (in a matrix game, the *gamemaster*) or team:
  - The Control Team is there to keep the game going in a direction that will allow the analysis team to gather enough data to use when attempting to answer the sponsor's questions. Keep in mind that these might resemble the one we posed earlier: "What are CINCPAC's options regarding future Freedom of Navigation (FON) transits in light of increasingly hostile rhetoric and aggressive behavior by the Chinese Coast Guard?" That means keeping the players out of any side streets or rabbit holes, and that sometimes means skewing the direction of the game.
  - The Adjudication Team takes the players' "orders," compares the two moves, looks for potential points of active conflict, adjudicates the results, and reports the new state of the game to both sides. The team works with the control team to ensure that moves or results will lead the game in an undesired direction.
  - The White Team might represent a separate body of intelligence advisors to assist players by answering questions not covered in the pre-game briefings.

Organizational Behaviors

Just as in the combat model we collected statistics on team behaviors, for a wargame, we need to understand something about how the other side tends to think and behave. As an illustration of this, let us go back to the VaTech vs. ODU game: Some columnists have suggested that Tech had not actually prepared for ODU, but had focused on the *next* game against Duke. As a result, they were not ready for an opponent that they had all but dismissed as an easy win. By contrast, based on the casualty predictions in 1990, the United States obviously over-prepared for Desert Storm One, and the subsequent fight was pretty much one-sided.[33]

In most wargames, organizational behaviors are derived from a variety of sources:

- Observations of the prospective opponent's behavior in exercises and drills, and/or in actual combat: This is usually the job of the intelligence departments,

---

33 I will admit that, as the father of two sons who were involved in Iraq and Afghanistan (Army), and two in the Persian Gulf (Navy), I'm all for over-preparation. . .!

and they usually do a good job, but there are gaps when the opponent is not considered likely (as in the case of Iraq), or is a completely unknown quantity (as with Al Qaida). Based on these observations, the players will be briefed on what we have seen them do, both recently and historically, and something about how they think. Those same intelligence professionals will often be part of the game control group, so as to provide additional information during the game, or may be assigned to different Blue and Red command groups for more direct advising.

• Defectors and ex-patriots from the opponent's military or country: Not all non-Western peoples think alike, nor do they hold the same values as we do. For them, suicide missions, the use of women and children as soldiers or bomb carriers, hiding their forces in hospitals and/or churches, all of these can be seen in opponents we find ourselves dealing with today.

Issue in Wargames (and Combat Models)

At this point, we might digress a bit and look at some of the issues that can affect both combat models and wargames. Some are ego-related; others reflect a lack of either sufficient analysis, or accurate data:

• Pet rocks: On at least two occasions, I have seen games where the implicit intent of the sponsor (in both cases, a flag officer) was to demonstrate the validity or military utility of a favorite "pet rock." In both cases, the failures were spectacular, and unfortunately, very public (in other words, there were a lot of participant witnesses). This is always a danger, as all of us wargamers and military officers share a common trait: Somewhere, perhaps buried deeply, we have an innate belief that we have a bit of unrecognized military genius just waiting to be recognized. My only advice would be to start small: Try your idea in a small, private venue first, and with at least one player who can be trusted to tell the truth, and not just endorse the idea because you promulgated it!

• Inadequate analysis: Whether this is due to a lack of data collection (no rapporteurs or surveys), lack of time ("I want the Hot Wash-Up out on Friday afternoon!"), or lack of interest, the result is the same: *learning the wrong lessons!* If one is to learn from a wargame, one must be willing to allow for an honest (and often brutally honest!) post-game discussion and analysis. Yes, the good guys might have "won," but. . . Did they all but ignore logistics by claiming that they had rearmed the guided missile cruiser completely in one hour? Did they overturn some of the game results by refloating the two carriers the Americans sank in the pre-operation Midway game? Did they hand wave the "minor issues" that resulted in an embarrassing outcome? Whatever it was, in the end, a lot of junior participants walk away with a warped sense of what makes good tactical and/or strategic sense and may use that erroneous knowledge at some point in the future!

- Inadequate understanding of the opponent: This is hard enough for wargamers; it is even harder for combat modelers. In most cases, governments and leaders tend to be modeled as individual entities (usually called "agents") that interact with each other and the model environment and make decisions (change things) based on a comparison of their individual values and the current model environment. How? you ask.

An agent-based model is based on a framework that lists some number (10?) of opposing traits:

- Religion vs. country
- Power vs. justice
- Family vs. society
- Order vs. chaos
- Etc.

The same intelligence analyst who briefed the wargamer now advises the modeler who now assigns a value (think of a set of slider bars); the modeler sets each trait based on the assessment of the analyst. If you will recall Jack Ryan in "Hunt for Red October," he had studied Captain Ramius over a number of years, and could predict many of his decisions, and so could likely do a reasonably good job of advising the modeler in the creation of a Marko Ramius agent. On the other hand, after the 2016 election, analysts throughout the world, both Allies and rivals, found themselves caught flat-footed: They had spent the previous months advising their leaderships on how to deal with the expected winner: Hillary Clinton, and had done little or no homework about the all-but-ignored Donald Trump!

yyyyn

We have looked at both combat models and wargames. We have shown how the former can assist us in understanding the potential odds of a conflict (as long as we have sufficient statistics!), and in demonstrating different ways by which we might be able to alter the odds. Similarly, we have examined the role of wargames in exploring issues where human decision-makers are the key to the problem, and where trying to model their behavior would only hide their logic and thought processes behind a computer screen. In the end, then, the question, "Which is the right tool?," gets turned around, and the real question becomes "What are you trying to answer?"

Part III

Wargaming and Operations Research

4

An Analysis-Centric View of Wargaming, Modeling, Simulation, and Analysis

Paul K. Davis

RAND Corporation, Santa Monica, California, USA

Background and Structure

This paper evolved from a presentation in a 2016 workshop on wargaming sponsored by the Deputy Secretary of Defense as part of an initiative to use wargaming to help foster innovation (Pournelle 2017). The presentation was to a working group charged with addressing how wargaming can support the Department of Defense (DoDo) larger analytic process. Doing so has always been a challenge. Analysis organizations often look askance at wargaming because of its lack of rigor and reproducibility. Wargamers often see themselves as doing something distinct from modeling and analysis, although acknowledging a larger cycle of research that includes analysis and experiments as discussed in the classic reference on wargaming by Perla (1990). A long awaited book by Matthew Caffrey describes important applications of wargaming over the centuries, including significant contributions to both military planning and force development (Caffrey 2019). An interesting literature also exists on recreational wargaming (e.g. Dunnigan 2003; Herman and Frost 2009).

   The chapter is structured as follows.[1] Section 2 makes distinctions and defines some terms. Section 3 describes the model-game-model paradigm – an analysis-centric view that is only one way to connect gaming to other analytic processes, but an important one. Section 4 describes a 2016 application that illustrates use of the approach. Section 5 draws implications for the building of simulations and using wargaming as part of doing so.

---

1 The chapter builds on and extends an earlier working paper (Davis 2017).

Simulation and Wargaming, First Edition. Edited by Charles Turnitsa,
Curtis Blais, and Andreas Tolk.

Relationships, Definitions, and Distinctions

Different Purposes for Wargaming

Wargaming (by which in this paper I mean wargaming with human players and include what are better referred to as exercises)[2] takes many forms and has many purposes, such as those in Table 1 (Perla 1990; Caffrey 2019; Pournelle 2017).

In this paper, I focus on the italicized portion of Table 1, using gaming to "broaden and enrich analysis-centric models." This includes understanding the adversary's potential reasoning and strategies.

Backdrop

A Common Critique of M&S

As backdrop let me elaborate on some all-too-common perceptions about the relationship between modeling and simulation (M&S) on the one hand, and wargaming on the other. The perception stems in part from how large campaign models were used in DoD's analysis for more than two decades.

Table 1 Some different purposes for wargaming and related exercises.

Tightening	Fill in details, as when working out and representing concepts of operations or estimating delays in human processes
	Test and strengthen tentative plans – gaming through a plan to find its flaws and ways to make plans more robust
Creating and enriching	Create ideas, options, concepts of operations, or strategies
	Understand possible futures
	Broaden and enrich understanding objectives, values, factors affecting achievement of objectives, and criteria for assessing options
Sensitizing	Raise consciousness about problems and potential consequences
Communication	Communicate and socialize ideas, resulting in education, shared understanding, and personal relationships

---

2 A wargame is "a warfare model or simulation whose operation does not involve the activities of actual military forces, and whose sequence of events affects and is, in turn, affected by the decisions made by players representing the opposing sides" (Perla 1990). Human exercises may not have separate teams; may have discussion but not moves; and may address a sequence of situations (vignettes) unaffected by decisions from earlier discussion.

Table 2 Relative attributes of models and games.

Attribute	As often perceived		Potential
	M&S	Games	Better M&S families
Quantitative versus qualitative	Quantitative to a fault	Qualitative	Quantitative and qualitative
Rigorous	Yes	No	Yes (careful, logical, systematic, and repeatable)
Authoritative	Yes	No	Yes, if authorities want uncertainty analysis
Scope	Kinetic	PMESII	PMESII
Character	Big, complex, opaque	Easy to understand	Varied with a mix of simple and more detailed M&S
Creative, forward-looking	No	Yes	Yes
Adaptive	No	Yes	Yes (no scripted models)
Able to address human issues and foibles	No	Yes	Yes, with both human interaction (gaming) and artificial intelligence agents
Interesting, compelling, good for team building	No	Yes	Yes
Man in loop	No	Yes	Yes, optionally
Clear and persuasive in communication	No	Yes	Variable: sometimes, wargaming will remain the method of choice; other times, its fruits can be communicated via M&S

PMESII: political, military, economic, social, information, infrastructure.

My depiction of these perceptions is shown in the first two columns of Table 2, which reflects interviews of defense officials (Davis 2016), comments made at the 2016 workshop (Pournelle 2017), and my previous work (Davis 2014). In this perception, M&S is seen as highly quantitative and allegedly rigorous and authoritative (i.e. bureaucratically approved); limited to kinetic phenomena; being complex, ponderous, and opaque; and as dealing poorly with uncertainty. Many senior figures have seen such institutionalized model-driven analysis as being backward-looking (focused on old scenarios), unresponsive to their questions, nonadaptive, mechanical (i.e. exercises in number-crunching), and not very helpful. The perception is unfair and arguable, but not without basis (Davis 2016). Wargaming, by contrast, is seen as more creative and forward-looking, more adaptive to modern problems, and more interesting and stimulating. And, because wargaming is so human centric, it is perceived as more realistic in some respects.

Moving now to what should be, the shaded column of Table 2 characterizes M&S as it can be. We can see much wargaming as a kind of modeling or, perhaps more apt, as an important activity within the process of building and refining models. The insights from wargaming should be incorporated in M&S and M&S should be a good vehicle for accomplishing many (not all) of the functions wargamers associate with wargaming.

In this view, we in the analytic community should routinely draw on a broad range of tools as suggested in Figure 1, my attempt to highlight how the strengths and weaknesses of the various tools varies (Davis2014). The reader should not nitpick the particular evaluations, which depend on unexplained assumptions and context. The larger story, however, is valid and – I believe – widely agreed. That said, most people in the analytic community focus exclusively on only one or a few of the tools noted.

In such a rich toolbox, some elements should be relatively simple; some should be at the campaign level and relatively detailed in some respects; and some should be far more detailed. Similarly, some M&S should focus on the more readily quantified issues (e.g. the resources needed to accomplish functions with nominal concepts of operations in a range of circumstances), but other M&S should represent the "softer" factors of crisis and conflict, including issues of deterrence, counterterrorism, the human and frictional complexities of actual operations, and – ideally – the use of combined political, military, economic, social, and informational instruments (Davis and O'Mahony 2013; Davis and O'Mahony 2017). Such M&S should truly embrace the relevant social sciences and their uncertainties rather than simplistic versions. Such broader M&S is likely to be more qualitative and game-like than an exercise in number-crunching.

This more ambitious approach to M&S is feasible but not easy. Merely as examples hinting at feasibility, the US Marines led an interesting study (JIWAB) for DoD that employed social-science methods and had many lessons for the analytic community (Wong et al. 2017). Several other articles in the same special issue of *Journal of Defense Modeling and Simulation*, 8[2], 2011) brought to bear social-science methods in addressing national security issues. A British simulation, the Peace Support Operations Model (PSOM) addressed political-military-economic issues at the campaign level in Afghanistan with heavy senior-leader participation (Body and Marston 2011; Connable et al. 2014). A system dynamics model has been used to represent – more qualitatively than not – the complex relationships involved in implementing counterinsurgency doctrine (Pierson et al. 2008). The value of the models is often in structuring, integrating, and posing issues, rather than predicting outcomes (Connable et al. 2014). As discussed in a recent book reviewing social-behavioral modeling, such models are often most useful when

| Instrument | Resolution | Relative Strength | | | | | | |
| | | Strategic-level functionality | | | | Physical phenomena | Human phenomena | Empirical cautions |
		Agility	Breadth	Strategic decision-aiding	Strategic integration			
Simple analytical[a]	Low	5	1	5	1	1	1	N.A.
Seminar-level human wargaming	Low	5	4	3	1	1	4	3
Red-teaming on capabilities and operations	Varied	5	3	3	1	3	5	5
Qualitative factor trees	Low	5	5	5	5	1	3	
Human wargaming	Medium	1	5	5	5	3	3	5
Campaign simulation (usual)	Medium	2	5	2	4	2	1	N.A.
+agents, political-economic factors, and exploratory analysis[b]	Medium	3	5	4	5	2	3	N.A.
Mission-level adaptive models, exploratory analysis	Medium	3	3	3	1	5	3	N.A.
High-fidelity simulation[c]	High	1	1	1	1	5	1	3
Historical case studies	Varied	1	1	1	1	3	5	5
Historical data analysis	Low	1	1	1	1	1	1	5
Field experiments and war data	Varied	1	1	1	1	5	5	5

NOTES: Ratings are 1 (very poor) to 5 (very good), with red, orange, yellow, light green, and green corresponding to 1, 2, 3, 4, and 5, respectively. Scores depend on assumptions.
[a]Examples include closed-form models and spreadsheet-level computer models.
[b]Exploratory analysis examines the effect of simultaneous variations of all important assumptions, not mere sensitivity analysis on the margin.
[c]In some instances, high-fidelity simulation can be a primary and reliable source of what can be considered to be empirical information. It is simply not feasible to obtain the equivalent information with physical testing.

Figure 1 A portfolio of methods and tools (all can be seen as M&S). *Source:* Davis (2014, p.24).

employed in settings that assist in the education of senior leaders, leaders who have no illusions about the models' ability to predict the future in detail. My own conclusion was (Davis et al. 2019, p. 921)

> One of the more robust conclusions from numerous examples of higher-level decision aiding has been to see model-supported decision aiding as being a process of educating and preparing decision-makers to understand their problem domain so that they can artfully construct and choose among options and later monitor progress and adjust (Rouse 2019; Thompson et al. 2019).

Such "educating" may occur, for example, with leaders strolling around a simulation facility (akin to a flight simulator in Rouse's discussion) and engaging with analysts who discuss selected issues, respond in real time to queries, and show parametrically how various interventions may influence the system's operations. Alternatively, it may occur in a wargaming setting or a type of brief-out that uses comprehensible parametrics to pre-emptively address "beyond what-if questions" (Davis et al. 2019, p. 914ff) and uses walk-through scenarios illustrating issues relating to the narratives that are so often crucial (Paul 2019).

Humans and M&S

If models are to help us understand phenomena, they should allow for human participation that reveals factors, options, and phenomena that can later be represented better in the model. In my view, the history of DoD simulation for analysis has long had four major flaws:

- Exclusive emphasis on "closed" models (i.e. models without human in the loop).
- Avoiding "soft" variables, i.e. qualitative variables that may be uncertain and difficult to measure or define.
- Largely avoiding adaptive logic by insisting on data-driven simulation (e.g. using "scripted models" rather than embedding intelligent agents).
- Largely avoiding uncertainty analysis by instead imputing to much value to base-case assumptions, even when doing so is absurd (e.g. imagining that we know details of future conflict scenarios, including warning time and attacker strategy).

Concerns on these matters are long-standing as I have noted elsewhere (Davis 2014). Great progress has been made, but old habits are difficult to overcome.

Distinctions

Another problem has been modelers moving too quickly to code cutting (writing the computer program) without having a solid conceptual model, something

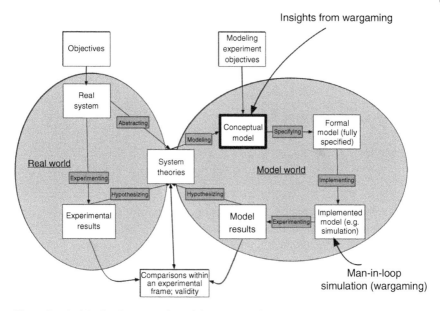

Figure 2 An idealized system view of theory, modeling, and experimentation.

lamented for years as reviewed in Tolk (2012). Figure 2 notes key distinctions and relationships among the real system, empirical experiments, theories, conceptual models, formal models, implemented models (usually simulations), and computational experiments (Davis et al. 2018). Here I use "model" to refer to the "conceptual model of Figure 2. I regard wargaming as an important source of insights for the conceptual model. Also, I see great value in man-in-the-loop simulations, a form of wargaming.[3]

More generally, the appropriate models may be qualitative or quantitative; simple or complex; rough and exploratory, or rigorous and predictive; imaginative, or constrained by conventional wisdom; and open or closed (i.e. with or without human interactions. Further, the models may be valid for only some of very different functions: (i) description, (ii) explanation (iii) postdiction, (iv) exploratory analysis of possibilities, or (v) prediction (Davis and O'Mahony 2017; Davis et al. 2019).

Many types of wargaming exist. I focus here on political-military seminar-style wargaming because the example in Section 4 is about analysis to prepare for and assist in crisis decision-making. I also do not discuss formal models. Elsewhere,

---

3 Many precedents for this. From the Cold War. I recall extensive man-in-the-loop wargaming in the US Joint Staff (what is now J-8), Germany's IABG, and NATO's Shape Technical Center. A more recent example is the PSOM work in Afghanistan mentioned earlier (Body and Marston 2011; Connable et al. 2014).

I describe how using a high-level visual-programming language closely tied to the conceptual model can be a good way to develop a fully specified but comprehensible model, after which it may or may not be desirable to spin off a production-model version in a more powerful or standard language (Davis and O'Mahony 2013).

A Model-Game-Model Paradigm

The Core Idea

Let us now move to the core of this paper, a model-game-model paradigm. The purpose of the paradigm, which I think of as classic rather than a new invention, is to improve models being used for analysis and other aspects of decision-aiding, such as helping to shape leaders' mental models.

Figure 3 sketches the paradigm, which is akin to the theory-test-theory version of the scientific method. Figure 3 uses the metaphor of learning a game such as chess when contemplating a class of possible future crisis. In this metaphor, we study a problem by seeing it as a game between adversaries. We study the game's rules, the game board, the nature of our adversary (or adversaries), the sides' possible strategies and tactics, and so on. We then build models (perhaps only mental models) to represent our understanding.

What can we do next? Physical scientists can go to the laboratory or to nature and run controlled experiments. Economists can craft "natural experiments" by exploiting variations in how otherwise similar groups of people approach particular problems. Often, however, we do not enjoy such luxuries. Thus, we employ

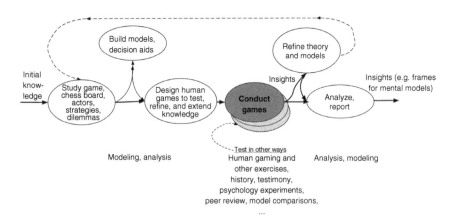

Figure 3   Model-test-model paradigm for using human gaming.

other methods, one of which is to run games and think about them – with cautions discussed below – as experiments.

Often, the primary value of wargaming is to leaders who directly participate, learn "the chess game" rapidly, and achieve common understanding with other participants. Another purpose is testing with a walk through, a crucial albeit insufficient exercise. If instead the purpose is analysis, then the games should probe the edges or illuminate matters not already understood.

Such games may extend knowledge or discover omissions or other errors in previous thinking. For example, we may gain insights about additional tactics and strategies. We might then do some analysis and report tentative conclusions, but we would also want to re-enter the cycle of research (Perla 1990). The cycle should continue until –if ever – we arrive at settled theory. That time may never come because the world keeps changing as do competitors and adversaries.

Can Human Gaming Truly Serve as "Testing"?

Can human gaming be used to test anything? A game, after all, is not a controlled laboratory experiment. The players are usually not decision-makers, although decision-makers have sometimes participated (Bracken 2012). Even if they are, they may not be the people who will make decisions if the crisis or conflict being gamed actually arises. The game's adjudication of moves may not represent well how real-world actions would play out or what effects they would have. And the conditions posited for the game are merely illustrative. The shortcomings of gaming are many if the measure is reproducing the real world or accurately predicting the future. How, then, can they be used meaningfully?[4]

Not surprisingly, Thomas Schelling had important comments on the matter (Schelling 1987). He forthrightly asked "Isn't a game a unique story that may never be repeated?" and "Is there not a danger that participants will be so carried away in this vicarious experience that they identify the game too much with the real thing and learn 'lessons,' perhaps overlearn them, that will prejudice their judgment in the future?" He went on to make a decisive point.

---

4 Yuna Wong discusses how gaming can and cannot help test a theory (Wong 2016); Stephen Downes-Martin discusses how Control Teams are often dominant de facto players, skewing results (Downes-Martin 2013); Robert Rubel describes the strength and weaknesses of wargaming broadly (Rubel 2006) and, in doing so, draws on 1991 dissertation work of John Hanley at Yale, which describes wargaming as a weakly structured tool appropriate to weakly structured problems – one that generates weakly structured knowledge that is conditional and subject to judgment in application (p. 112). In a recent paper Hanley lists numerous virtues of wargaming, but none are what would normally be regarded as proving validity or calibrating parameters (Hanley 2017). From the very old literature, a 1964 debate among Robert Levine, Thomas Schelling, and William Jones is still interesting (Levine et al. 1991).

The answer is that games, in this regard, are not different from real experience. . .

Ideas are so hard to come by that one should be ready to take them anywhere one can find them.

As with tentative "insights" from real-world events, we must ponder and assess whether the insights are generally valid or merely peculiar to the situation.

In addition to providing insights, wargames can sometimes

- Falsify a model by demonstrating that it misrepresents human processes or cognition.
- Enrich a mode by noting additional factors that it should contain because they are important to decision-makers, such as "What is driving the adversary's behavior? Internal dissension? Fear? Or does he believe that we will capitulate under pressure?"
- Creatively change the character of the problem by noting additional tactics and strategies. Perhaps these recognize vulnerabilities that "should not" logically exist (and which the model therefore did not anticipate), but that exist in reality (e.g. the edges of the Maginot Line). Or perhaps they will exploit rule breaking, which gamers are good at.
- Identify "frictions" such as decision delays caused by the necessity of communication and discussion with allies, or by peacetime rules and procedures that cannot suddenly be put aside. Another class of examples relate to differences between nominal and real capabilities, as when a mission capability exists in principle, but not in reality because "Oh, the unit that could do that hasn't trained or exercised for that mission in years. It will take weeks of preparation before it could execute it again."

Most such contributions from human gaming are qualitative rather than quantitative.

We should be skeptical about using wargames as evidence, although there may be more evidentiary value than has been appreciated. Recent debates discuss using online games as a source of empirical data. They yield fascinating information (Reddie et al. 2018; Lakkaraju et al. 2019) but views differ on what conclusions can be drawn (Oberholtzer et al. 2019; Reddie et al. 2019). For further work on online gaming, see Guarino et al. (2019). Other research exploits agent-based models, recreational-style simulations, and traditional social-science research to study ways to influence human behavior (Miller et al. 2019).

Let us now ask how to think about human wargaming in a model-centric approach, Table 3 suggests some of the appropriate and inappropriate questions to ask of human gaming. The second column indicates questions that are

Table 3 Questions to ask and not ask of human gaming.

Good questions	Bad questions
What strategies might the adversary plausibly employ?	Cut to the chase. What *will* he do?
What considerations (factors) have we omitted that need to be included?	What is the full set of factors? What are their precise values?
What could go wrong? What could be much less efficient or effective than assumed?	How precisely will things go?
How might the adversary be reasoning, again considering both likely and plausible?	How *will* he reason?

inappropriate because of inherent uncertainties. Or, to be more precise, it may be legitimate to ask the questions, but one should not imagine that the answers that come from gaming are reliable.

Case Study: Deterrence and Stability on the Korean Peninsula

A 2016 effort served as a case study of using the paradigm of Figure 3. In this effort, I collaborated with a team of analysts in the Korea Institute for Defense Analyses (KIDA) led by Dr. Jaehun Lee, under the general direction of Dr. Yuntae Kim.

Background

An earlier study with KIDA had reviewed Cold War deterrence theory critically and then applied concepts to the Korean peninsula. An important conclusion was that US-Korean extended deterrence had serious difficulties (Davis et al. 2016):

> Maintaining and strengthening the U.S. extended deterrence is no longer straightforward—no matter how fervent and sincere the official statements of assurance. Extended deterrence is inherently difficult when the adversary can strike the extending power with nuclear weapons. NATO faced up to the credibility issue in the 1970s. Something analogous is necessary for Korea.

The conclusion was provocative at the time because of previous complacency and because so much official US effort had offered repeated (and sincere) reassurances to South Korea. The conclusion may still be troubling, but warning signs are now abundant. Matters can change quickly in international relations. North and South Korea are well aware of President Trump's doubts about the wisdom of having the United States still be defending rich, modern nations such as South Korea and Japan (Landler 2018).

The problem is that extended deterrence is easily possible when the deterrer is secure and can severely punish the nation to be deterred with impunity. It is not so simple when the question arises "Would the United States really trade Washington or Los Angeles for Seoul?" This question is not very different from "Would the United States really trade San Francisco for Paris?," a question that NATO faced and dealt with in the 1960s (Delpech 2012). The United States and South Korea have not done so as yet.

Foreseeing challenges and crises ahead for extended deterrence as North Korea's nuclear and Intercontinental Ballistic Missile (ICBM) programs proceeded, the project team chose a focus problem for collaborative work in 2016:

How should the Republic of Korea (ROK) and the United States prepare for possible major crisis in which war—even nuclear war—is plausible?

By major crisis, we meant a crisis significantly worse than those caused by usual North Korean provocations, a crisis in which thinking the previously unthinkable could not be avoided. Our work would apply the model-game-model paradigm. Subsequent sections deal, respectively, with (i) model building and game design and (ii) designing and executing a human wargame and learning from it.

Model Building

Ideal Methods and Practical Expedients

Ideally, we would have used a sophisticated game-structured simulation as in Figure 4, which shows the architecture of the RAND Strategy Assessment System (RSAS) from the 1980s (Davis and Winnefeld 1983; Davis 1989). It had a multi theater combat model at its core (center of the diagram), which covered the spectrum from conventional war to general nuclear war, and which promoted uncertainty analysis. That combat-model portion evolved into the Joint Integrated Contingency Model (JICM) campaign model (Fox and Jones 1998). The rectangles indicate Red, Blue, and Green agents, but human teams were interchangeable with agents, a core innovation.[5] The RSAS had alternative agents because of

---

5 Documentation exists on the Red and Blue agents (Davis et al. 1986), the Green Agent (Shlapak et al. 1986), the military-level agents (Schwabe 1992), and on lessons learned (Davis 1989).

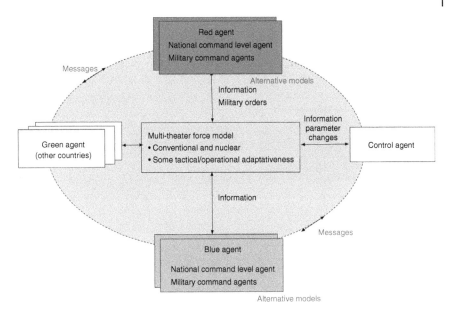

Figure 4 A precedent: the RAND strategy assessment system.

uncertainties about how real-world Soviet/Warsaw Pact and US/NATO leaders would reason. The Green Agent consisted of simpler rule-based models for third countries. France, for example, could make independent decisions about its own nuclear weapons. This mattered a great deal in the simulations, a surprise to many of us at the time because of our superpower-centric mindsets.

The RSAS embodied knowledge from political science, psychology, and organizational theory among others. Its agents were unlike the ones used today in agent-based modeling (ABM). They Red and Blue agents used structured forward-chaining rule-based modeling at the national-command level and a version of slotted scripts at the level of theater commanders. Even if the models were good, the input uncertainties were enormous. Thus, we introduced "multiscenario analysis" across different assumptions about the thinking of Red and the myriad uncertainties of military operations. This came to be called exploratory analysis (Davis 2014), closely linked to Robust Decision-making (RDM) (Lempert et al. 2003).

As the 2016 project with KIDA began, neither the RSAS nor anything like it still existed. So how could we proceed with the vision of Figure 4? We used a poor man's version of the RSAS concept: no real combat model (just some estimates of damage), no treatment of most countries, and only very simple representations of Red (the Democratic People's Republic of Korea (DPRK)), Blue (the Republic of Korea and United States, ROKUS), and China. We included, however, one of the

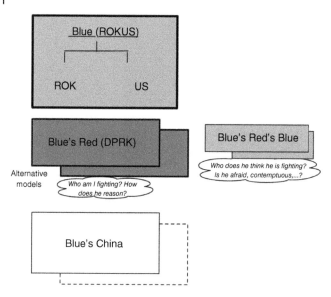

Figure 5 A highly simplified pol-mil model.

most important aspects of the earlier work (Figure 5): a Blue Agent model with a model of (Blue's Red), which in turn reflected Blue's notion of how Red perceived and would respond to Blue (i.e. a primitive Blue's Red's Blue).[6]

To actually build models even for our limited purposes, it was necessary to address a number of issues described in what follows. The challenges were how to organize thinking about levels of conflict and escalation possibilities, structure cognitive models, identify the options that the models should consider, identify the considerations (factors) that should inform decisions, and estimate the consequences of executing the options.

Modernizing the Escalation Ladder

Our structuring concept adapted Herman Kahn's escalation ladder (Kahn 1960; Kahn 1965). A primary theme was that nuclear war is not like a binary switch. It is not as if one either has peace or general nuclear war. Aggression can be small or large, narrow or broad, subtle or crude. How does one deter aggression generally when the threat of massive across-the-board nuclear attack would make no sense and has no credibility? This is the "problem of limited war" that Henry Kissinger

---

6 The astute reader will note the possibility of infinite recursion, but it is seldom worthwhile to go beyond Blue's Red's Blue or perhaps one Blue's Red's Blue's Red. As an example: "Blue: I think Red is very aggressive, that he sees us as indecisive and prone to imagining that we (Red) will settle for a peaceful compromise)."

Table 4 Modern complications in an escalation ladder concept.

Issue	Comment
Space and cyberspace	Mostly new
Chemical and biological weapons	The United States sees them as weapons of mass destruction
Precision conventional weapons	Devastatingly effective in conventional warfare
Strategic cyberattack	Role in decapitation, infrastructure attack, and counterforce attack (e.g. disruption of communications)
Electromagnetic pulse (EMP)	New implications (effects on command and control and electronic systems)
Inherent weakness of extended deterrence	Not new. As for NATO in 1970s
Little green men and hybrid warfare	Not really new
Multiparty crises	Not really new

wrote about as he began his career (Kissinger 1957). Today, it is common to read virtuous-sounding claims that "There can be no victors in nuclear war," but in fact there can be the problem of limited war remains as discussed in Chapter 2 of a national-panel study to which I admittedly contributed (National Research Council 2014).

Escalation ladders today are more complicated because of the items indicated in Table 4: the "new domains" of space and cyberspace, the asymmetric situation in which some US adversaries have chemical and biological weapons while the United States has profound capabilities for precision strikes. So also, strategic cyberattack and electromagnetic pulse (EMP) attacks have added new dimensions. The last three rows show complications that only seem new. As mentioned earlier, extended-deterrence problems were evident in the 1950s. So also, history has numerous examples of analogues to "little green men" and "hybrid warfare" (van Creveld 1991).

Glossing over some subtleties, Table 5 shows a simplified Korea-specialized escalation ladder with separate ladders for Red and Blue. North Korea might find it necessary to use chemical and even biological weapons because of the US advantage in precision weapons. So also, it would not likely treat US space systems as in sanctuary (see also [Bennett 2013]). The ladder shown also assumes that the North Korean leaders might – in extremis – prefer to go out with a dramatic last-gasp destruction of its enemies, rather than capitulation. Hitler reportedly entertained such notions and part of North Korean lore is "World Without

Table 5 Simplified escalation ladders for North Korea (Red) and for South Korea and the United States (Blue) (bullets indicate rungs available to Red or Blue).

Option	Red	Blue
Wait and see	•	•
Demonstrative use of nuclear weapons	•	•
Limited conventional strike	•	•
Decapitation	NA	•
Conventional war	•	•
Counterforce against nuclear threat	NA	•
Use of chemical or biological weapons	•	NA
Use of tactical nuclear weapons in ground operations	•	•
Limited strategic-nuclear use	•	•
All-out war	•	•
Countervalue attack	•	•
Extended countervalue attack (against US homeland)	•	NA

North Korea Need Not Survive," as discussed in a review of Cold War lessons that convinced me that irrationality of leaders is not something to dismiss lightly, especially under duress (Davis et al. 2016). That said, we did not predict such reasoning and we saw Kim as probably rational (something more evident after Kim's artful international activities in 2018).

Cognitive Decision Models

The next challenge was creating simple "cognitive decision models" roughly akin to the agents of the RSAS described earlier. The approach was to create a structure allowing for rational decision-making, but also allowing major departures. Here a rational structure is one with multiple options, assessment criteria, and multiple factors to consider.

Real people cannot do the related calculations because they lack the information for the complex and uncertain probabilistic mathematics. Because of such bounded rationality, they use heuristics (Simon 1978). A heuristic that I have used in representing limited rationality is to have the model assess the outcomes of each option using perceived best-estimate, best-case, and worst-case assumptions.[7] This is perhaps the best that can actually be achieved in most real-world situations.

---

7 This approach was first used for Saddam Hussein before and during his invasion of Kuwait in 1990 (Davis and Arquilla 1991b; Davis and Arquilla 1991a; National Research Council and Naval Studies Board 1996).

Table 6 Generic top-level structure of cognitive decision model.

Cell values (not shown) give scaled outcomes or net assessments (e.g. 0–10).								
Option	Most likely outcome		Best-case outcome		Worst-case outcome		Net assessment	
Model	A	B	A	B	A	B	A	B
1								
2								
3								

Memoirs of past decisions by presidents and top advisors show evidence of their attempting this level of reasoning. To be sure, the assessments will be flawed due both to misperceptions, miscalculations, and cognitive biases (Kahneman 2011; Jervis et al. 1985). Another key element of the approach is to demand alternative adversary models because we do not know how adversaries (or even our own national leaders) will think. Analysis should recognize this, rather than focusing on best estimates. The "tyranny of the best estimate" is an affliction constantly besetting decision-makers (Davis et al. 2005).

Top-Level Structure

Table 6 shows the essence of the top-level concept, simplified to consider only three options and two models of the adversary, Model A and Model B.

In the simplest computational version, the net assessments are linear weighted sums of quantified versions of the option assessments. The models may have different weighting factors as when a risk-taking model with glorious objectives gives little weight to worst-case possibilities and a risk-averse model is concerned primarily with avoiding disaster. The models may also different assessments of outcomes. The "leanings" with respect to weights and with respect to option assessments will often be correlated as when risk-takers are also more optimistic. The net assessment can be nonlinear, as when a model rules out an option, rather than just rating it a bit lower. This is a familiar strategy in real-world decision-making.

Lower Level Structure

The next level of sophistication is to characterize how the models estimate option outcomes given perceptual errors in perceptions, including perceptions of adversaries. Where do the cell values in Table 6 come from? Figure 6 shows the basic concept. As above, an option is assessed using "most-likely," "best-case," and "worst-case" assumptions. These assessments depend on various factors (the Fs)

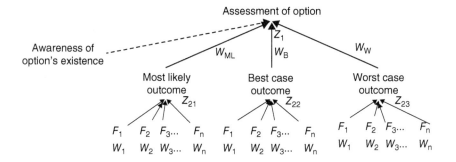

Z_1 = function of personality, pscyhological domain, desperation, optimism...

Z_{2i} = function of utility function, if any; (beliefs, perceptions, personality, psychological domain, desperation, optimism, other cognitive biases...)
W_i: weight of factor i

Figure 6 Multilevel factor tree assessment of options.

and assumptions about factor values. Precisely how the factors combine to generate assessments (represented by the Z functions) could be as simple as linear weighted sums, or something more complex.

Despite the simplicity of this cognitive-model concept, implementing it is not trivial. Figure 7 illustrates one element of the difficulty. Suppose a model is contemplating a preventive-war or pre-emptive option, i.e. an option to escalate before the adversary does. The model should consider each preventive/pre-emptive option, estimate what the adversary will do in response (recognizing uncertainties in how the adversary thinks), and characterize the outcome and uncertainty. He should then compare results to those allowing the adversary to go first. That said, if he is not certain of the adversary's intentions, then he should account for the probability of mis-estimating them. After all, the negative consequences of mis-estimating would be catastrophic.

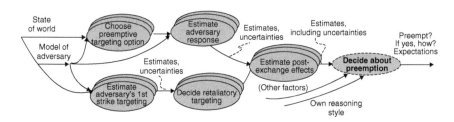

Figure 7 Features of even a "Simple" cognitive model considering pre-emption.

These ideas were implemented in an uncertainty-sensitive model, Crisis Options, written in Analytica®, a product of Lumina Decision Systems. The options that Blue considers, for himself and for his image of Red, correspond to the various levels shown in Table 5. The model was drastically simplified relative to an operational-level simulation or war game in which ground, air, and naval forces would be executing strategies. The premise for our work was that we wanted to focus on weapons-of-mass-destruction-level issues. Details of conventional operations were not important for our purposes even though they might involve major battles and substantial loss of life (as would occur with a major artillery barrage of Seoul).

Continuing, Figure 8 indicates schematically what we sought initially to include in the model. My KIDA colleagues and I used the graphic for discussion and debate before the model was refined and implemented. Although the graphic is qualitative (a kind of mind map generated by analysts trying to be both creative and comprehensive), concepts were mapped to numbers so as to generate charts. Nonetheless, the figure is still approximate, soft, and squishy – as with reality. Although Figure 8 shows an along-the-way representation, a more mature version would be what is called a "factor tree" as described with examples elsewhere (Davis and O'Mahony 2017).

Even in a simplified depiction, it was necessary to characterize the consequences of executing various options and receiving Red's response. For simplicity and because it is what matters most, the Crisis Options model focused on damage to Blue. As shown in Figure 9, the model used a 0 to −100 scale for the various types of attacks. The values were subjective, informed by rough-cut background knowledge. They could have been based on detailed studies and could have been converted to casualties or losses of economic value, but we did not see that as important for our purpose.

At this point, we have seen the essence of the model used: its cognitive structure, the options considered, the factors affecting evaluation of options, and crudely estimated consequences ascribed to various options. Although more qualitative than not (despite using numbers), the model was useful for thinking. It seemed to incorporate a good deal of knowledge. But could human wargaming test or enrich the model?

Designing and Executing a Human Game

Having developed a model that incorporated a good deal of strategic theory, what did we want to accomplish with a human game? Consistent with the ideas described earlier, we sought primarily to look at qualitative matters, such as:

- Strategies and options (for both Blue and Red): Would teams go beyond those we had identified?

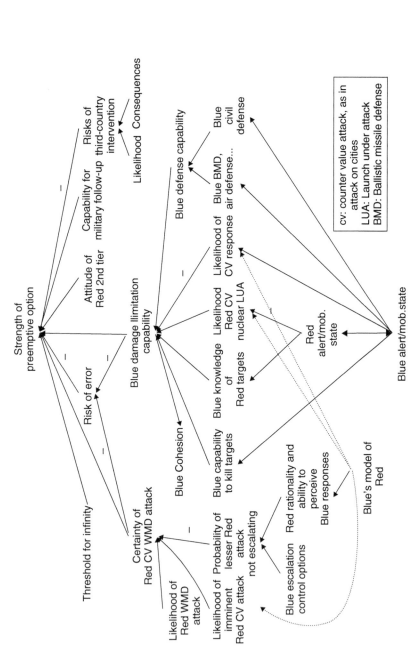

Figure 8 Modeling the military value of a pre-emptive option.

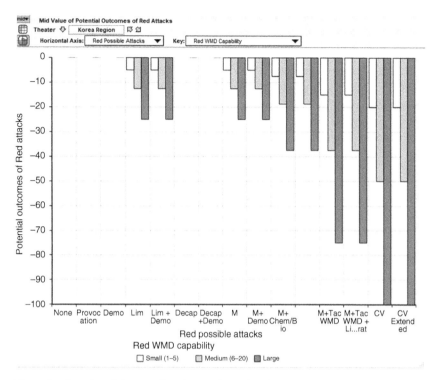

Figure 9 Postulated damage data from red attacks. *Note:* the notional values are sufficient to make qualitative distinctions visual. Abbreviations: demo, demonstrative nuclear use; decap, decapitation attack; M, substantial military attack; chem/bio, use of chemical and biological weapons; Tac WMD, tactical use of weapons of mass destruction (not just demonstrative use), CV, countervalue attack (e.g. against cities); CV Extended, countervalue attack extended to the United States.

- Criteria for evaluating options, such as short-term or long-term benefits and costs, both military and political: Would teams use different or additional criteria?
- Factors in evaluating options: Would teams use different or additional factors?
- Mindsets: Would teams reason in ways understandable with the model or in very different ways?

With these constructs in mind, we designed the human war game with some specific features:

- Use an alarming initial scenario unlike "normal" DPRK provocations, one making it difficult to proceed with normal, complacent thinking. Make war, even nuclear war, appear truly possible (not "unthinkable").

- Have separate teams for Blue (ROK and United States), Red (DPRK), and China. Give special instructions to teams (not necessarily consistent) to sensitize and goad them, but with substantial uncertainties, raising war alarms and dilemmas.
- Instruct teams to (i) sketch options, (ii) sketch alternative models of their opponent, (iii) consider most-likely, best-case, and worst-case outcomes for each option; and (iv) explain their reasoning.
- Have two rounds of play, with the second round's conditions ratcheting up tensions.

The exercise was held in June 2016 at KIDA headquarters in Seoul. KIDA organized and conducted the game. I advised and observed. The game was a day in duration, had no computer support, and had minimal supporting materials. It was closer to a seminar exercise than an operational-level wargame addressing maps, details of maneuver and logistics, etc. Participants were both civilian and military, largely associated with think tanks, government policy offices, or with military political-military divisions. The game was conducted in Korean, with translation for English speakers.

The game reflected "thinking the unthinkable." The teams recognized that, in such a crisis, the US and South Korean leaders would find it necessary to consider preventive-war or pre-emptive options and to recognize uncertainties about subsequent actions by North Korea, China, and other states. The China team had to consider uncomfortable options.

Dealing with these matters was prescient. Reportedly, in the real world, both the Obama and Trump administrations later examined options for preventive-war strikes on North Korea because of the grave threat that North Korean nuclear weapons and ICBMs posed for the United States (Sanger and Broad 2017). As in the game, such options proved unattractive because of intelligence shortcomings and a likely DPRK response destroying Seoul. However, cyberattacks were apparently authorized to delay DPRK developments.

Reflections and Conclusions

Since the game was an experiment, we prepared grade cards. Table 7 shows an assessment informed by a post-game survey of players conducted by KIDA. The game itself went well with players reporting favorably on the day's activities. Our biggest disappointment was that players did not ask for or demand much information, as would surely occur in a real-world crisis environment. A next kind of grade (Table 8) compared discussion in the game to the model developed beforehand. A primary reason for using human wargaming is that models are often too simplistic. Human players can quickly point out additional factors and considerations. They can often be more creative. In this case, however, the pre-existing

Table 7 Assessment of game experiment.

Issue: Did teams	Result
Accept scenario?	Yes
Develop options?	Yes
Define options adequately?	Yes
Discuss in depth?	Yes, at the political level
Have alternative models of adversary?	Yes, sketchy but meaningful
Most-likely, best-case, and worst-case assessments?	Yes
Have disagreements?	Yes, but polite
Provide explanations?	Yes
Take their role seriously?	Yes
Consider contingent options?	Implicitly at best
Request information?	No

Table 8 Important factors in model or game.

Model variable (hypothesis)	Discussed by teams?
Adversary model (Blue's Red, Red's Blue)	Yes, but best estimate prevailed
DPRK alert status	Minimal
ROK-US alert status	Built-in; shallow recognition
"Other" indicators of attack and WMD use	Acknowledged, but not assimilated
Indicators of irrationality	Acknowledged, but not assimilated
Certainty about Red intentions	Irrationality was discounted
Option feasibility govem alerts, intelligence, etc.	No
Blue cohesion	No
ROK damage-limitation capability	No
DPRK launch under attack and delegation possibilities	No
Potential of Chinese intervention	Yes, but little discussion of consequences

theory and model were richer in some respects than the game discussion. The human game tended to confirm the prior theory and model. As intended, however, human teams also added options and elaborated reasoning in ways that the original model has not anticipated.

We were surprised that few of the important factors identified in the model (left column of Table 8) were dealt with by the human teams. This reflected the choice of people recruited and the game's political-military (rather than military-political) nature. The players were outstanding, but reflected their day jobs. An additional factor may have been that the game included both Koreans and Americans. Finally, a big factor was that the model constructed ahead of the game reflected my own years of experience with nuclear-crisis analysis during the Cold War era, experience that the game participants did not have. That experience included a mix of modeling, simulation, wargaming, and analysis. Thus, the case described here was actually an iteration and specialization, rather than a first pass through the process of Figure 3. For a more typical analysis of possible crisis, prior thinking might be more primitive with no reasonable starting model. Early human wargaming might be especially valuable. One notable example, the lessons from which were not heeded at the time, was Red Teaming of the US Millennium Challenge wargame in 2002. The Red Team immediately found vulnerabilities in the US concept of operations, which it wanted to exploit. That would have disrupted the exercise, so the Red Team was ignored and its leader, Lt. General Paul Van Riper (USMC), stepped down.[8]

Based on subsequent discussion with other Korean and American figures, including some who have been senior officials or military officers, it is clear that in the same game but with different players, some players would have paid close attention to issues such as "what do you mean by the state of alert; who has done what so far; and which of our options are still feasible?" Also, there would have been discussion of ballistic-missile defense, a sensitive subject in South Korea. Discussion of Chinese options and preferences would have gone differently. KIDA conducted additional games on its own, in which I did not participate.

For subsequent work, we considered mechanisms to enhance games of this type. These include simple decision aids to remind players of factors they might otherwise not think about, charging the teams with much more demanding questions, and more "Control Team" interventions in the midst of team deliberations.

Some conclusions about using wargaming in a model-game-model paradigm are

- It is crucial to plan sets of games, whether parallel, sequential, or both, and to define the games, recruit players, and establish details so as to probe different corners of the possibility space. Regrettably, we did not have the time or resources to do so.

---

8 The most dramatic account appears in a larger book (Gladwell 2005). A fuller and more balanced account appeared subsequently (Zenko 2015) but reinforced the value of Red Teaming.

- The insights from a particular game will sharpen understanding of some analytic aspects of the problem, but may contribute nothing on others.
- In using a model-game-model approach, a thoughtful pre-existing model will sometimes be more insightful, in some respects, than a human game.

All of this supports another long-standing principle, which has been expressed in diverse ways:

- Valid analysis will only seldom follow directly from human gaming, and results of human games should rarely be considered as "evidence" akin to empirical data. Analysts are responsible for the logic and flow of their analysis, which should be understandable and compelling in a way that transcends the particulars of the human gaming.

Implications for Simulation

In this paper, I have focused on relating gaming to the conceptual model and a highly simplified computational model. How should gaming be related to more ambitious simulations, by which I mean computational models generating behaviors over time? My own advice on the matter is as follows:

- Assure separate development of a conceptual model that is not constrained by the practical issues that often limit what is included in full simulations.
- Use gaming to enrich that conceptual model, focusing primarily on fundamental issues such as identifying the appropriate factors, perspectives, and dilemmas. Accept that these will often be qualitative.
- Subsequently, develop simulations that incorporate as much of the richness as appears suitable to the application. Do not omit important matters because of an imagined requirement for precise quantification or because of confusing that with rigor.
- Assure that the simulation is designed from the outset so as to support extensive exploratory analysis across uncertainties and disagreements. Sound analysis requires this.
- Use model-supported analysis to pre-emptively address a vast range of what-ifs, presenting outcomes in intelligible parametric displays that communicate well the basic truths (as we understand them) without claiming precision or certainty.

Some of these admonitions are long-standing, consistent with the past literature on capabilities analysis and the preferences of analytically savvy decision-makers who are rightfully skeptical about the more precise point calculations sometimes favored by the modeling-and-analysis community.

References

Bennett, B.W. (2013). *The Challenge of North Korean Biological Weapons*. Santa Monica, CA: RAND Corporation.

Body, H. and Marston, C. (2011). The peace support operations model: Origin, development, philosophy and use. *Journal and Simulation* 8 (4): 69–77.

Bracken, P. (2012). *The Second Nuclear Age: Strategy, Danger, and the New Power Politics*. New York: Times Books.

Caffrey, M.B. (2019). *On Wargaming: How Wargames Have Shaped History and How They May Shape the Future*. Newport, RI: Naval War College Press.

Connable, B., Perry, W.L., Doll, A., Lander, N., and Madden, D. (2014). *Modeling, Simulation, and Operations Analysis in Afghanistan and Iraq: Operational Vignettes, Lessons Learned, and a Survey of Selected Efforts, RR-382-OSD*. Santa Monica, CA: RAND Corporation.

Davis, P.K. (1989). *Some Lessons Learned from Building Red Agents in the RAND Strategy Assessment System, N-3003-OSD*. Santa Monica, CA: RAND Corporation.

Davis, P.K. (2014). *Analysis to Inform Defense Planning Despite Austerity*. Santa Monica, CA: RAND Corporation.

Davis, P.K. (2016). *Capabilities for Joint Analysis in the Department of Defense: Rethinking Support for Strategic Analysis*. Santa Monica, CA: RAND Corporation.

Davis, P.K. (2017). *Experimenting with a Model-Game-model Paradigm for Using Human Wargames in Analysis. Working Paper WR-1179*. Santa Monica, CA: RAND Corporation Corp.

Davis, P.K. and Arquilla, J. (1991a). *Deterring or Coercing Opponents in Crisis: Lessons from the War with Saddam Hussein*. Santa Monica, CA: RAND Corporation.

Davis, P.K. and Arquilla, J. (1991b). *Thinking about Opponent Behavior in Crisis and Conflict: A Generic Model for Analysis and Group Discussion*. Santa Monica, CA: RAND Corporation.

Davis, P.K., Bankes, S.C., and Kahan, J.P. (1986). *A New Methodology for Modeling National Command Level Decisionmaking in War Games and Simulations*. Santa Monica, CA: RAND Corporation.

Davis, P.K., Wilson, P., Kim, J., and Park, J. (2016). Deterrence and stability for the Korean Peninsula. *Korean Journal of Defense Analyses* 28 (1): 1–23.

Davis, P.K., Kulick, J., and Egner, M. (2005). *Implications of Modern Decision Science for Military Decision Support Systems*. Santa Monica, CA: RAND Corporation.

Davis, P.K. and O'Mahony, A. (2013). *A Computational Model of Public Support for Insurgency and Terrorism: A Prototype for More General Social-Science Modeling, TR-1220*. Santa Monica, CA: RAND Corporation.

Davis, P.K. and O'Mahony, A. (2017). Representing qualitative social science in computational models to aid reasoning under uncertainty: National security examples. *Journal of Defense Modeling and Simulation* 14 (1): 1–22.

Davis, P.K., O'Mahony, A., and Pfautz, J. (eds.) (2019). *Social-Behavioral Modeling for Complex Systems*. Hoboken, NJ: John Wiley & Sons.

Davis, P.K., O'Mahony, A., Gulden, T.R., Osoba, O.A., and Sieck, K. (2018). *Priority Challenges for Social-Behavioral Research and Its Modeling*. In: *Santa Monica*. Calif.: RAND Corporation.

Davis, P.K. and Winnefeld, J.A. (1983). *The RAND Corporation Strategy Assessment Center, R-3535-NA*. Santa Monica, CA: RAND Corporation.

Delpech, T. (2012). *Nuclear Deterrence in the 21st Century: Lessons from the Cold War for a New Era of Strategic Piracy*. Santa Monica, CA: RAND Corporation.

Downes-Martin, S. (2013). Adjudication: The diabolus in machina of war gaming. *Naval War College Review* 66 (3): 67–80.

Dunnigan, J.F. (2003). *How to Make War: A Comprehensive Guide to Modern Warfare in the Twenty-First Century*, 4the. New York: Harper Perennial.

Fox, D.B. and Jones, C.M. (1998). *JICM 3.0: Documentation and Tutorial, DRU-1824-OSD*. Santa Monica, CA: RAND Corporation.

Gladwell, M. (2005). *Blink: The Power of Thinking Without Thinking*. New York: Little, Brown.

Guarino, S., Eusebi, L., Bracken, B., and Jenkins, M. (2019). Using sociocultural data from online gaming and game communities. In: *Social-Behavioral Modeling for Complex Systems* (eds. P.K. Davis, A. O'Mahony and J. Pfautz), 407–473. Hoboken, NJ: John Wiley & Sons.

Hanley, J.T. (2017). Changing DoD's analyst paradigm: The science of war gaming and combat/campaign simulation. *Naval War College Review* 70 (1): 64–103.

Herman, M.L. and Frost, M.D. (2009). *Wargaming for Leaders: Strategic Decision Making from the Battlefield to the Boardroom*. McLean, VA: Booz Allen Hamilton Inc.

Jervis, R., Lebow, R.N., and Stein, J.G. (1985). *Psychology and Deterrence*. Baltimore, MD: Johns Hopkins University Press.

Kahn, H. (1960). *On Thermonuclear War*. Princeton, NJ: Princeton University Press.

Kahn, H. (1965). *On Escalation: Metaphors and Scenarios, Westport, CT*. Greenwood Press.

Kahneman, D. (2011). *Thinking, Fast and Slow*. New York: Farrar, Straus and Giroux.

Kissinger, H. (1957). *Nuclear Weapons and Foreign Policy*. New York: Council on Foreign Relations.

Lakkaraju, K., Epifanovskaya, L., Stites, M., Letchford, J., Reinhardt, J., and Whetzel J. (2019). "Online games for studying behavior," In: *Social-Behavioral Modeling for Complex Systems* (ed. P. K. Davis, A. O'Mahony, and J. Pfautz), 387–405. Hoboken, NJ: John Wiley & Sons.

Landler, M. (2018). U.S. considers reducing force in South Korea. *New York Times*),3 May: A1.

Lempert, R. J., S .W.Popper, and S. C.Bankes (2003). *Shaping the Next One Hundred Years: New Methods for Quantitative Long-Term Policy Analysis*.Santa Monica, CA: RAND Corporation.

Levine, R., Schelling, T.C., and Jones, W. (1991). *Crisis Games 27 Years Later: Plus C'est Déjà Vu, P-7119*. Santa Monica, CA: RAND Corporation.

Miller, L.C., Wang, L., Jeong, D.C., and Gillig, T.K. (2019). Bringing the 'real world' into the experimental lab: Technology-enabling transformative designs. In: *Social-Behavioral Modeling for Complex Systems* (eds. P.K. Davis, A. O'Mahony and J. Pfautz), 359–380. Hoboken, NJ: John Wiley & Sons.

National Research Council (2014). *U.S. Air Force Strategic Deterrence Analytic Capabilities: An Assessment of Methods, Tools, and Approaches for the 21st Century Security Environment*. Washington, DC: National Academies Press.

National Research Council and Naval Studies Board (1996). *Post-Cold War Conflict Deterrence*. Washington, DC: National Academy Press.

Oberholtzer, J., Doll, A., Frelinger, D., Mueller, K., and Pettyjohn, S. (2019). Applying war-games to real-world problems. *Science and Justice* 363 (6434): 1406.

Paul, C. (2019). Homo narratus (the storytelling species): The challenge (and importance) of modeling narrative in human understanding. In: *Social-behavioral modeling for complex systems* (eds. P.K. Davis, A. O'Mahony and J. Pfautz), 849–864. Hoboken, NJ: John Wiley & Sons.

Perla, P.P. (1990). *The Art of Wargaming: A Guide for Professionals and Hobbyists*. Annapolis, MD: US Naval Institute Press.

Pierson, B., Barge, W., and Crane, C. (2008). The hairball that stabilized Iraq: Modeling FM 3-24. In: *The Human Social Cultural Modeling Workshop* (eds. A. Woocock, M. Baranick and A. Sciaretta), 362–372. Boca Raton, FL: CRC Press.

Pournelle, P. (ed.) (2017). "MORS wargaming special meeting, October 2016: Final report." https://www.mors.org/Portals/23/Docs/Events/2016/Wargaming/MORS%20Wargaming%20Workshop%20Report.pdf?ver=2017-03-01-151418-980.

Reddie, A.W., Goldblum, B.L., Lakkaraju, K., Reinhardt, J., Nacht, M., and Epifanovskaya, L. (2018). Next-generation war-games. *Science* 362 (6421): 1362–1364.

Reddie, A.W., Goldblum, B.L., Reinhardt, J., Lakkaraju, K., Epifanovskaya, L., and Nacht, M. (2019). Applying war-games to real-world policies: Response. *Science* 363 (6434): 1406–1407.

Rouse, W.B. (2019). Human-centered design of model-based decision support for policy and investment decisions. In: *Social-Behavioral Modeling for Complex Systems* (eds. P.K. Davis, A. O'Mahony and J. Pfautz). Hoboken, NJ: John Wiley and Son.

Rubel, R.C. (2006). The epistemology of war gaming. *Naval War College Review* 59 (2): 108–128.

Sanger, D.E. and Broad, W.J. (2017). Trump inherits a secret cyberwar against North Korean Missiles. In: *New York Times, Asia Pacific (4 March)*.

Schelling, T.C. (1987). The role of war games and exercises. In: *Managing Nuclear Operations* (eds. A.B. Carter, J.D. Steinbruner and C.A. Zraket), 426–444. Washington, DC: Brookings.

Schwabe, W.L. (1992). *Analytic War Plans: Adaptive Force-Employment Logic in the RAND CorporationStrategy Assessment System (RSAS)*. In: *N-3051-OSD*. Santa Monica, CA: RAND Corporation.

Shlapak, D.A., Schwabe, W.L., Lorell, M.A., and Ben-Horin, Y. (1986). *The RAND Strategy Assessment System's Green Agent Model of Third-country Behavior in Superpower Crises and Conflict*. Santa Monica, CA: RAND Corporation.

Simon, H. A. (1978). "Nobel Prize lecture: Rational decision-making in business organizations."NobelPrize.org.

Thompson, J., McClure, R., and DeSilva, A. (2019). A complex systems approach for understanding the effect of policy and management interventions on health system performance. In: *Social-Behavioral Modeling for Complex Systems* (eds. P.K. Davis, A. O'Mahony and J. Pfautz). Hoboken, NJ: John Wiley and Sons.

Tolk, A. (2012). Verification and validation. In: *Engineering Principles of Combat Modeling and Distributed Simulation* (ed. A. Tolk). Hoboken, NJ: John Wiley and Son.

van Creveld, M. (1991). *The Transformation of War: The Most Radical Reinterpretation of Armed Conflict Since Clausewitz*. New York: Free Press.

Wong, Y.H. (2016), How Can Gaming Help Test Your Theory, PAXsims, https://paxsims.wordpress.com/2016/05/18/how-can-gaming-help-test-your-theory/

Wong, Y.H., Bailey, M., Grattan, K., Stephens, C.S., Sheldon, R., and Inserra, W. (2017). The use of multiple methods in the joint irregular warfare analytic baseline (JIWAB) study. *Journal of Defense Modeling and Simulation* 14 (1): 45–55.

Zenko, M. (2015). Millennium challenge: The real story of a corrupted military exercise and its legacy. *War on the Rocks* 5 November.

5

Wargaming, Automation, and Military Experimentation to Quantitatively and Qualitatively Inform Decision-Making

Jan Hodicky[1] and Alejandro S. Hernandez[2]

[1] *NATO HQ SACT, Norfolk, Virginia, USA*
[2] *Naval Postgraduate School, Monterey, California, USA*

Introduction

The tug-of-war between military credibility and mathematical rigor to inform decision-making is continuous. A complex world, made even more complex in multidimensional operations, diversity of forces, idiosyncrasies of leadership personalities, considerations for global policies, and ubiquity of information, reinvigorates the tremendous need to reduce Clausewitzian friction (Watts 1996). A means for mitigating this friction is by increasing quantitative analysis to support decisions. Today's commanders demand traceable, credible, quantifiable information.

The number of tools that are available for collecting information steadily grows throughout the military and commercial domains. However, leaders are unsatisfied with just more information; they seek *new knowledge*. Commanders require the ability to examine and re-examine problems to obtain a *grand innovation* that results in a decisive comparative advantage over an adversary (Worley 1999). Such innovation comes from discovered knowledge (Box, Hunter, and Hunter 2005). A need to balance the objectives of acquiring information and developing insights is not new. Gaining information is critical to decision-makers (Kaplow and Shavell 2004; Raiffa 1970), but the need to learn about previously unseen paths to victory is the goal of any commander. New approaches for attaining knowledge are at the nexus of technology and research methodologies.

Simulation and Wargaming, First Edition. Edited by Charles Turnitsa, Curtis Blais, and Andreas Tolk.
© 2022 John Wiley & Sons, Inc. Published 2022 by John Wiley & Sons, Inc.

A first level classification of ways to examine a problem shows the dichotomy between qualitative and quantitative methods (Creswell 2014). Quantitative methods are processes for collecting, analyzing, and interpreting resultant data that stem from samples gained through closed-ended inquiry about populations of interest. The types of data from quantitative methods are numerical in nature and generally come from surveys and experiments to help *explain* an event. On the other hand, qualitative approaches for investigation involve open-ended questions and personal definitions or interpretations in order to *describe* an event (Hamalainen, Sormunen, Rantapelkonen, and Nikkarila 2014). Qualitative data may be in several forms, to include textual, visual, anecdotal, or tabular. Whether qualitative or quantitative, each method entails specific techniques to obtain information about the problem. This chapter frames a unique application of selected techniques to provide leaders with militarily credible knowledge that can withstand scientific scrutiny. The result is a combination of automated computer-aided wargames (CAWs) and stage-wise experimentation that we illustrate later.

Military Methods to Knowledge Discovery

In discussing wargames, we specifically mean professional games, versus hobby games (Dunnigan 1992). Professional games focus on analysis of outcomes for the purposes of education and research (Perla and Curry 2011). Historically, wargames are a military staple. In the fifth century B.C., Sun Tzu introduced *Wei Hai* (encirclement) as a fundamental tool for understanding concepts in the *Art of War*, replacing Sumerian and Egyptian games of war. From the era of SunTzu to medieval times, *Chaturanga* and its cousin, *Chess*, dominated games for the strategic study of war. These high-level games, though educational, were difficult to translate for practical use. In 1664, Christopher Weikhmann presented a hands-on application of wargames through *Koenigspiel*, the King's Game. From *Koenigspiel* sprang a series of similar games under the title of "military chess." The most lasting impression came from Lieutenant George Heinrich Rudolph Johann von Reisswitz's version, *Kriegsspiel* or wargame (Perla and Curry 2011). In recent years, the US military has rediscovered wargames as an effective way to explore increasingly difficult problems upon which traditional decision-making tools are less effective.

Military commanders often have a narrow view of wargames. The military decision-making process employs wargames solely to analyze courses of action (DA – Department of the Army 2012). This human-in-the-loop (HITL), turn-by-turn execution of an existing military plan serves as a means to scrutinize courses of action while educating commanders and the staff. Analysis consists of reverse engineering the series of decisions that led to game outcome(s). Experimentation in this context consists of investigating different decisions. This limited venue fits the needs of the Commander, but not necessarily the conditions for knowledge discovery.

In his tenure at the Center for Naval Analysis, Peter Perla published a series of papers that distinctively described the nature and utility of wargames. The eventual book that emerged from these papers, *The Art of Wargaming*, edited by Curry (2011), is a rarity in its combined comprehensiveness and directness for defining wargames and wargaming. A primary discussion of the text is a comparison of wargames with other methods of inquiry that are available to military leaders. Details of these comparisons are in the following paragraphs. Most keen in these observations is the investigative use of wargames (Perla and Curry 2011). The power to learn about a future situation is a form of scenario methodologies to obtain knowledge (Lindren and Bandhold 2009, von Reibnitz 1988). Kress and Washburn (2009) further recognize that at times, the only means to address inherent complexities in truly competitive situations may be with a wargame.

Unlike wargames, which involve no live forces, military exercises consist of actual organizations, equipment, and environments in which operations occur. The main purpose of an exercise is to practice activities and tasks necessary to achieve a given mission. Exercises execute plans to determine the efficacy of the plan and to familiarize and train the forces. The scale of an exercise is dependent on the mission, the forces involved, and the intent. For instance, a live exercise is typically a tactical training event to drill the military members and staff on their respective roles in the plan. Command post exercises prepare the headquarters elements, including commanders at appropriate levels, on the command and control of forces (Cayirci and Marincic 2009). Because the goal of an exercise is to educate forces on a given plan and to examine that plan, knowledge acquisition is not a major objective, but can occur, although not by design.

Campaign analysis is an approach to examine a scenario and manipulate factors in order to achieve a desired outcome, in many cases, the success of a force of interest. Comparatively, wargames elicit information about future situations that may result in many outcomes, including failure of a mission. Wargames promote dynamic, continuous play; turn-by-turn, players consider new situations and make decisions until the game ends. Campaign analysis has the option to stop gameplay, study potentially undesired results, reset the situation after manipulating factors, and then continue playing. Iterative play enables campaign analysis to shape conditions in which the force of interest is successful. While the focus of campaign analysis is on examining the results through quantitative insights, wargames study the processes, such as preceding events and decisions that lead to the outcomes (Perla and Curry 2011). The knowledge that comes from campaign analysis is the formulated relationships among the factors of interest and the desired outcome(s).

Experiment is a test or series of tests that relate purposeful changes to input values with the observed variations in the output or response (Montgomery 2005). In a general sense, experiment has commonality with campaign analysis. The main difference from campaign analysis is that experiments deliberately plan and systematically manipulate the set of variables as a whole prior to beginning the game or run. Moreover, similar to wargames, experimental events are uninterrupted once the variable values are set. The experimental design (variables to manipulate and combinations of ways to do so) determines the analyses that may be borne from the resultant data. Campaign analysis is a "walk" though the possible factor settings to achieve a desired end. In contrast, experiment sets the paths to explore without inclination toward any outcome.

Box, Hunter, and Hunter (2005) emphasize that experiment is critical to learning and essential in the scientific process. Continued discussion from Box et al. shows a perspective for continued discovery through experimentation. Richard Feynman (1995) is even more adamant about the power of experimentation and contends that experiment is the test of all knowledge, and *is itself (experiment) the source of knowledge*. A series of efforts since the early 2000s has made experiment a focal point for knowledge discovery. The Naval Postgraduate School Simulation Experiments and Efficient Designs Center has an international reputation for customized design of experiments to support decision-makers (https://my.nps.edu/web/seed/). Kleijnen, Sanchez, Lucas, and Cioppa (2005) cover approaches for designing simulation experiments for specific problem sets. Coupled with the drive for analytically rigorous methods from which national leaders can confidently make decisions (Hase, Matos, and Styer 2014), these statements contain the end product that leaders are seeking, discovered knowledge and a potential path to grand innovation (Worley 1999).

We must be clear in our discussion of experiments in this chapter to ensure that further exchange is in terms that serve our intended use. Experimentation from a military standpoint is a method of inquiry in three areas: development or examination of operational plans, understanding of new technologies, and a study of military behavior or success in new environments (Hernandez, Ouelett, and McDonald 2015). For our purposes, the working definition of *military experiments* is the "process of exploring innovative methods of operation, especially to assess their feasibility, evaluate their utility, or determine their limits" (Worley 1999, p. 11). Although this definition was developed many years ago, its suitability for the narrative in our thesis is near exact. As Worley (1999) outlines, the quantum purpose of military experimentation is the acquisition of knowledge about military operations.

Technology: Knowledge Enablers

Technology emerges from need, curiosity, inventiveness, creativity, innovation, and application. Opportunity to shape raw resources into a new or undiffused product is met with an intuition to ease previous efforts, or to realize new achievements (Diamond 2005, Schumpeter 1934). These circumstances frequently explain how technologies sprout from required activities in an organization, and in turn mark the implementation of that invention. Consequently, our discussion of technology is narrowly focused on those inventions that support the military leader's desire to acquire new knowledge: modeling and simulation (M&S) or computer simulation is a principal tool in the commander's cache and is at the core of our discussion.

Simulation models are specific technologies developed from mathematical and logical relationships to gain some understanding of a system of interest (Law 2015). However, connecting the appropriate M&S tool with the situation in question is a challenge, especially as the complexity of the problem increases. Connable, Perry, Doll, Lander, and Madden (2014) studied the collection of M&S tools that were developed to support decision-making in Iraq and Afghanistan. Their conclusions include a substantial description of the limitations for M&S applications as the sole means to address complex problems. A major finding details the successes and failures for identifying the proper M&S application for the given situation and associated research questions. In the operational landscape of Iraq and Afghanistan, the appropriate M&S tools were rare and limited to relatively deterministic problems (Connable et al. 2014). Subsequently, results from a single M&S application seldom illuminates greater understanding about the problem. Determining how to precisely couple methods and simulation models is one of the objectives of this chapter.

A logical progression of innovative application of simulation models is in support of wargames and exercises (Hernandez, et al. 2015, Cayirci and Marincic 2009). Benefits of CAW and computer-assisted exercises (CAEs) are evident. After front-end investments to develop the computer event, the infrastructure is reusable, thereby saving time and costs (Kress and Washburn 2009). Automated adjudication of interactions further enables free play. Additionally, authoritative sources establish a baseline for common understanding of the problem. Both the CAW and CAE involve HITL interactions, which require decisions that are singularly in the realm of training audiences and role players. Incorporating knowledgeable humans in computer events helps establish the credibility of CAW and CAE outcomes (Hamalainen, et al. 2014).

Computer simulation experiments systematically explore the decision space of military operations in a virtual environment. Analysts frequently turn to computer simulation to overcome challenges that physical experimentations encounter in safety, monetary, time, or resource issues (Gianni, D'Ambrogio, and Tolk 2015). The ability of computers to simulate increasingly complex problems provides analysts with great potential to assist decision-makers (Page 2016). Studies in human behavior and biomimetics often use computer simulations and rely on efficient designs of experiments that allow analysts to explore a broader range of possible innovations (Booker 2005). Agent-based models have enabled the United States Marine Corps Warfighting Lab to study and gain insights about nontraditional combat through the aggregate study of large numbers of nonlinear actors. The United Kingdom's Defense Science and Technology Laboratory developed the Peace Support Operations Model (PSOM) as a response to socioeconomic issues at the strategic level (Body and Marston 2011). For more than 35 years, Joint Forces have used the Joint Theater Level Simulation to train military and civilian audiences on contingency plans. However, it is possible to gain more than training and education from traditional exercise and wargame applications. Hernandez, Hatch, Pollman, and Upton (2018) demonstrate employment of experiments to these types of computer simulation to elicit new knowledge. In this chapter, we discuss several case studies that offer opportunities to develop an overall framework for knowledge discovery through military experimentation that open a space for the new classification of CAW.

Wargaming Automation Challenges in M&S Perspective

In this section, wargaming is defined and its relation to modeling and simulation is explained. Further, basic wargaming elements are explained followed by a discussion on the automation challenges where constructive simulation is the main tool to support wargaming.

Wargaming Relation to M&S

To get the meaning of wargaming and its relation to M&S, comparisons of several selected definitions of wargame are explicitly cited.

Perla (2011) defines a wargame as "A warfare model or simulation, using rules, data, and procedures, not involving actual military forces, and in which the flow of events is affected by, and in turn affects, decisions made during the course of those events by players representing the opposing sides."

NATO (2015) defines a wargame as "A simulation of a military operation, by whatever means, using specific rules, data, methods and procedures."

Red Teaming Guide (2013) defines a wargame as "A scenario-based warfare model in which the outcomes and the sequence of events affect, and are affected by, the decisions made by the players."

All definitions are putting stress on different elements of the wargaming; however, all together they create a scope of all the basic wargaming elements needed for its successful execution. These elements are further discussed in the section "Wargaming Elements". What these three cited definitions have in common is the use of models or simulation. Very fundamental for any wargame is a replica of the operational environment as close to the real situation as possible. The replica is expressed as a model and its behavior is scrutinized in simulation over the execution of the model in time. Simulation creates stimuli to the Players who are firstly forced to employ their creativity in the wargame planning phase and secondly, learning in the cognitive phase of wargame, both by wargaming environment.

The value of M&S approaches is well defined and experienced in the wargaming community, mainly in the training domain. Constructive simulations are mainly used in training and bring computer assistance to the wargaming (Cayirci, 2009). The chapter puts stress on the experimentation domain with constructive simulation support.

Wargaming Elements

Wargame can be seen from two main points of view. Firstly, wargame has its own life cycle. In this sense, wargame goes through five phases. The first phase of a wargame is known as Design. In this phase, mainly the objective, outputs and all the resources and roles needed to carry out the wargame are specified. It is followed by game Development. In this phase, all the wargame elements are defined and implemented. The implemented wargame elements are then verified and integrated into the whole wargame. Next phase, called Execution, stimulates Players to make plans and decisions that affect the flow of the wargame. Execution phase is usually composed of cycles in which the initial scenario is introduced, blue units make their plans, red units make their planning as well and a cognitive sub-phase reveals results of blue and red plans. This sequence may differ based on a type and a form of the wargame. Validation/Analytical phase is just after Execution and it serves as an evaluation of the result of the performed event. It is the main resource of knowledge of the analytical wargame and a pool of experience for future similar events. The wargame life cycle must be concluded by the Refine phase where the main findings are incorporated into the Design phase of a new wargame or a new event under the same Experimentation or Training Campaign. For further reading on a wargame life cycle, refer to Wargaming Handbook (2017).

Secondly, a wargame can be seen as a system, and therefore, it can be decomposed into the following key elements.

- Scenario is the scene opener for the Players. It is an exhaustive description of the scene and it contains geopolitical information of the area of operation, describing Political, Military, Economic, Social, Information, and Infrastructure (PMESII) factors. It contains minimum information needed to start the planning activity of the Players.

- Order of Battle (ORBAT) contains capability of your own forces and its characteristics. It is found in the hierarchy approach with described personnel, equipment and their status.
- Objectives of the wargame are critical points for successful design and implementation of the game. It contains declaration of the expected wargaming outcome. Solely execution of the wargame cannot be an objective of the event.
- Maps and Charts create the interface between the Scenario and the Players. Quality of the map resources and the way it is visualized, in two or three dimensions, significantly influence the overall quality of the wargaming life cycle (Hodicky, 2011). It is an extra add-on to the Scenario, emerging the Players into the operational environment.
- Clock is the driver of the flow of the game. It should reflect needs of the Objectives and the Players. Time elapsed during the wargame does not need to correspond with real time of the operation. Some vignettes might go faster or slower than in reality to fit better the needs of the Players and the rhythm of the wargame.
- Rules and Data create boundaries for the Players' moves and the decision-making process. Umpire enforces it during the game. Even the Umpire behavior is driven by the Rules and Data.
- Players create dynamicity of the wargame. They manage their planning and decision-making process. Typically, there are two teams of Players, Blue and Red.
- Analysts read the results of the wargame execution and make arguments to support the Umpire in his decision. He formulates the main findings of the whole wargame and makes proposal for the refinement of the upcoming cycles if needed.
- Facilitator supports the execution of the game to achieve the objectives of the wargame (WG). He neutrally supports the Players to adhere to the pace of the wargame. He has to stimulate the wargaming audience to be proactive and to use the provided wargaming time effectively.
- Umpire makes decision over the executed plans of the Players. He is the main enabler of the successful cognitive sub-phase of the wargame. He is in charge of formulating the achievements, drawbacks, and limitations resulting from the execution of a single wargaming cycle.
- Operators operate analytical tools if the form and the structure of the wargame allows so. It may include a discrete or stochastic model with a static or dynamic description of its behavior. Players instruct Operators to exploit their analytical tools making their planning procedures better and faster. Umpire instructs Operators to see the results of the executed plans of the Players to support their sub-cognitive phase. Analysts instruct Operators to employ tools to see the details of the executed planes. Operators should not make any initiatives to influence Players, Umpire, or Analysts. They are just following instructions.

Constructive Simulation Building Blocks

Very fundamental classification of M&S makes the difference between Live, Virtual, and Constructive simulation (Hodson 2018). The essential difference is based on how the aspects of reality are included, specifically from people or a system of interest. Constructive simulation, the main interest of the paper, is qualified as a simulation involving simulated people operating simulated systems with simulated effects over a simulated environment. Real people – operators in constructive simulation make inputs to such simulations. Following part describes basic block building in the constructive simulation, generalized from the existing constructive simulations portfolio, and used mainly for training purposes:

- Scenario block is used to design and implement scenario including ORBAT with a structure of their own forces and enemy forces.
- Executable Models block makes calculations of replicated entities' behavior in the simulation, following defined rules applied to data.
- DB Parameters block contains inputs for Executable Models block. Probabilities of hit and kill or a weapon characteristic are the examples of these parameters.
- Graphical User Interface (GUI) provides a communication between the operator of the Constructive simulation and the Executable Models.
- Event Planner makes a plan of activities with the entities represented in the simulation. There are simulated entities instructed to follow the prebuild orders in the extent of its integrated autonomy.
- Visualization ensures that common operational picture is visualized in 2D or 3D environment. Visualization is used not only to represent real time execution of the implemented models, but it serves also to After Action Review (AAR) block.
- AAR demonstrates the flow of events and the status of all the entities represented in the synthetic environment in the logical time.
- Clock drives the execution of models in accordance with the logical time of the simulated scenario.
- External Communication follows the principles of interoperability by enforcing standards for the distributed simulation and the command and control systems.
- Terrain contains geographical data used to model the environment over which the simulated entities are represented.

Wargaming Elements Not Supported by Constructive Simulation

There are two main categories of wargame elements. The first category are elements representing human beings in the game and the second category are technical elements of the game. Unified constructive simulations building blocks, mentioned in previous section, are not covering all the elements from both categories.

From the category containing Scenario, ORBAT, Objectives of wargame, Map and Charts, Clock and Rules and Data, only Objectives of wargame is not directly supported by the constructive simulation. The others are supported with limitations given by the training purpose of the constructive simulation.

The second category contains Player, Analyst, Umpire, Operator, and Facilitator. Among them, Player and Umpire are represented, to some extent, in the wargame with limitations given by the training character of the wargame.

Umpire is represented by the executable models and by the ability to express the effects and results of the executed plans over the battlefield.

Players are represented by automata that replicates behavior of the commander and doctrine, applied at an appropriate level of the command. Therefore, aggregation and disaggregation mechanism of the replicated entities in the simulation represents, to some extent, the Players. However, representation of a Player that makes a plan, as a reflection of a current situation and as an effect of the wargame life cycle cognition phase, has not been fully introduced yet.

Challenges to Combined Methodologies for Knowledge Discovery

Wargaming literature is rife with arguments against using research wargames for statistical analysis. As CAW became common practice, the prevalent argument was that such events do not facilitate classical analysis (Perla and Barrett 1984). Perla and Curry (2011) and Rubel (2003) maintain that the very nature of wargames prevents the application of statistical methods on data collected from a game. However, Rubel (2003) does eventually find an approach for extracting credible statistics from wargames, which we discuss later in this section. Dynamic play and the temperament of human beings guarantee that no two traditionally played wargames are the same. In fact, the expense in time, people, and resources dissuade efforts to repeat the same game with the same people (Kress and Washburn 2009). Consequently, the data from these games prove unusable for statistical inferences.

Game theory has lineage with wargames and potential for providing valid analytics. Yet, game theory suffers from the same gaps in information that compels the need to conduct wargames. Originating from the study of military tactics in World War II, game theory evolved as a discipline centered on an organized approach for strategic decision-making (Kaplow and Shavell 2004). The data requirements for a formal game is demanding. The set of outcomes for each decision must be known, along with the probabilities for following a decision branch, and the ultimate "payoff" value of the outcome at the end of the branch. As Kress and Washburn (2009) explain, the values for an outcome result from the study of multiple actors in similar situations, which enable the development of utility functions. Formal games can then be summarized as matrices for analyses. This point of view has extremes. For instance, De Mesquita (1981) formulates an expected utility function for *global war*, which requires essentially unattainable data, i.e. the perception of anticipated changes in world views between actors *i* and *j*. Of course, the circuitous dilemma that the situation presents is how to make such computations when the dearth of information about the problem is precisely the reason for the investigation.

Appleget and Cameron (2015) proclaim misplaced confidence in the results of computer-aided wargames and computer simulations to address complex problems. Concerns about the validity of the simulation outcomes are raised from the absence, or limited presence, of human decisions that automated decision rules (or matrices) have replaced. Appleget and Cameron are not alone in their assertions that some computer simulations are mischaracterized as wargames. Koopman (1956) had recognized that there is much more to the study of operations than mathematics and the experimental sciences. A MITRE study shows that many classes of experiments that use M&S are also being viewed as wargames (Page 2016). Yet, these simulation experiments do not have the advantages of scenario methodologies that value the human element (Mietzner and Reger 2005). Hamalainen et al. (2014) further argue that the prominent role of experienced players in research wargames contributes to the value and credibility of the method, and its absence, the opposite effect.

Kaplow and Shavell (2004) present some options to capture realism or human factors by constructing the wargame to have a manageable set of branches that would reference a credible set of player decisions. However, such an approach constrains the exploratory nature of wargames. More importantly, analysts must deal with the difficult prospect of explaining how the sparse data set represents the actual decision space. Since large-scale wargames may have a near infinite number of endings, a handful of outcomes is unlikely to define the behavior of any output of interest from the game. Structuring a research wargame to explore the decision space in a manner that produces credible data for scientific analysis is the crux of the problem.

Constructive Simulation Constrains in the Context of Automation and Wargaming

Current constructive simulations are mainly used in training domain. Very high investments into constructive simulation are not needed. Therefore, the reuse of existing building blocks and its appropriate update, to serve the experimental domain better, is preferable and more effective than implementing new simulation tools to support the experiment from scratch. If the wargame is implemented for experimentation purposes, the need to reproduce results with the same inputs variation is a must. Replication is enabled by a maximum automation, implemented in the wargame life cycle and in the wargame elements. The following discussion focuses on main challenges and constrains in the sense of the wargame being supported by constructive simulation with a maximum automation for experimentation purposes (Hodicky, 2019) (Hodicky, 2017).

Current approach for describing a **scenario** in constructive simulation is based on Military Scenario Description Language (MSDL) and Coalition Battle Management Language (CBML) formalisms. MSDL is a language that was designed to enable different real entities or simulation entities to share their scenario description files and to reduce the time for a recreation of the scenario for every single entity based on the proprietary formats when integrated into one executable environment (Blais, 2009). CBML is designed to facilitate exchange of orders and reports among the simulation applications, C2 systems and autonomous systems (Blais, 2012). Latest effort merges these two formalisms into a single framework called C2SIM and brings interoperability between the simulation and the real battlefield domain. This framework has a potential to bring an automation into the processes of scenario design and implementation in the constructive simulation at all levels of command and control. However, wargames at a strategic level have different scenario wording than wargames at the operational and tactical levels. There is a limitation of the CBML and MSDL vocabulary, in the context of a scenario description at a strategic level. These two formalisms are very tactically oriented.

Constructive simulation works at a defined level of the command and control, e.g. it is suitable only for an operational level, at the most it might be used at two levels – operational and tactical. If the wargame is supported by such constructive simulation, then, there is a need to transform the scenario description from a higher level to a lower level. Even if the wargame is played at a higher level, the scenario transformation mechanism might allow using the constructive simulation originally designated for one level below. Now, there are no best practices or best automation for the translation from the strategic level down to the operational level, and from the operational level down to the tactical level. MSDL and CBML may be good candidates for extending their vocabulary with more elements, allowing semiautomated translation of the scenario between the different levels of the command and control. Moreover, MSDL and CBML allow design and implementation of the scenario and there are already proposals to translate the scenario into the executable form readable by the constructive simulation (Khimeche, 2016).

Design and implementation of a scenario is a time- and resource-consuming effort. Constructive simulation has a feature allowing implementation of the scenario into the simulation environment. However, it is done manually, without any automation support. Ideally, the constructive simulation package would generate a scenario close to the most likely realistic situation, respecting the objective of the wargame. That would require building interfaces between the constructive simulation and the databases containing relevant real-time military data. JANE'S database can be taken as an example (Janes, 2019). Designing such a programmable interface would also serve as a scenario validation tool.

All in all, these steps would bring more realism into the wargame and it would increase its reliability coming from experimentation. One promising approach is to design a general part of the scenario called Settings, similar to the one used in CAX events, that are valid and reused for more events. Modifications that reflect the current security and safety environment and the objective of the experiment are updated faster and easier than building the whole scenario from scratch. These Settings should be maintained in the DB that is interconnected with the constructive simulation.

Order of Battle in the wargame is generally an identical copy of already existing structure of a training event. However, for experimentation, ORBAT can be even the subject of the experiment. The optimum structure of its own forces for defined mission can be taken as an example. In this case, the validity of the hierarchically organized systems, weapons and their parameters must be checked against the existing databases (Janes, 2019) and it has to be done in accordance with the national doctrine. Now, there is no such a tool with any level of automation.

An **objective** of a wargame, in the experimentation domain, is transformed and expressed by the constructs in the form of hypotheses, metrics, and indicators. NATO environment defines three types of experiments and not all of them require using all the constructs. The first type is Discovery Experiment with the objective of getting new insights that were not known before. It can be used as a mean to generate the hypothesis. The second type is Hypothesis Testing with the only objective to prove or disapprove the selected hypothesis in the context of the defined inputs, outputs, and condition of the experiment. The last type is Validation Experiment with the objective of demonstrating that the military capability, proved by testing hypothesis, fulfilled the requested set of requirements. Only in Discovery Experiment, there is no need to have hypothesis. Otherwise, the hypothesis is constructed to follow the objective of the experiment.

Firstly, it is defined, at the very abstract strategic or operational level, to be understandable to the operational and the key decision personnel. Afterward, this hypothesis is transformed into the levels where it can be finally proved or disapproved, by employing metrics and indictors. The following examples illustrate the process that is explained in detail in the Guide for Understanding and Implementing Defense Experimentation (2006).

For military-capability-oriented experiment, the hypothesis is formed using the following template:

- New capability covers operational task(s).

As an example, a new Intelligence Surveillance and Reconnaissance (ISR) capability can be taken:

- Robust ISR eliminates safe places of the opponent.

The hypothesis is transformed into one or more hypotheses at the experimental level:

- ISR version 1.D ensures continuous monitoring of the opponent units and system.

Afterward, the metrics and indicators are defined to prove the hypothesis. Potential Metrics can be Monitoring Effectiveness and it is formed by the following indicators:

- Number of Identified Units, Number of Real Units, Number of Identified Systems, Number of Real Systems, Time to Indicate, Area of Operation, and Area of Detection.

If the experiment is supported by a constructive simulation, the indicators are the inputs and outputs of the Executable Models implemented in the constructive simulation. If the selected indicator is not already implemented, the implementation has to be performed so that the automation of the experiment can be enabled.

The process of designing and connecting the hypothesis, metrics, indicators, inputs, and outputs of the Executable Models is not supported in the current constructive simulation tools. A database containing the already designed and executed experiments and their related scenarios would be a good starting point for validation of the hypothesis, metrics, and indicators against the objective of the wargame, and would be also useful for a further automation of the experiment design phase. Otherwise, there are potential sources of errors coming from the biased view of the game and the experiment designer.

Having a **map** in the wargame is a critical requirement. The wargame objective drives the required resolution of the visualized operation area based on the available terrain data. A common misleading practice is to provide as detailed information to the Players as possible (Hodicky, 2016). On the contrary, a good practice is to automatically define the requested resolution based on the scenario, the size of the operational area, and the objective of the wargame.

There are experiments for getting a real territory replicated. Getting products like 2D or 3D visualized terrain database is still a time- and resource-consuming effort. There are some promising approaches to fully automate the generation process of 3D data on demand, in almost real time from satellite, vector, and relief data (Hodicky, 2011).

On the other hand, there are some cases when the terrain database must be created as a faked terrain. Therefore, there must not be any similarity to the real-world geo features as it could cause political problems. Thus, automation of a fake

Table 1 Commonalities and differences of constructive simulation in the experimentation and training domain.

Selected wargame elements with employed constructive simulation	Training	Experimentation
Environment where Players act	As close as possible to the situation in the battlefield	Not critical
Execution	Simulation time is equal to the real time	Simulation time is as fast as possible
Resolution (Rules and Data)	Not critical	As close as possible to the situation in the battlefield
Design	Same techniques applied	Same techniques applied
Analysis	AAR applied, fast process	Statistical modeling applied, slow process

terrain generation process is among hot topics and challenges in the experimentation and training domain.

Current constructive simulations are complex systems or even a system of systems. The complexity depends on the system architecture, level of aggregation, and the number of mutually communicating executable models describing behavior of the battlefield entities. Current description of the **Rules**, implied in the constructive simulation, is not done in the most natural way to be understandable to the operational people. It results in a very complicated and, in some cases, even impossible validation process of the constructive simulation that is used for experimentation purposes. Constructive simulation cannot be taken from the training domain as is and be reused in the experimentation. They have some features in common, however, there are some differences too; it is summarized in Table 1.

Therefore, unified and native description of the rules and data used in the constructive simulation, mentioned in third line of Table 1 (Resolution of analytical tools), is a critical and needed activity from the validation point of view. Omnipresent Service-Oriented Approaches, applied in the simulation domain, seem to be a promising way to standardize this description (Prochazka, 2017) and to bring it closer to the operational people by interpreting it into an operational schema. Rules and Data might be described accordingly to XML schema and their associations can be expressed in an operational schema form.

In a wargame, **Players** are the main advantage and disadvantage at the same time. If the wargame is used for training purposes, then creativity of the Players is requested, and it fulfils their unique role representing human factors in the game. On the other hand, if the wargame is used for experimentation purposes, human factors are going against the idea of reproducibility of the experiment results. If the same wargame is repeated by another group of operational people, the results would be most likely different. Therefore, the current approach in constructive simulation is to aggregate human being(s) together into a single automata with system characteristics, trying to mimic human factors that influence the decision-making process. Results coming from such approach are reproducible as it follows defined algorithms. The main challenge is to replicate the Player by automata at all the decision levels that are applied in the wargame. Player automata validation techniques are needed and currently there are not any in the constructive simulation.

Another approach is building an expert system that puts together similar wargame scenarios and similar plans that have been already implemented as the reaction to the scenario. When the expert system is mature, then the appropriate plan is selected and executed in the experiment based on the scenario category. Further step may be to implement machine-learning techniques to train the expert system using current plans and to predict the most likely outcome as inputs for the execution phase of the wargame. This further step in the automation would introduce a reduction of the Players bias when making their plans, by putting them into the plausible plan limits.

Analyst creates the interface between the gathered results of the experiment and the key decision-maker who will then use these results to support his decision. This role is critical for the credit of the wargame and it is requested not to have any Analyst bias. This requirement is not easy to fulfil as there are attempts to interpret the results in a way that fits better to the expected or needed outcomes of the experiment. However, in a constructive simulation, the role of Analyst is neglected. It is only represented by an AAR module that makes basic statistical summaries over the experimental results. Missing is also implementation of the automation indicators into the set of easy to understand set of metrics for the decision-maker. There is no implemented tool/approach in the constructive simulation that would allow automatically proving or disapproving the defined hypothesis of the experiment. Automation of Analyst is the most challenging and provocative objective in the wargaming automation.

There is still a problem of mixing rigid and open wargames to serve their objectives better. An open wargame gives possibility to the **Umpire** to decide over the situations that are not yet described by rules; however, at the same time it brings Umpire bias into the wargame. On the other hand, a rigid wargame implements the Umpire in the form of the wargame internal rules. Current constructive simulation allows to mix both approaches only in a very naive way. If the results of the executable models are not corresponding to the idea of the Umpire, then he may modify it. However, this approach is not recommended as it degrades the value and validity of the constructive simulation. One promising way to solve this problem is to implement a learning phase of the executable models. It would allow the

Umpire to modify the implemented rules on the fly in the constructive simulation within the provided limits. Such modification cannot be done without a proper validation process. However, it can be executed automatically using statistical analysis to be sure that the generated results are still reasonable.

Umpire makes his decision about the results of the planned activities of both sides based on his experience and support from his analytical team. Analytical tools give him needed assistance to increase the quality of the cognitive phase of the wargame. However, current constructive simulators are too complex and discrete in their character. They are simply not fast enough to be used in a wargame as a support to critical time decisions. Therefore, new approaches are needed to be implemented into the constructive simulation to get the recommendation to the Umpire in the cognitive phase as fast as possible. One promising approach is to implement a new layer over the constructive simulation using continuous modeling or system dynamic paradigms to describe the effects of planned activities in the form that is independent of a particular scenario. Therefore, effects are not described by the numbers of destroyed units and systems, but they are summarized in the general parameters like resiliency, deterrence, elimination, reaction. The challenging part is to transform the ORBATS and Plans of both sides into the inputs of the structure of the expert system.

The following section contains latest examples where the automation of the selected wargaming elements in experimentation has already proved its benefits.

Stage-Wise Experimentation in CAW

Our research posits that there are combinations of qualitative and quantitative methods that can reliably lead to knowledge discovery. The ensuing methodology that we describe automates a CAW and innovatively implements a form of experimentation, designated as stage-wise experimentation. This new approach for computer experimentation employs human expertise, accounts for peculiarities of leader traits, and seizes upon the benefits of tremendous computational power. Other conveniences that computer simulations offer include realism through random play, rapid adjudication of game time decisions, and variable game speed (Gianni, et al. 2015). Blending methods and technologies produces a unique approach that incorporates credible warfighting input, while producing data that can withstand scientific scrutiny.

A Progression of Mixed Methods to Grand Innovation

The approach for developing a blended research approach to military experimentation leverages how M&S frequently provides the environment for studying a problem. The ubiquity of wargames involving computer simulations and past studies for improving wargame analyses provide the basis for linking CAW and military experimentation. Automation and improvements in agent-based models make it possible to replicate the trace of a wargame outcome. In the following paragraphs, we discuss the progressive efforts of other researchers to automate wargames. Ideas from these studies lead to a novel approach for transforming a CAW into an experimental environment, and to which a process for stage-wise experimentation can preserve the value of the human element in the resulting wargame. We designate this new approach as stage-wise experimentation of CAW (SWE-CAW). The data from this new approach is traceable to human-based decisions and can support conventional analysis techniques.

In 1982, RAND researchers, Davis and Williams, developed wargames with the intent of using decisions that they captured from background studies of human expertise. The results are heuristics for applying specific combat models for particular scenarios, player decision rules, and a menu of strategies. Different combinations of these factors create instantiations of a new wargame. Establishing factor settings of each instance creates specific branches of sequels that describe the entire game. As such, the wargame can be run many times to collect data for scientific analyses. Davis and Williams applied rigorous discipline in the assumptions and rationale that they made for the rules in each decision branch. Unique to this early effort to automate human decisions was coding logic from experienced officers to replace HITL in common operations of the game. Davis and Williams (1982) implemented a simple decision tree to follow common operations in the game. The levels of the tree include the decision to invade and fight, and methods to use. We recognize that current M&S languages and sophisticated agent-based modeling techniques can easily incorporate such decisions. When unusual events occurred, the RAND team employed a Control Team, and then updated the decision rules later to accommodate the special situation in the following runs. These early efforts to automate a HITL wargame still inform techniques to create decision rules in a wargame (Hernandez, et al. 2015).

Rubel (2003) concedes the difficulty in generating realism and rigor in experimentation, but argues that by setting adequate controls, a wargame can play a significant role in experimentation. Recognizing that command and control (C2) could be the key element for developing an analytically rigorous wargame, Rubel focused on player behavior. By meticulously examining the player C2 of a CAW, Rubel isolates the audience from the simulation play. Analysis of player performance and behavior leads to a reference of game time decisions. The recorded CAW can then be replayed using the simulation environment as the backdrop for experimentation. Asserting that these decisions in the wargame are no longer simulated, Rubel (2003) claims that they are real human decisions because they are what the players would actually do in the given situation. Therefore, valid experiments on the decisions (as the variables) could be run for analyses.

Testing this approach in a naval war college wargame for hunting Scuds (mobile missile launchers), Rubel (2003) used a Latin Square experimental design to examine two factors that had impact on the accuracy of locating a Scud. In this experiment, the factors are the command style for decision-making of the team and the way that the common operational picture (COP) is developed for the team to view. The treatments for command were one of three styles: one player is in charge, players followed a given search plan, and free play by each individual in the command cell (self-synchronization). Treatments for the second factor were a display of the true outputs of the various intelligence, surveillance, and reconnaissance assets, which included false positives and negatives, and a COP that players populated based on their perception of the actual situation. The response variable in the experiment was the team's accuracy of Scud location. A full factorial experiment has six total combinations of potential wargames. In this context, each wargame is relatively simple, but the resultant data has the requisite rigor for statistical analysis. Rubel (2003) further states that the experiment was valid. It was correctly structured, involved sufficient numbers of replications, and could be independently repeated, with a reasonable expectation of producing similar results.

Recognizing that the complexities of the wargame structure make an analytically derived solution impossible, Zhuang, Hsu, Newell, and Ross (2012) resolved to use M&S to gain understanding of the form that a solution may take. In their study of decision-making for insurgency operations, Zhuang et al. sought to improve the development of decision-making heuristics. This effort began with an examination of a six-player, paper-based seminar game. The team transformed the seminar game into a computer wargame in which HITL decisions are automated. Applying Matlab programming, the team developed the Interactive and Automatic Wargames Training System, along with a graphical user interface (Zhuang et al. 2012).

The objective of the team's efforts was to determine the decision rules that would result in the most advantageous situation for each player at the conclusion of the game. Through experimentation, the team would determine an appropriate decision that each player should take for a given situation. The team began with very few automated rules (two to four) for each of the six players. The rules determined how a player may implement options based on PMESII: Political (P), Military (M), Economic (E), Social (S), Infrastructure (I_1), and Information (I_2). The automated players could only use the set of heuristics that the team had programmed (Zhuang, et al. 2012). For each combination of decision rules for each player, the team automatically ran 1,000 replications of the game to determine the frequency of "wining" situations for each player. Through iterations of modifying rules, adding or removing rules, and injecting human experience in the heuristics the team improved the set of decision rules for each player. In this methodology, Zhuang et al. (2012) implement experimentation to improve decision heuristics that an automated version of their wargame can reliably credit human behavior for the game outcomes.

Hernandez, McDonald, and Ouellet (2015) used a combination of the previous efforts to automate a HITL, computer-aided wargame and then to apply experimentation and modified decision heuristics at specified "turns" of the wargame. In the context that Hernandez et al. developed the analytic framework for combining an automated CAW (ACAW) and experimentation, the team called it postwargame experimentation and analysis (PWEA).

Operational commanders continually search for means to improve understanding of the battlefield in which they must preside. In the period 2011 through 2013, the Commanding General of the International Security Assistance Force (ISAF) – Afghanistan directed the Afghanistan Assessment Group (AAG) for more innovative ways to use available tools and processes to identify solutions for the ISAF's most immediate problems, such as examining force mix issues in preparation of US departure from theater. The ISAF staff chose wargaming for its research method. However, the resultant data and insights from the in-theater, HITL, CAW using the Peace Support Operations Model (PSOM) were insufficient to address the Commander's needs. The AAG required a better approach to wargame analysis and a means to extract more answers from an executed wargame. Hernandez et al. (2015) partnered with AAG to develop PWEA.

Though model-agnostic, the team incorporated the PSOM in the analytical framework. The process transforms a HITL, CAW, including human decision elements, into an automated, closed-form operational simulation model that can be methodically explored using advanced experimental designs. Analysts "farm" the simulation for data in areas of the decision space where before information was unavailable. Data analyses enabled new insights supported by scientific rigor and traceable human behavior. In these experiments, the factors were specific types of units located in specified regions of the operational area. The response variable was the level of violence as computed from PSOM indicator variables.

The PWEA effort was unique in its compilation of all previous methods to automate human. While the team developed initial decision rules much as Zhuang et al. (2012) and Davis and Williams (1982) had done, the team acknowledged that the heuristics would need to adapt to the dynamics of the situation and the players involved. The team derived the initial decision heuristics from the actual HITL CAW and verified the rules with the ISAF staff. The significance of the PWEA approach was in the innovative approach that the team followed for improving the decision heuristics in the ACAW.

Using the initial decision heuristics, the team and the ISAF staff ran the PWEA to determine if the results were reasonably similar to the actual CAW. The ISAF staff confirmed the similarity of the PSOM outcomes in the game. However, the ISAF staff also recognized significant changes in the operational output during execution of the original wargame and in PWEA. It was evident that the changes occurred after strategic and operational decisions at turns #2 and #9 of the game. To investigate the impact of these strategic decisions on violence, the ISAF staff developed new decision matrices with the team and injected them at the specified turns. Therefore, the analysis team performed the planned experiments up to the specified turns. At that point, the team applied the new decision heuristics, treating the turn (#2) as the beginning of the game and with "initial" conditions. Meantime, the original set of experiments would continue using the actual game start. The change in decision heuristics would again change at the other significant turn (#9). Because the original branches and sequels of the design point were still run to completion, they could serve as a basis for comparison. Therefore, the differences of outcomes could be attributed to the new decision rules that were implemented from that decision point. In this manner, stage-wise experimentation focuses on a wargame's strong point – analysis of human decisions made during the game. The results of these phased decision heuristics is stage-wise experimentation that permitted the staff to use a different set of rules than the original wargame.

Figure 1 is an abstract of a wargame's decision tree and a visual for explaining stage-wise experimentation. If the execution of a single wargame is viewed as a

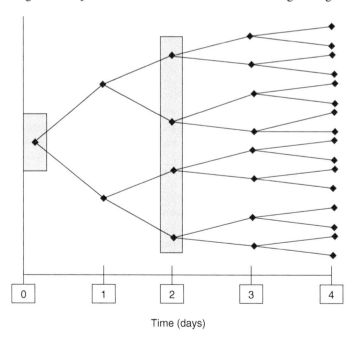

Time (days)

Figure 1 Illustrating stage-wise experimentation through a decision tree (Hernandez, et al. 2015).

decision tree, we can consider performing an experiment on selected portions of that tree. In Figure 1, Day 0 represents investigation of variations from the starting conditions. Alternatively, one may choose to isolate a specific portion of the wargame tree, and perform variations at that point. On Day 2, variations (in decision rules) are propagated forward, while not changing decisions and actions that occurred prior to Day 2. In short, performing experiments at different portions of the overall wargame tree allows for comparison of risks and opportunities associated with new strategic considerations such as deviations from planned force reductions, major shifts of combat power or strategy, or level of Red activity (Hernandez, et al. 2015).

Stage-wise experimentation (SWE) adds significantly to the complexity of automating the wargame. Partitioning the experiment into "stages" requires new considerations and internal controls to modify the scenario to correspond with the players' new decision matrix. Mapping new decisions to factor settings in the simulation model requires investments in time and effort. However, once complete, the analysis team has a closed-form experimentation venue that can accept experimental designs to explore different questions about the scenario. The results have credible human input that is traceable to game outcomes. The procedure from implementing stage-wise experimentation follows the approach that Rubel (2003) had described. Automation of the CAW should also be narrowed to a specific period because of the exponential rise in the number of interactions that each human decision will influence over time.

A Complete Application of ACAW and SWE for Future Capability Insights

Naval leaders are in the midst of redesigning the architecture of its forces. The US Navy Warfare Systems Directorate (N9) examines future capabilities that support the strategic vision for maritime forces. As such, the directorate seeks a self-contained, experimentation environment to develop insights regarding the impact of projected naval capabilities on military success. Implementation of ACAW and SWE to an executed exercise addresses the N9 need and illustrates the means to enable knowledge discovery.

The backdrop for this recent effort is a semiannual, multinational, Joint Forces exercise, Cobra Gold 2018 (CG18). Interagency and nongovernmental agencies also played major roles in the operational scenario. Operations occurred in a fictional landmass where multinational force used combined air, land, and sea power to meet its mission and perform a number of key tasks. These tasks included achieving air superiority, peace enforcement, maritime interdiction, counter piracy, and humanitarian assistance. The CG18 scenario has the necessary elements for an analyst to investigate the performance of current, new, or envisioned systems. From this scenario, specific vignettes make it possible to stress a system and visualize its relationship within a network of systems and physical surroundings.

To address the questions in this exercise, the exercise command group chose to use the joint theater level simulation – Global Operations (JTLS – GO) or JTLS for quick reference. This simulation is an interactive, internet-enabled application. As discussed earlier, the choice of the correct simulation for the problem to be addressed requires planning and understanding of the underlying issues (Connable et al. 2014). The JTLS has supported Unified Commands in training exercises for over 35 years and is recognized as a suitable choice for this exercise (https://www.rolands.com/jtls/j_over.php).

Upon completion of CG18, the team developed the ACAW in two phases. In a HITL CAW, the game is split into finite decision intervals as defined by a "turn." In a wargame, "turns" are specific intervals of time in which player decisions are adjudicated so that new conditions may be presented to the training audience for a new round of interactions and decisions (Sabin 2014). Automation of CAW decisions is only needed at each turn. However, an exercise is continuous. Decisions can occur at any instance of the game. Figure 2 shows the development of events in an exercise (Hernandez, et al. 2018). The link between the training audience, the computer simulation, and the response cells that input audience decisions enables a constant communication and activity.

Transforming the exercise into an ACAW first required shaping the exercise in terms of a CAW. Automating a player for an extended period is a significant challenge for a programmer. The team scoped the problem so that automated player decisions and actions involved only a slice of the CG18 scenario. Specifically, a continuous 24-hour period, which was the most active period for naval operations in the exercise

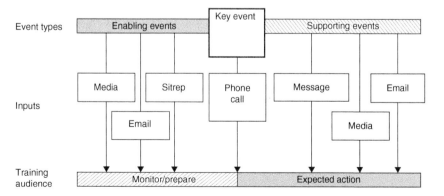

Figure 2 Event development in an exercise consists of constant communication between players, simulation, and training audience (Hernandez et al. 2018).

fit the project's purpose. Studying the activities of the exercise, the team designated 8 or 12 periods as turn. In these turns, the conditions could be considered and a decision matrix could be referenced. The decisions are then adjudicated in JTLS. Automating JTLS play in this segment required specific software development to:

- Create a game from a scenario.
- Configure the game.
- Start web services (the primary JTLS communication mode).
- Start the JTLS Combat Events Program.
- Send a start game order to the JTLS Order Management Authority.
- Systematically and periodically collect and parse messages for metrics of interest, anecdotal information, and simulation time.
- Post-process game output from JTLS messages and convert pertinent information into a usable data format.

While the software development for JTLS is specific, the approach is similar for other simulation models. Accordingly, we introduce a corresponding structure for transforming a CAE into an ACAW (Figure 3).

The steps in Figure 3 begin after considering a CAE and the associated simulation software. Step 1 examines the computer simulation to determine if its architecture is open enough to accept orders from external routines. A check to ensure that the scenario and operations have utility for the customer makes up Step 2. The third step is simply data collection. The team that transforms the CAE must understand the context of the orders that are recorded in the simulation, as well as the exercise results. Step 4 consists of identifying a period that captures the activities that most meet customer objectives. Additionally, the team determines the time intervals for the game "Turns." This step is critical because it provides the opportunity for where a HITL can submit orders in the first run of the exercise. These orders are later submitted without a HITL in Step 7. In Step 5, the team models the new systems and\or capability that will be incorporated in the CAE for study. Step 6 corresponds with Step 4; the team creates the vignettes and manual orders that the new system executes at each Turn. Step 7 automates the game setup, which supports a replication of a simulation run. This step also automates Player submission of orders. Finally, the complete ACAW and SWE must collect simulation data to support analyses that address the research objectives. While we will not attempt it in this paper, this abbreviated explanation for creating an automated CAW from a HITL CAE could be the start of a taxonomy for automated wargames.

Constructing an experimentation environment was the next step in the effort. Figure 4 shows that the first step in experimentation is identifying the factors of interest. These are variables that analysts or study sponsors believe have an impact on relevant metrics of the game. Based on the number of factors, the team constructs a design of experiments (DOE) that may be ingested into the JTLS

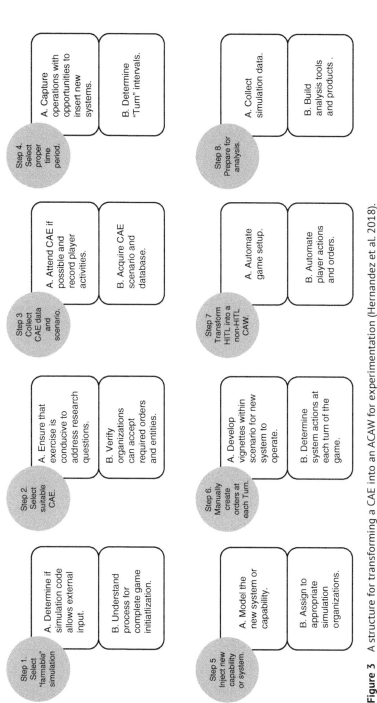

Figure 3 A structure for transforming a CAE into an ACAW for experimentation (Hernandez et al. 2018).

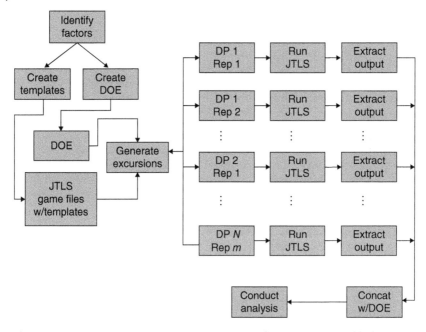

Figure 4 Incorporating design of experiments in JTLS (Hernandez et al. 2018).

architecture in much the same way a database becomes part of the exercise configuration. A software development included a function to initiate the DOE in JTLS at the start of the game. Software (jtlsfarmer) to extract relevant data from each run produces a sample of observations for analysts to study (Hernandez et al. 2018). In this fashion, stage-wise experimentation is applied as the dynamics of the exercise, the game, or the research questions require.

With the advent of M&S tools, coupled with the exponential rise in computing technologies, a wargame can be transformed into an experimentation environment. Researchers recognize the power of computers to explore nonlinear decision spaces. Computer simulation experiments permit analysis of the relevant decision space for the given operational environment. Study of the operation through experimentation enables the emergence of new knowledge as a result of this innovative method of inquiry.

Computer-Assisted Wargaming Classification

Previous discussion about wargaming elements, its automation challenges, missing features in the constructive simulation, and Use Case examples of a creative computer assistance to a wargame creates an environment for making a new CAW classification based on the level of automation of its elements.

Red Teaming Guide (2013) manual divides wargames on the bases of the application purpose. It puts the wargames into two big families. The first family belongs to a decision support domain and the second family belongs to a training/education domain. In the second family, the decision-makers are trained in teams; in the first family, the decision-makers are provided with information that supports their decision-making process. The decision support domain is further divided into operational analysis, experimentation and operational planning (COA) support application sub-domains. Classification into these two families is unclear and it does not have explicit borders as you use the same tenants of the wargame, and moreover, you use the decision-making process even when creating the decision-making support information.

To demonstrate the different wargaming classification approaches, a web reference of LBS company was used (LBS, 2019). It states that wargames have different variations starting from Command Post Exercises (CPX), Computer-Assisted Exercises (CAX), Seminar Wargames, Decision Games, Crisis Management Exercises (CMX), Course of Action Wargames and Experimentation.

The Guide for Understanding and Implementation of Defense Experimentation (2006) introduces experiment classification based on the form of the involved simulation type. It forms four groups of methods applied in the experiment. The first method, an analytical wargame, requires employment of its commander and his staff planning their course of actions and reacting to the opponent activities. The second method, an experiment with constructive simulation, is based on the idea of replication of some aspects of the operational environment and human behavior in the form of models in the constructive simulation. The third method, Human-in-the-Loop simulation, is based on the idea of operators formulating inputs and outputs of the simulation. The last method, life simulation, is in fact an exercise with a human in a field. This classification of experiment does not bring any clarity to the wargaming and to its potential future automation as the elements that form the wargame can be used in all four groups and the level of the element automation is hidden in the definition. This classification is only experimentation domain oriented.

This short excursion into the current wargaming classification approaches demonstrated none-unified and none-explicit classification in this domain that complicates fundamental understanding of the wargaming for military purposes.

The new proposed CAW classification is domain and purpose independent. Common denominator of the wargaming classification is the level of automation of the wargame elements. Table 2 shows abbreviations and basic hierarchy of the special terms.

The entry point and the very top of the classification hierarchy is the wargame. Any definition of a wargame, as proposed in the earlier section on **Wargaming Relation to M&S**, is valid. The wargame is then divided into two main classes,

Table 2 Wargaming classification terms and their hierarchy.

Abbreviation		Proposed term
WG		Wargame
MWG	CAWG	Manual Wargame / Computer-Assisted Wargame
	CAWGHIL	Computer-Assisted Wargame Human in the Loop
	CAWGHOL	Computer-Assisted Wargame Human on the Loop
	CAWGHOLAO	Computer-Assisted Wargame Human on the Loop Automated Operator
	CAWGHOUTL	Computer-Assisted Wargame Human out of the Loop
	CAWGHOUTLAA	Computer-Assisted Wargame Human out of the Loop Automated Analyst

the first class is Manual Wargame and the second class is Computer-Assisted Wargame. MW does not use any computer assistance at any phases of the wargame. CAWG is a wargame with specific wargaming human behavior elements automation capability. This is further divided, in this family, into the following subclasses:

- CAWGHIL where any human behavior wargaming element is being automated. It means that Player, Umpire, Technical and Operator Controller and Analyst are in the wargame represented by a human being.
- CAWGHOL with Player, Technical and Operator Controller and Analyst being represented by a human being. Umpire is, in this class, represented by automata.
- CAWGHOLAO with Player and Analyst being represented by a human being and Umpire and Technical and Operator Controller are played by automata.
- CAWGHOUTL with Analyst represented by a human being and automata taking care of the roles of Umpire, Player, Technical and Operator Controller.
- CAWGHOUTLAA with all the human behavior wargaming elements played by Automata.

Military experiment can be described by the level of experiment control, time needed to go through all the phases of the wargame, and the level of fidelity provided by the experiment environment. As defined earlier in CAW classification, Figure 5 describes these wargaming classes in relation to the level of experiment control, and fidelity of the environment where the experiment is carried out in the time needed to successfully perform the experiment.

Figure 5 demonstrates that experiments with some level of automation of human behavior wargaming elements bring better control and decrease time

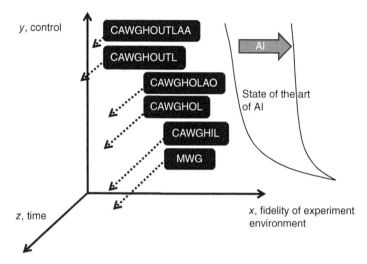

Figure 5   Attributes of military experiment related to CAW.

needed to perform all the phases of the experiment. On the other hand, such computer assistance strongly depends on the current state of the art of the Autonomous domain and artificial intelligence field. Nevertheless, the level of environment fidelity replicated by the experiment is smaller for CAW with its defined level of WG element automation than in nonautomated wargames.

Conclusion

The management of violence and curbing rampant chaos are primary directives of any military commander. This grave responsibility requires the discovery of new knowledge. Leaders thirst for means to understand their environment and project plans that achieve the mission while taking into account all the factors that affect tactical and strategic decision-making. Knowledge discovery is critical for today's generals, or any commander that makes life and death decisions. This knowledge must be based on expert judgement and scientific underpinnings from credible data.

Wargaming has been in the military fold for thousands of years. This approach to uncover new information about combat is imbued in commanders when planning war, "...the general who wins a battle makes many calculations in his temple ere the battle is fought..." (Giles 2014, p.6). Whether Sun Tzu meant "temple" as the commander's mind, or institutions with throngs of advisors and staffers, or both, the need to develop insights from the operational situation is imperative. Computer-aided wargames offer incredible advantage for adjudicating decisions and enabling a visual of virtual operations of modeled units in realistic environments. Decision-makers have other tools in their kit, including campaign analysis, unit and command post exercises, and simulation models. These methods provide useful information, but not necessarily with the analytical rigor that commanders need.

Stochastic computer simulations, wargames, and experimentation offer a means to provide information that meet the pedigree that commanders demand. Computer-aided wargames incorporate the experience of battle-tested professionals who inject their expertise at each turn of the game. Considering the game situation, actual humans determine the appropriate response(s) to the conditions of the wargame. However, any wargame is singular, and therefore, the resultant data cannot support traditional analysis techniques. The ability to reproduce the exact same branches and sequels of a game has been impossible until the recent past. Practitioners have developed approaches for creating closed-loop wargames that use computer simulations. This chapter describes a progression of research efforts that have furthered these efforts. From these studies, a common theme emerges; an essential element for transforming a human-in-the-loop computer-aided wargames into a closed-loop wargame is automating the human decisions in the game.

There are many challenges to automating humans in a computer-aided wargames. Researchers have found that scoping the scenario, simplifying the actual decisions that are automated, then allowing the inherent decision-making features of the computer simulation to take over, just as they do in the nonautomated wargame, enables the creation of an experimentation environment. This experimental venue allows the analysts to run the same wargame many times. The stochastic nature of the computer simulation produces variances in the results. Through different programming techniques, the computer simulation can inject a design of experiments. Each design point in the experiment is a new instantiation of the game. Replications of the design point reduces errors in estimated values from the experiments. However, the data from the results can withstand scientific scrutiny. Coupled with the military basis for decisions in the game and the data resulting from experimentation, analysts can present a balanced set of information for the commander to consider for upcoming decisions.

This new approach to knowledge discovery and the expected progress in the autonomous domain opens possibility to make new proposed CAW classification. It is domain and purpose independent founded in the level of automation of wargame elements. In the close future the quality of replicated human behavior wargaming elements will be at such level that experimentation using wargaming with no human actors will reach the same results credibility like Human in the Loop wargaming.

References

Appleget, J.A. and Cameron, F. (2015). Analytic wargaming on the rise. *Phalanx* 48 (1): 28–32.

Booker, L. (2005). Learning from nature: Applying biometric approaches to Military tactics and concepts. *The Edge: Biotechnology Issue* 9 (1): 10–11.

Body, H. and Marston, C. (2011). The peace support operations model: Origins, development, philosophy and use. *The Journal of Defense Modeling and Simulation: Applications, Methodology, Technology* 8 (2): 69–77.

Box, G.E.P., Hunter, J.S., and Hunter, W.G. (2005). *Statistics for Experimenters: Design, Innovation, and Discovery.* Hoboken, Nrsey: John Wiley & Sons, Inc.

Blais, C. 2012. Strategies for aligning and employing the Coalition Battle Management Language (C-BML) and the Military Scenario Definition Language (MSDL). *Proceedings of Fall Simulation Interoperability Workshop 2012.*

Blais, C., Dodds, R., Pearman, J., Baez, F. 2009. Rapid Scenario Generation for multiple simulations: An application of the Military Scenario Definition Language (MSDL). *Proceedings of Simulation Interoperability Standards Organization - Spring Simulation Interoperability Workshop 2009.*

Cayirci, E. and Marincic, D. (2009). *Computer Assisted Exercises and Training: A Reference Guide.* Hoboken, New Jersey: John Wiley & Sons, Inc.

Connable, B., Perry, W.L., Doll, A. et al. (2014). *Modeling, Simulation, and Operations Analysis in Afghanistan and Iraq: Operational Vignettes, Lessons Learned, and a Survey of Selected Efforts, 76.* Santa Monica, CA: RAND.

Creswell, J.W. (2014). *Research Design, Qualitative, Quantitative, and Mixed Methods Approaches.* London, United Kingdom: Sage Publication, Inc.

Davis, P.K. and Williams, C. (1982). *Improving the Military Content of Strategy Analysis Using Automated Wargames: A Technical Approach and Agenda for Research.* Santa Monica, CA: RAND Corporation.

Department of the Army (DA) (2012). *Army Doctrine Reference Publication (ADRP) No. In: 5-0: The Operations Process.* Headquarters: Department of the Army.

De Mesquita, B.B. (1981). *The War Trap.* London, United Kingdom: Yale University Press.

Diamond, J. (2005). *Guns, Germs, and Steel.* New York, NY: W.W. Norton & Company, Inc.

Dunnigan, J.F. (1992). *The Complete Wargames Handbook, How to Play, Design, and Find Them.* New York, NY: William Morrow & Company, Inc.

Feynman, R.P. (1995). *Six Easy Pieces.* Reading, Massachusetts: Addison-Wesley Publishing Company.

Gianni, D., D'Ambrogio, A., and Tolk, A. (2015). *Modeling and Simulation Based Systems Engineering Handbook.* Boca Raton, Florida: Taylor and Francis Group.

Giles, L. (2014). *The Art of War, Sun Tzu*. San Digeo, CA: Canterbury Classics.

Guide for Understanding and Implementing Defense Experimentation. 2006. The Technical Cooperation Programme (TTCP).

Hamalainen, J., Sormunen, J., Rantapelkonen, J. and Nikkarila, J-P. 2014. Wargame as a Combined Method of Qualitative and Quantitative Studies. *The Journal of Slavic Military Studies, Special Issue*. May 2014.

Hase, C., Matos, R., and Styer, D. December 2004. Evidence-Based Decision Making, Techniques for Adding Rigor to Decision Support Processes in Complex Government and Industrial Organizations, *Phalanx*, 22-27.

Hodicky, J., Prochazka, D., and Prochazka, J. (2019). *Automation in Experimentation with Constructive Simulation. Proceedings of the Modelling and Simulation for Autonomous Systems*: 2018.

Hodicky, J., Melichar, J. 2017. Role and Place of Modelling and Simulation in Wargaming. *Proceedings of the MSG-149 Symposium on M&S Technologies and Standards for Enabling Alliance Interoperability and Pervasive M&S Applications.*

Hodicky, J. Castrogiovanni, R., Presti, A.L. 2016. Modelling and simulation challenges in the urbanized area. *Proceedings of the 17th International Conference on Mechatronics - Mechatronika, ME 2016.*

Hodicky, J. (2011). *Tactical data analysis for visualization in three dimensions.* Simulation and Communication: *Proceedings of the International Conference of Distance Learning.*

Hodson, D.D. 2011. Military simulation: A ubiquitous future. *Proceedings of the 2018 Winter Simulation Conference.*

Hernandez, A.S., Hatch, W.D., Pollman, A.G., and Upton, S.C. 2018. Computer Experimentation and Scenario Methodologies to Support Integration and Operations Phases of Mission Engineering and Analysis, *Proceedings of the 2018 Winter Simulation Conference.*

Hernandez, A.S., McDonald, M., and Ouellet, J. (2015). Post Wargame Experimentation and Analysis: Re-Examining Executed Computer Assisted Wargames for New Insights. *Military Operations Research Journal* 20 (4): 19–37.

Janes. 2019. http://www.janes.com/ accessed 22.04.2019.

Kaplow, L. and Shavell, S. (2004). *Decision Analysis, Game Theory, and Information.* New York, NY: Foundation Press.

Khimeche, L. (2016). *A disruptive approach for Scenario Generation: an agile reuse bridging the gap between Operational and Executable Scenario. Proceedings of the SIW*: 2016.

Kleijnen, J.P.C., Sanchez, S.M., Lucas, T.W., and Cioppa, T.M. (2005). A User's Guide to the Brave New World of Designing Simulation Experiments. *INFORMS Journal on Computing* 17 (3): 263–289.

Koopman, B. O. 1956. Fallacies in Operations Research. *Operations Research,* 4(4), pp. 422–430.

Law, A.M. (2015). *Simulation Modeling and Analysis*. New York, NY: McGraw-Hill Education.

LBS. 2019. http://lbsconsultancy.co.uk/our-approach/what-is-it/.accessed 22.04.2019.

Lindren, M. and Bandhold, H. (2009). *Scenario Planning: The Link between Future and Strategy*. United Kingdom: Palgrave McMillan.

Mietzner, D. and Reger, G. (2005). Advantages and Disadvantages of Scenario Approaches for Strategic Foresight. *International Journal of Technology Intelligence and Planning* 1 (2): 220–239.

Montgomery, D.C. (2005). *Design and Analysis of Experiments*, 6the. New York, NY: John Wiley & Sons, Inc.

NATO Glosssary of Terms and Definitions. 2015. Paris: NSO, https://nso.nato.int/natoterm/Web.mvc, accessed 22.04.2019.

Page, E.H. (2016). *Modeling and Simulation, Experimentation, and Wargaming – Assessing a Common Landscape, Case Number 16-2757*. The MITRE Corporation.

Perla, P.P. and Barrett, R.T. (November 1984). *Wargaming and Its Uses, Professional Paper 429*. Alexandria, VA: Center of Naval Analyses.

Perla, P.P. and Curry, J. (2011). *The Art of Wargaming: A Guide for Professionals and Hobbyists*. Annapolis: U.S. Naval Institute Press.

Prochazka, D., Hodicky, J. 2017. Modelling and simulation as a service and concept development and experimentation. *Proceedings of the ICMT 2017 - 6th International Conference on Military Technologies*.

Raiffa, H. (1970). *Decision Analysis, Introductory Letters on Choices and Uncertainty*. Reading, Massachusetts: Addison-Wesley Publishing Company.

Red Teaming Guide. 2013, 2nd Edition, UK Development Concepts and Doctrine Centre (DCDC).

ROLANDS & ASSOCIATES, Corporation. 2018. JTLS – GO Executive Overview. https://www.rolands.com/jtls/j_over.php, accessed 29.03.2018.

Rubel, R.C. (June 2003). *Using Wargames for Command and Control Experimentation*. Newport, RI: Naval War College, Wargaming Department, Center for Naval Warfare Studies.

Sabin, P. (2014). *Simulating War: Studying Conflict through Simulation Games*. New York, NY: Bloomsbury Academic.

Schumpeter, J.A. (1934). *The Fundamental Phenomenon of Economic Development, An Inquiry into Profits, Capital, Credit and Interest in the Business Cycle*, pp. In: *57-94, Cambridge*. Massachusetts: Harvard University Press.

Simulation Experiments and Efficient Designs (SEED) Center, Naval Postgraduate School. 2019. Welcome to SEED. https://my.nps.edu/web/seed/, accessed 22.04.2019.

Wargaming Handbook. 2017, 1st Edition, UK Development, Concepts and Doctrine Centre (DCDC).

Washburn, A. and Kress, M. (2009). *Combat Modeling*. New York, New York: Springer.

Watts, B.D. (1996). *Clausewitzian Friction and Future War*. Washington, DC: National Defense University.

Worley (1999). *Defining Military Experiments*. Alexandria, VA: Institute for Defense Analysis.

Zhuang, J., Hsu, W., Newell, E. A., and Ross, D. O. 2012. Heuristics, Optimization, and Equilibrium Analysis for Automated Wargames. Non-published Research Reports. Paper 174. Homeland Security Center.

6

Simulation and Artificial Intelligence Methods for Wargames

Case Study – "European Thread"

Andrzej Najgebauer[1], Sławomir Wojciechowski[2],
Ryszard Antkiewicz[1], and Dariusz Pierzchała[1]

[1] *Military University of Technology, Warsaw, Poland*
[2] *Multinational Corps North East, Szczecin, Poland*

Introduction

The chapter is based on our team's experience as active participants in a variety of wargames in player roles as well as from extensive support of wargaming environments with special simulation and artificial intelligence tools.

- better evaluation of possible moves during the game (course of action); or
- better and comprehensive planning probable network of events in phase of preparation the scenarios; or
- better analysis of whole process in phase "after action review" of the game.

We intend to describe simulation and analytical support of wargaming.

Authors of the study - the representatives of the Research Team for Modeling, Simulation and IT Decision Support in Conflict and Crisis Situations from the Military University of Technology (MUT) and the main operational expert gen. S. Wojciechowski - took part in a few war games so-called "Defending European Frontline." The games were successively organized in MUT, Polish MOD, Washington (the office of the American "think tank" Center for Strategic and Budgetary Assessments CSBA), and EU NATO Command (organized by RAND Corporation). The basic assumption of the games was to test certain operational concepts in the

Simulation and Wargaming, First Edition. Edited by Charles Turnitsa, Curtis Blais, and Andreas Tolk.
© 2022 John Wiley & Sons, Inc. Published 2022 by John Wiley & Sons, Inc.

event of a threat of escalation and then conflict in the Baltic states, members of NATO. The potential participation of Poland and NATO countries, including the United Sates, had to be checked with the assumption that they would use forces in future (10 and more years), equipped with different combat systems. The participation of the MUT team consisted in quantitative support for the "tabletop" game in these experiments. The proposed quantitative support of the team focused on statistical, optimization, computer simulation tools and also artificial intelligence techniques, based on probabilistic reasoning methods [11]. *In order to make good decisions under such conditions, it is beneficial to introduce probability to the reasoning process. Since working with the full probabilistic description of the world can be in many cases too difficult, the Bayesian models are used to simplify the reasoning process* [11]. Both the scope of quantitative methods and the selection of research tools as well as the methodology of using these methods and tools in the described experiments will be presented below. The presented scenario is different to the games mentioned above and is rather the generalization toward more universal approaches, however, based on European prerequisites. The basic experiment has been prepared by combining and successively applying the following components: game scenario on the strategic/operational level, mathematical modeling based on network complex activities model, an original software application for the analysis of the possible moves during the gaming, experiments with the game, fast analysis of course of action and evaluation defined important characteristics like the probability of success (the aim achievement), time of complex activities realization and the others (also presented in the chapter). purpose of method presentation, there are two aims. The first aim of the game is to test the decision-making process of the players. Secondary aim is the measurement of the requirement of the military potential needed. The game scenario addresses the land domain extended to chosen air operations.

We intend to describe simulation and analytical support of wargaming, which can be presented in the following steps:

1) Aims of game – main and additional.
2) Organization of the game:
   a) Location.
   b) Personnel.
   c) Structure and time of game.
   d) Budget.
   e) IT support (simulation and AI tools).
   f) Kick-of-meeting – time and range.
3) Description of game:
   a) Conflict situation (scenarios, taking into account the aims of game) – geopolitics.
   b) Sides of conflict (countries or organizations, forces engaged and reserves, possible moves with the division onto phases, etc.).

c) Initial situation (startex).
d) Description of game process – potential moves and expected results.
4) Wargaming:
   a) Realization of moves.
   b) Evaluation after the moves (who or what evaluates).
   c) Seminar after each move or after phase.
   d) End of the wargame (results) (endex).
   e) A summary of the wargame.
5) Final seminar – after action review – the evaluation of achievement of the aims of game.

Our analysis is focused on points 1, 3, 4, and 5 of the schema.

Assumptions and Research Tools

We use military joint operations in the present study as an example of a complex operation. The process of joint operation design is outlined in the following steps [6]:

1) Define the end state (in terms of desired strategic political-military outcomes).
2) Define the objectives describing the conditions necessary to meet the end state.
3) Define the desired effects that support the defined objectives.
4) Identify friendly and enemy center(s) of gravity (COGs) using a systems approach.
5) Identify decisive points (DPs) that allow the joint force to affect the enemy's COG and look for decisive points necessary to protect friendly COGs.
6) Identify lines of operation describing how decisive points are to be achieved and linked together in such a way as to overwhelm or disrupt the enemy's COG.
7) Identify how decisive points relate to phases in order to identify how operations are arranged as regards time, space, and effect. Also identify changes in phases, especially the critical transition from Phase III (Dominate) to Phase IV (Stabilize).

Complete the detailed synchronization and integration of forces and functions, tasks, targets, and effects centered on decisive points and phases to achieve unity of effort.

A very simple excerpt from a complex campaign planning concept is presented in Figure 1 [10].

The end state is a rectangle with incoming arcs (from preceding decisive points) and obviously without any outgoing arcs. It might be treated as a friendly or enemy center of gravity (COG). The three decisive points (triangular: 2, 3, and 4) are related to phase and identify how operations are arranged in time/space.

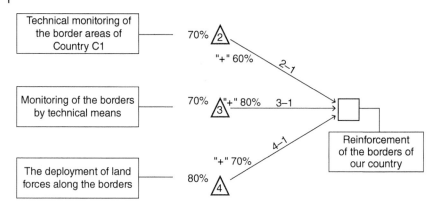

Figure 1 Simple excerpt from a campaign planning concept (course of action – COA) [10].

Moreover, their parameters as well as parameters of edges describe an effect of influencing decisive points on next node (in that case: end state).

The important issue is a quantitative evaluation of the prepared plan (COA). There are papers dealing with modeling and evaluation of COA based on Timed Influence Nets used in CAusal STrength (CAST) logic [4, 8, 16]. They are appropriate for modeling situations such as military situations, in which the estimate of the conditional probability is hard and subjective. So in plan recognition, the CAST logic could break the bottleneck of knowledge acquisition and give the uncertainty modeling a good interface. Falzon in the paper [7] describes a modeling framework based on the causal relationships among the critical capabilities and requirements for an operation. The framework is subsequently used as a basis for the construction, population, and analysis of Bayesian networks to support a rigorous and systematic approach to COG analysis. Authors of paper [3] show that problems of military operation planning can be successfully approached by real-time heuristic search algorithms, operating on a formulation of the problem as a Markov decision process.

The concept of CAST logic was presented for the first time in paper [2] and subsequently developed in many others [7, 12, 14, 15, 16].

Especially important is paper [15], which removes some shortcomings of earlier versions of CAST logic implementation in influence networks and takes account of time-dependency in complex operations. In this paper, we modeled time-dependency using a stochastic PERT model and a CAST logic model in compliance with paper [16]. An interesting proposal for modeling stochastic dependences between actions was given in the paper [5].

In this chapter, we present a method of quantitative evaluation of the COA based on stochastic model of the complex activities for the determined stage of wargame or real conflict. Another possible way is dynamic response of the opposite side actions in the wargame mode of the proposed toolset development.

Modeling of Complex Activities

Network Model of Complex Activities

The MCA software package for (M)odeling (C)omplex (A)ctivities analyzer with many embedded tools, based on a network model, is described below. The legacy version of the MCA is mentioned in [1, 10].

The MCA is a computer tool/application, which allows to evaluate some characteristics of complex activities (operation). We understand a complex activity or operation as such undertaking that contains many related actions. The succession relationships mean a sequence in time and reflect the impact of the success of one action on another. Military or political-military operations are good examples of complex activities.

We have assumed that complex activities can be defined as the following network [1, 10]:

$$S_p = \langle G, \Phi, \psi \rangle \tag{1}$$

where $G = \langle D, U \rangle$, G is the Berge's graph without loops, D is a set of graph nodes, where a node is related to one action of complex activities, $D = \left\{ x^p, 1, 2, ... \bar{d}, x^k \right\}$, U is a set of graph edges – edges define the time sequence of actions; $\Phi = \{f | f : D \to R\}$ – family of functions defined at graph nodes, $\Phi = \{ \tau_{min}, \tau_{max}, \tau_p, pb, F, pr, PR\}$; and $\Psi = \{f | f : U \backslash \{(d, k) : k = x^k\} \to [-1, 1]\}$ – family of functions defined at graph edges, $\Psi = \{h, g\}$.

We assume that values of the following functions are determined by the user of MCA during the modeling phase of planned process:

$\tau_{min}(d)$ – the shortest time needed to complete tasks connected with action d.

$\tau_{max}(d)$ – the longest time needed to complete tasks connected with action d.

$\tau_p(d)$ – the most probable time needed to complete tasks connected with action d.

$pb(d)$ – baseline probability of completing tasks connected with action d.

$h((d, k))$ – the value of this function reflects the influence of completing task of action d on the probability of completing task of action k. If the value of h((d, k))>0 then the influence of d on k is positive, while if the value of h((d, k)) < 0 it means that the completion of tasks connected to node d causes a negative effect on completing tasks of node k.

$g((d, k))$ – the value of this function reflects the influence of completing task of action d on the probability of completing tasks of action k. If value of $g((d, k)) < 0$, it means that failure in completing tasks of node d has negative influence on completing tasks of node k. If value of $g((d, k)) \geq 0$, it means that failure in completing tasks of node d has positive influence on completing tasks of node k.

We assume that values of $\tau_{min}, \tau_{max}, \tau_p, pb$, and $h((d, k)), g((d, k))$ are given by the analyst using MCA.

The presented model is a modification of MCA model in [1, 10]. The idea of the extension is presented below.

We divided nodes of the model into two types:

- *DA* actionable nodes – nodes connected with some tasks, which should be actually completed in order to obtain some effect, actionable nodes in the graph are nodes without incoming edges.
- *DS* state nodes – nodes with $\tau_{min}(d) = 0$, $\tau_{max}(d) = 0$, $\tau_p(d) = 0$. State nodes are influenced by actionable nodes and describe changes of some abstract states of complex operations.

Additionally, we define two types of actionable nodes:

- *DA_P* – actionable nodes with positive influence on state node.
- *DA_N* – actionable nodes with negative influence on state node.

We will further show the method of evaluating three important quantitative characteristics of complex activities in conflict progress:

$F(d, t)$ – probability of execution of tasks connected with node d in a time shorter than t, execution of the above tasks does not guarantee positively completing tasks of node d.

$pr(d)$ – probability of positively completing tasks of node d in unlimited time.

$PR(d, t)$ – probability of positively completing tasks of node d in a time shorter than t.

The idea of the $pr(d)$ evaluation is based on CAST logic described in paper [1] and some extensions in Ref. [10]. According to CAST algorithm, we should set the following values as input data:

Let us denote $Pred(k) = \Gamma^{-1}(k)$.

If $\Gamma^{-1}(k) = \{d\}$, then it is assumed that:

- If decisive point d is achieved then:

$$P\left(k \mid d\right) = \begin{cases} pb_k + h\left(d,k\right) \cdot \left(1 - pb_k\right), if\ h\left(d,k\right) \geq 0 \\ pb_k + h\left(d,k\right) \cdot pb_k, if\ h\left(d,k\right) < 0 \end{cases}$$

$$(2)$$

- If decisive point d is not achieved then:

$$P\left(k \mid \sim d\right) = \begin{cases} pb_k + g\left(d,k\right)\cdot\left(1 - pb_k\right), if\ g\left(d,k\right) \geq 0 \\ pb_k + g\left(d,k\right)\cdot pb_k, if\ g\left(d,k\right) < 0 \end{cases} \tag{3}$$

where $P(k \mid d)$ means the conditional probability of achieving point k, given that point d was achieved ($P(k \mid \sim d)$ d not achieved).

In case $|\Gamma^{-1}(k)| > 1$, we have defined additional quantities

$$y^k = \left(y_d\right)_{d \in \Gamma^{-1}(k)}, \tag{4}$$

$$Y^k = \left\{ y^k = \left(y_d\right)_{d \in \Gamma^{-1}(k)} : y_d \in \left\{0,1\right\} \right\} \tag{5}$$

where

$$y_d = \begin{cases} 1 & \text{if point } d \text{ achieved} \\ 0 & \text{otherwise} \end{cases} \tag{6}$$

$$h_d^k\left(y_d\right) = \begin{cases} h\left(d,k\right) & y_d = 1 \\ g\left(d,k\right) & y_d = 0 \end{cases} \tag{7}$$

$$h\left(y^k\right) = \left[\prod_{d \in \Gamma^{-1}(k):h_d^k(y_d)<0} \left(1 - \left|h_d^k\left(y_d\right)\right|\right) - \prod_{d \in \Gamma^{-1}(k):h_d^k(y_d)\geq0} \left(1 - \left|h_d^k\left(y_d\right)\right|\right) \right] /$$

$$/ \max\left[\prod_{d \in \Gamma^{-1}(k):h_d^k(y_d)<0} \left(1 - \left|h_d^k\left(y_d\right)\right|\right), \prod_{d \in \Gamma^{-1}(k):h_d^k(y_d)\geq0} \left(1 - \left|h_d^k\left(y_d\right)\right|\right) \right]. \tag{8}$$

Now we can calculate conditional probability $P(k \mid y^k)$ using the following formula:

$$P\left(k \mid y^k\right) = \begin{cases} pb_k + h\left(y^k\right)\cdot\left(1 - pb_k\right), if\ h\left(y^k\right) \geq 0 \\ pb_k + h\left(y^k\right)\cdot pb_k, if\ h\left(y^k\right) < 0 \end{cases}. \tag{9}$$

The unconditional probability $pr(k)$ of achieving the decisive point k is computed via the expression:

$$pr\left(k\right) = \sum_{y^k \in Y^k} P\left(k \mid y^k\right)\cdot \prod_{d \in \Gamma^{-1}(k)} pr\left(d\right)^{y_d} \cdot\left(1 - pr\left(d\right)\right)^{1-y_d}. \tag{10}$$

The unconditional probability $pr(k)$ of positively completed tasks of node k is computed via the expression:

$$pr\left(k\right) = \sum_{y^k \in Y^k} P\left(k \mid y^k\right) \cdot \prod_{d \in \Gamma^{-1}\left(k\right)} pr\left(d\right)^{y_d} \cdot \left(1 - pr\left(d\right)\right)^{1-y_d}$$

(11)

In order to compute value of $F(k, t)$ we should find the longest path from the actionable states, with positive influence on state nodes to the k.

Let us denote by $\bar{\mu}\left(i, j\right)$ a certain path leading in the graph G from node i to node j. The length of this path will be determined by the sum of times $\tau_E(d)$ for subsequent nodes of the path:

$$Dl\left(\bar{\mu}\left(\left(i, j\right)\right)\right) = \sum_{d \in \bar{\mu}\left(i, j\right)} \tau_E\left(d\right)$$

(12)

where $\tau_E\left(d\right) = \dfrac{\tau_{\min}\left(d\right) + 4 \cdot \tau_p\left(d\right) + \tau_{\max}\left(d\right)}{6}$.

We denote $\mu_{max}(i, j)$ as the longest path between nodes i, j. So:

- The earliest possible starting point for the performance of the tasks leading to the node d – $t_{min}(d)$ – we determine using the formula

$$t_{\min}\left(d\right) = Dl\left(\mu_{\max}\left(x^p, d\right) - \tau_E\left(d\right)\right) = \max_{k \in \Gamma^{-1}\left(d\right)}\left(t_{\min}\left(k\right) + \tau_E\left(k\right)\right)$$

(13)

The latest possible starting point for the performance of the tasks leading to the node d, without delaying the realization of the entire operation – $t_{max}(d)$, we determine using the formula

$$t_{\max}\left(d\right) = t^* - Dl\left(\mu_{\max}\left(d, x^k\right)\right) = \min_{l \in \Gamma\left(d\right)}\left(t_{\max}\left(l\right) - \tau_E\left(d\right)\right)$$

(14)

where

$$t^* = Dl\left(\mu_{\max}\left(x^p, x^k\right)\right)$$

(15)

Then,

$T_{min}(d)$ – the earliest possible moment of completing the tasks leading to the achievement of the node d – we determine using the formula

$$T_{\min}\left(d\right) = Dl\left(\mu_{\max}\left(x^p, d\right)\right) = \max_{k \in \Gamma^{-1}\left(d\right)}\left(t_{\min}\left(k\right) + \tau_E\left(k\right) + \tau_E\left(d\right)\right) = t_{\min}\left(d\right) + \tau_E\left(d\right)$$

(16)

$T_{max}(d)$ – the latest possible moment of completing the tasks leading to the achievement of the node d, not causing a delay in the realization of the entire operation, is determined using the formula

$$T_{\max}\left(d\right)=t^{*}-Dl\left(\mu_{\max}\left(d,x^{k}\right)+\tau_{E}\left(d\right)\right)=\min_{l\in\Gamma(d)}\left(t_{\max}\left(l\right)\right)=t_{\max}\left(d\right)+\tau_{E}\left(d\right)\tag{17}$$

In order to determine *F(k,t)* let us denote by

τ_k a random variable, the value of which represents the time needed to complete the tasks necessary to reach node *k*.

We are assuming that τ_k distribution is $N(\tau_E(k),\ \sigma^2(k))$, where

$$\sigma^{2}\left(k\right)=\left(\frac{\tau_{\min}\left(k\right)-\tau_{\max}\left(k\right)}{6}\right)^{2}.$$

We further assume that the time needed to reach node *k* is the time taken for passing the critical path (the longest distance) from the starting point x^p to node *k*, so:

$$F\left(k,t\right)=P\left\{\sum_{d\in\mu_{\max}\left(x^{p},k\right)}\tau_{d}<t\right\}\sim N\left(\sum_{d\in\mu_{\max}\left(x^{p},k\right)}\tau_{E}\left(k\right),\sum_{d\in\mu_{\max}\left(x^{p},k\right)}\sigma^{2}\left(k\right)\right)\tag{18}$$

In order to take into account the influence of actionable nodes with negative influence on state node k, we do the following:

- We find all actionable nodes *d* with negative influence such that:

$$DA_{N}\left(k\right)=\left\{d\in DA_{N}:\exists\overline{\mu}\left(d,k\right)\right\}\tag{20}$$

- We find $d^{*}\in DA_{N}:\tau_{E}\left(d^{*}\right)=\max_{r\in DA_{N}\left(k\right)}\tau_{E}\left(r\right)$ (21)

- We find all actionable nodes *b* with positive influence such that:

$$DA_{P}\left(k\right)=\left\{b\in DA:\exists\overline{\mu}\left(b,k\right)\right\}\tag{22}$$

- We find $b^{*}\in DA_{N}:\tau_{E}\left(b^{*}\right)=\max_{r\in DA_{N}\left(k\right)}\tau_{E}\left(r\right)$ (23)

We assume that the time of completing tasks connected with nodes d^{*} and b^{*} is a random value with normal distribution $N(\tau_E(d^{*}),\sigma^2(d^{*}))$ and $N(\tau_E(b^{*}),\sigma^2(b^{*}))$. We can now evaluate the probability that:

$$P\left\{\tau_{b^{*}}<\tau_{d^{*}}\right\}=\int_{-\infty}^{+\infty}F_{\tau_{b^{*}}}\left(y\right)dF_{d^{*}}\left(y\right)\tag{24}$$

Applying formulas for *pr(k)* and *F(k,t)* and $P\left\{\tau_{b^{*}}<\tau_{d^{*}}\right\}$, we can compute the probability of positively completed tasks of node *k* in a time shorter than *t* using the expression:

$$PR(k,t) = F(k,t) \cdot pr(k) \cdot P\{\tau_{b'} < \tau_{d'}\}.$$

(25)

For the end node x^k, which corresponds to ES(End State), we define a random variable $SR(x^k)$ whose value indicates the degree to which the end state has been achieved. In order to determine the expected value $SR(x^k)$ we introduce additional symbols:

$$\forall d \in D \setminus \{x^k\} I(d) = \begin{cases} 1 - \text{a point d has been reached} \\ 0 - \text{a point d not been reached} \end{cases}$$

(26)

- It follows from the previous assumptions and designations that $P\{I(d)=1\}=pr(d)$. Using the above notations, we obtain:

$$E\{SR(x^k)\} = \sum_{d \in \Gamma^{-1}(x^k)} pr(d) \cdot w(d)$$

(27)

where $w(d)$ denotes the importance of reaching node d for degree of end state achievement

$$\sum_{d \in \Gamma^{-1}(x^k)} w(d) = 1, w(d) > 1.$$

(28)

Taking time into account, we could define the time-dependent indicator of end state achievement:

$$E\{SR(x^k,t)\} = \sum_{d \in \Gamma^{-1}(x^k)} PR(k,t) \cdot w(d)$$

(29)

The MCA Software Package for Wargaming

The (M)odeling (C)omplex (A)ctivities software package was implemented with recommendations based on object-oriented methodologies. The main programming language is Java. There are two versions of the software package. The first version was prepared as the standalone application for ease of use also for wargamers. The second version was implemented for web-based use. Its key architectural objective was to ensure ease of reconfiguration and further development.

Figure 2 and 3 show the architecture of a package. It is compliant with multi-layer approach. The Net package contains classes responsible for determining graph/network parameters – the main ones are:

- **Node** – stores data describing graph nodes – represents one of the activities in the network.
- **Edge** – stores data describing graph arcs – represents transitions between actions, creating their sequences and cause and effect sequences.

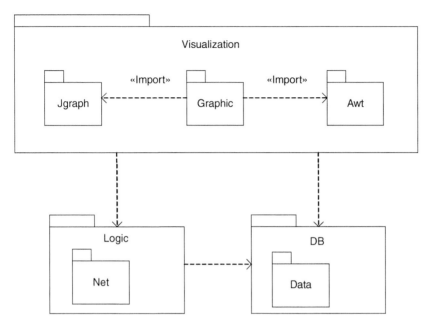

Figure 2 MCA package class layers.

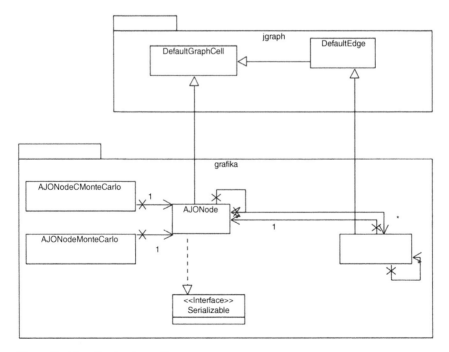

Figure 3 Visualization Layer Classes.

- **Pert** – stores graph structure data and provides methods to calculate the values of defined network characteristics.
- **SimGenerator** – is used to generate sequences of pseudo-random numbers.

The PERT class is the basic class responsible for the creation and calculation of network functions/characteristics (described in nodes and arcs). Characteristics are positive real numbers representing time and probability. It contains methods dedicated to, among others, detecting cycles in the graph, searching for the shortest and longest paths, calculating the time of starting and ending the activity, and estimating the probability of successful completion of the activity (defined as a node). Both known algorithms (Dijkstra algorithm, Depth-first Traversal) as well as heuristic algorithms, which are our own elaboration, were used. One of them – MonteCarlo – implements a method that calculates the probability of the operation in a given time (the argument is the time of completion). For the activity defined by the node for which the calculations are performed, the following sequence of calculations is performed:

- For each node, the action time is generated by the pseudo-random number generator (PRNG).
- Then the execution time for the network is calculated (from the point of view of the given node).
- This operation is repeated N times (N is the number of attempts).
- The result is the ratio of the number of successes (when the end time at the node is less than or equal to the set time) to the number of all attempts.

Another algorithm is C-Monte Carlo, which is based on a sequence of calculations for Monte Carlo, although some analytical calculations are also performed.

The following parameters are set for each decisive point:

- Name.
- Description.
- The shortest time of completion of tasks.
- The longest time of completion of tasks.
- The most probable time of completion of tasks.
- Baseline probability of completion of tasks - values in [0..1].

For each decisive point (node), the following characteristics are estimated:

- Expected duration of the tasks leading to the achievement of the decision point.
- The earliest possible moment of starting the implementation of tasks leading to the achievement of the decision point (maximum time distance of the decision point from the starting point).

- The earliest possible moment of starting the implementation of the tasks leading to the achievement of the decision point without delaying the implementation of the entire operation (the shortest time distance from the starting point).
- The earliest possible moment of completing the tasks leading to the achievement of the decision point.
- The latest possible moment of completing the tasks leading to the achievement of the decision point, not causing a delay in the implementation of the entire operation.
- Probability of reaching the decision point in less than the "worst" time, taking into account the impact of achieving the previous decision points (calculated for the worst time of the implementation of tasks leading to the point).
- Probability of achieving a decision point in less than "t," taking into account the impact of achieving the preceding decision points.
- Probability of reaching the decision point, taking into account the impact of achieving the preceding decision points.
- Variety of random variable of time needed to complete tasks necessary to reach the decision point (normal distribution).

The transitions (arcs) between nodes are described by the following parameters:

- Preceding decision point.
- Another decision point.
- The name of the transition, identifying the relation of the sequence of decisive points.
- Additional description (e.g. transition conditions).
- The coefficient of increase in the probability of reaching the decision point, provided that the decisive point is not reached.
- Factor for reducing the probability of reaching the decision point, provided that the preceding decision point is reached.

Since all information about the structure of the graph and the network is maintained in the Pert class, the images can be produced using different graphic packages. The MCA package has a graphical interface implemented using JavaFX 2 framework.

Wargame – Course of Action Evaluation

Assumptions

To present "in action" the developed model, method, and software, an experiment regarding a fictional military conflict and operations was carried out. The scenario of the game covers an unrealistic region in which are located countries A, B, C, D that are in possible conflict with fictional countries E, F. The main purpose of the game is the examination of the decision-making process of the player, who defends A, B, C, D. Secondary game objective is the measurement of the requirement of the level of the military potential needed for successful conduct of the operation. The game focuses on operations and actions of the ground forces with limited employment of close air support assets.

Situation

Y Forces

States A, B, C, D belong to the security block Y. Their respective military defense plans are synchronized and coordinated by the block command and control authorities. Military potential of the countries is based on standardized unit structures and they respectively consist of the following states:

- State A has two infantry brigades, in which they have air defense, artillery, engineer battalions, as well as recce company[1].
- State B has one infantry brigade.
- State C has two infantry brigades.
- State D in the vicinity of the border with state C has one mechanized division that consists of one armored brigade and two mechanized brigades[2].

Additionally in the region, the Y block deployed one international battlegroup subunit and is capable of deploying VHRF (Very High Readiness Force) brigade-sized unit.

X Forces

Countries E and F belong to block X, which is opposing the forces of block Y. Block X has the ability to employ combined army that consists of one tank division, one mechanized division, one parachute division, two independent mechanized brigades, brigade-sized air defense, artillery, engineers, electronic warfare, and reconnaissance units. Ground operations can be reinforced by aviation brigade with attack helicopter regiment and close air support fixed wing regiment. This size of the forces can be utilized after preliminary preparations. However, block X in the region is organized and prepared to engage 15 battlegroup-sized subunits without preliminary preparations (SHORT NOTICE).

---

1 This unified structure applies to all brigade-level units in the region.
2 Mechanized brigade consists of three mechanized battalions and a recce company, combat service, and combat support elements. Armored brigade consists of two armored battalions and one infantry battalion.

Intents of the Opposing Sides

A) X block operational objectives are focused on defeating A, B, C, D combat forces. In the alliance, country E plays supportive/enabling role in the ground domain. Its capabilities are not to be offensively employed. In the air domain, it plays a key role for the X alliance due to the ability to integrate with the air defense system. Country F intends to operate in the full-depth range. Its operations will be mainly ground focused due to lack of superior maritime forces. X intends to achieve its objectives with a speed and simultaneity focusing its attacking effort on the weakest adversary.

Block X sequence of actions[3]:

a) Course of action I – PROLONGED PREPARATIONS – will consist of following decisive points:

1) READINESS[4] – Increased level of readiness of the armed forces.
2) DEPLOYMENT – deployment to forward positions and assembly areas in order to pose pressure on Y block forces (24/72/96 hrs).
3) MOBILIZATION (5/10/15 days) – selective mobilization of the main offensive capabilities with intent to gradually finish all mobilization within an additional month.
4) ISTAR INFILTRATION – deployment of deep reconnaissance and intelligence (12/24/48 hrs) immediately after the decision of armed offensive actions.
5) OFFENSIVE (24/48/72 hrs) – conducted with main effort (ME) focused on state B or C. In the other directions limited offensive actions will be blocking or screening. In the case of an attack on state B, blocking actions will be executed against states C and D. If attack on the state C – block will be executed against states A and D.

Attack on ME will be run with 2/3 ratio of all advancing forces. To achieve shock and tempo, no preparatory fires or CAS operations will be run longer than 3/6/9 hours. In this option, attacking forces will be reinforcing their actions by employing second tactical echelons successively, 12/24/48 hrs (the greater the success the faster).

b) Course of action II – SHORT NOTICE – will have the decisive point MOBILIZATION as the last one. The operational objectives of offense will be limited to the defeat of the forces of the country B and seizing the capital when gaining maximum terrain elsewhere. The force ratio on LoE (level of effort) of ME for state B - compared to the other lines of effort - will be 2/3. The success

---

3 These actions are taken before actions of block Y due to fact that block Y is defensive by principle.
4 Name of the decisive point.

of the second course of action depends on as early as possible decision for attack with comparison to block Y preparation measures.

B) Y block operational objectives are to employ deterrence measures to prevent possible aggression conducted by block X. Its operation is foreseen to be divided into several phases that encompass: deterrence (preparation), defense divided into sub-phases delay and hold, and offence. The operation is to be conducted along lines of effort that reflect four countries in the region. Y block identifies X block operational center of gravity as the ability to maintain high tactical tempo with the operational objective to defeat X forces. Deterrence (preparation) phase is divided into decisive points:

1) Mobilization (2/5/10[5] days).
2) Deployment and readiness of forward deployed force (FDF) battlegroups: (12/24/48 hrs).
3) Reception of VHRF (5/10/15 days).
4) VHRF integration (3/5/7 days).
5) Y Alliance C2 establishment (2/7/- days).
6) Counter-mobility (7/14/21[6] days).
7) Land LOC maintained[7].
8) Deception (3/5/7 days).
9) Joint fires enabled (5/7/12 days).
10) Follow-on forces (20/25/30 days[8]).

If deterrence is not effective and X block attacks, Y defending own territory in two separate sub-phases (DELAY, HOLD) is going to create conditions to employ VJTF/follow-on forces to attack to defeat the aggressor and regain territory (if lost). In the DELAY sub-phase decisive points are:

11) X forces diverted[9] (12/24/48 hrs).
12) Reconnaissance capabilities denied (12/24/48 hrs).
13) X LOS[10] disturbed (24/48/72 hrs).
14) X combat capabilities reduced[11].
15) Enemy advance stopped.

---

5 Numbers reflect minimum/medium/maximum time to prepare. The more time – the higher is probability of success.

6 This includes engineering obstacles, demolitions, etc. and reflects minimum/medium/maximum time for preparation.

7 LOC – lines of communications.

8 All DPs apply to countries A, B, C, D. States A, B, C have to rely on allied reinforcements.

9 Preparations succeeded and X forces advance into Y forces defense main effort (ME): fires / own maneuver/local counterattacks/blocking positions.

10 LOS – lines of logistic supplies

11 On Main Effort (ME) the potential is down to 65%, on other lines 70%.

In the phase HOLD[12] decisive points are:

16) X second operation echelon denied.
17) Own forces resupplied.[13]

Additional Comments

Conduct of the game and the results are dependent on the choice of the enemy course of action. If we assume that DETERRANCE (PREPARATION) phase was completed, the game is about to check how efficient would be the enemy forces along specific lines of operation. In that case, it is assumed that the enemy would operate with forces that completed fully the processes of preparation (Combined Arms Army size offensive). If the X block will not allow defenders to prepare and create favorable conditions, it will act against Y block being in minimum and/or medium level of preparation (along respective lines of effort). It will most probably be in the medium level of preparations (defenders successfully read initial symptoms and interpreted enemy intents). In that situation, effects of DELAY sub-phase would be affected by progress of DPs in the DETERRENCE/PREPARATION phase.

To achieve these objectives, each side of the conflict can make moves described in the model presented in point 2. The influence of these moves is expressed by the evolution of the unconditional probability of achieving a decisive point (in other words, the move's success).

Model of Operation

The next Figure 4 shows a single first phase of complex operation. Each triangular symbol addresses decisive points – for Y block simple triangles and for X reverse triangles accordingly. A rectangle is reserved for end state symbol.

A Collection of Values of the Function *h(g)*

Decisive impact: $h=1$; Very strong impact: $h=0.9$; Strong impact: $h=0.75$; Average impact: $h=0.5$; Weak impact: $h=0.25$; Very weak impact: $h=0.1$; No impact: $h=0$.

1) Time of implementation steps:
    a) Times are expressed in days.
    b) The nodes corresponding to the countries, and not the operations, assign lead times $c=[0, 0, 0]$.

---

12 Phase is not to be played.
13 Up to 80%.

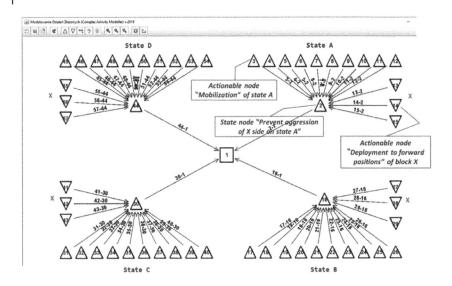

Figure 4 Graphical model of an analyzed conflict situation.

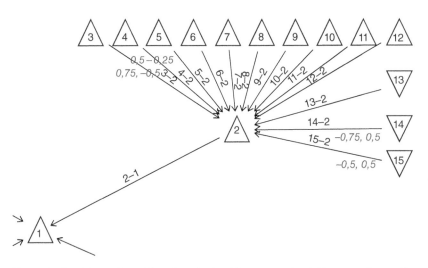

Figure 5 The illustration of functions f, g (positive or negative influences of preceding decisive points on the conflict escalation) related to edges: 3-2, 4-2, 14-2, 15-2.

2) *PB* – the baseline probability of completing tasks without the influence of any other events.

Time is described in vector $c = [$*The shortest time of completion of tasks, The most probable time of completion of tasks, The longest time of completion of tasks*$]$.

Deterrence Phase

Parameters Value – Deterrence Phase

1) End state: prevent aggression of X side on state A – node 1
   Parameters: Pb=0,1;
2) End state: prevent aggression of X side on state B – node 16
   Parameters: Pb=0,1;
3) End state: prevent aggression of X side on state C – node 30
   Parameters: Pb=0,1;
4) End state: prevent aggression of X side on state D - node 30
   Parameters: Pb=0,1;

Table 1 Parameters of decisive points of state A in deterrence phase.

Node no.	Decisive points	Side	Influence	PBA	h	g	C_{min}	C_{med}	C_{max}	Units of time
1	COG									
2	Prevent aggression of X side on state A			0,1						
3	Mobilization	A	positive	0,6	0,75	-0,5	14	24	42	[days]
4	Deployment and readiness of FDF BG	A	positive	0,8	0,5	-0,25	0,5	1	2	[days]
5	Reception of VHRF	A	positive	0,7	0,9	-0,75	5	10	15	[days]
6	VHRF integration	A	positive	0,9	0,9	-0,75	5	10	16	[days]
7	Y Alliance C2 establishment	A	positive	0,9	0,75	-0,5	2	7	10	[days]
8	Counter-mobility	A	positive	0,9	0,75	-0,75	7	14	21	[days]
9	Land LOC maintained	A	positive	0,8	0,9	-0,75	2	4	8	[days]
10	Deception	A	positive	0,8	0,5	-0,25	3	5	7	[days]
11	Joint fires enabled	A	positive	0,8	0,5	-0,25	5	7	12	[days]
12	Follow-on forces	A	positive	0,8	0,9	-0,75	20	25	30	[days]
13	Raise of forces' readiness status – READINESS	X	negative	0,9	-0,75	0,5	2	5	7	[days]
14	Deployment to forward positions and assembly areas – DEPLOYMENT	X	negative	0,9	-0,75	0,5	1	3	4	[days]
15	Mobilization	X	negative	0,9	-0,5	0,5	5	10	15	[days]

Table 2 Parameters of decisive points of state B in deterrence phase.

Node no.	Decisive points	Side	Influence	PBA	h	g	C_{min}	C_{med}	C_{max}	Units of time
1	COG			0,1						
16	Prevent aggression of X side on state B									
17	Mobilization	B	positive	0,9	0,75	-0,5	2	5	10	[days]
18	Deployment and readiness of FDF BG	B	positive	0,8	0,5	-0,25	0,5	1	2	[days]
19	Reception of VHRF	B	positive	0,7	0,9	-0,75	5	10	15	[days]
20	VHRF integration	B	positive	0,9	0,9	-0,75	5	10	15	[days]
21	Y Alliance C2 establishment	B	positive	0,9	0,75	-0,5	2	7	10	[days]
22	Counter-mobility	B	positive	0,8	0,5	-0,5	7	14	21	[days]
23	Land LOC maintained	B	positive	0,7	0,9	-0,75	6	8	12	[days]
24	Deception	B	positive	0,7	0,5	-0,25	3	5	7	[days]
25	Joint fires enabled	B	positive	0,5	0,5	-0,25	5	7	12	[days]
26	Follow-on forces	B	positive	0,8	0,9	-0,75	20	25	30	[days]
27	Raise of forces readiness status – READINESS	X	negative	0,9	-0,75	0,5	2	5	7	[days]
28	Deployment to forward positions and assembly areas – DEPLOYMENT	X	negative	0,9	-0,75	0,5	1	3	4	[days]
29	Mobilization	X	negative	0,9	-0,5	0,5	5	10	15	[days]

Table 3 Parameters of decisive points of state C in deterrence phase.

Node no.	Decisive points	Side	Influence	PBA	h	g	C_min	C_med	C_max	Units of time
1	COG									
30	Prevent aggression of X side on state C			0,1						
31	Mobilization	C	positive	0,8	0,75	-0,5	2	5	10	[days]
32	Deployment and readiness of FDF BG	C	positive	0,8	0,5	-0,25	0,5	1	2	[days]
33	Reception of VHRF	C	positive	0,7	0,9	-0,75	5	10	15	[days]
34	VHRF integration	C	positive	0,9	0,9	-0,75	5	10	15	[days]
35	Y Alliance C2 establishment	C	positive	0,9	0,75	-0,5	2	7	10	[days]
36	Counter-mobility	C	positive	0,5	0,75	-0,75	7	14	21	[days]
37	Land LOC maintained	C	positive	0,6	0,9	-0,75	5	7	10	[days]
38	Deception	C	positive	0,7	0,5	-0,25	3	5	7	[days]
39	Joint fires enabled	C	positive	0,8	0,5	-0,25	5	7	12	[days]
40	Follow-on forces	C	positive	0,9	0,9	-0,75	20	25	30	[days]
41	Raise of forces readiness status – READINESS	X	negative	0,9	-0,75	0,5	2	5	7	[days]
42	Deployment to forward positions and assembly areas – DEPLOYMENT	X	negative	0,9	-0,75	0,5	1	3	4	[days]
43	Mobilization	X	negative	0,9	-0,5	0,5	5	10	15	[days]

Table 4 Parameters of decisive points of state D in deterrence phase.

Node no.	Decisive points	Side	Influence	PBA	h	g	C_{min}	C_{med}	C_{max}	Units of time
1	COG									
44	Prevent aggression of X side on state D			0,1						
45	Mobilization	D	positive	0,9	0,75	-0,5	5	10	15	[days]
46	Deployment and readiness of FDF BG	D	positive	0,9	0,75	-0,25	0,5	1	2	[days]
47	Reception of VHRF	D	positive	0,8	0,8	-0,5	5	10	15	[days]
48	VHRF integration	D	positive	0,9	0,8	-0,5	5	10	15	[days]
49	Y Alliance C2 establishment	D	positive	0,9	0,75	-0,5	2	7	10	[days]
50	Counter-mobility	D	positive	0,9	0,75	-0,75	7	14	21	[days]
51	Land LOC maintained	D	positive	0,9	0,9	-0,75	5	7	10	[days]
52	Deception	D	positive	0,8	0,5	-0,25	3	5	7	[days]
53	Joint fires enabled	D	positive	0,8	0,9	-0,75	5	7	12	[days]
54	Follow-on forces	D	positive	0,8	0,9	-0,75	20	25	30	[days]
55	Raise of forces readiness status – READINESS	X	negative	0,9	-0,5	0,25	2	5	7	[days]
56	Deployment to forward positions and assembly areas – DEPLOYMENT	X	negative	0,9	-0,5	0,25	1	3	4	[days]
57	Mobilization	X	negative	0,9	-0,5	0,5	5	10	15	[days]

COA Evaluation

The analytical process is illustrated by the following graphs – Fig. 6.

We can see four state nodes: 2, 16, 30, 44. For each state node we can show the actionable nodes d with negative and positive influence:

$$DA_P(2) = \{3\}$$

$$DA_N(2) = \{15\}$$

$$DA_P(16) = \{26\}$$

$$DA_N(16) = \{29\}$$

$$DA_P(30) = \{41\}$$

$$DA_N(30) = \{43\}$$

$$DA_P(44) = \{54\}$$

$$DA_N(44) = \{57\}$$

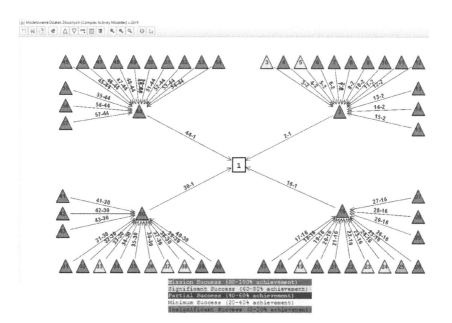

Figure 6 The sample analysis of conflict escalation.

Table 5 The main probabilistic characteristics of the COA- according to the MCA.

State Node No.	$P\{\tau_{b^*} < \tau_{d^*}\}$	pr(k)	F(k,tmin)	F(k,tmax)	F(k,1.5tE)	PR(k,tmin)	PR(k,tmax)	PR(k,1.5tE)
2	0.00035	0.960	0.11239	0.96296	0.91260	0.00004	0.00032	0.00030
16	0.04734	0.942	0.27425	0.72575	0.93319	0.01223	0.03236	0.04161
30	0.04734	0.926	0.27425	0.72575	0.93319	0.01202	0.03181	0.04090
44	0.04734	0.996	0.27425	0.72575	0.93319	0.01293	0.03422	0.04400

Now, we can show the probability of positively completed tasks of node k in a time shorter than t using the expression: $PR(k,t) = F(k,t) \cdot pr(k) \cdot P\{\tau_{b^*} < \tau_{d^*}\}$, according to the model defined in "Network model of complex activities," where:

for nodes 16, 30, 44

$tmin = 20$

$tmax = 30$

$tE = 25$

and for node 2

$tmin = 14$

$tmax = 42$

$tE = 25.33.$

Summary

The proposed method of modeling and analyzing complex projects combines the advantages of two approaches known from the literature: influence networks with the CAST logic method and the PERT method. The use of this combination allows to assess the possibility of completing a complex project in a specific time. The possibility of realization is determined by the probability of performing the operation in a specific time. This probability depends on the time distribution of the realization of individual activities and the relationship between the success of the realization of a specific activity and the success in the realization of preceding activities.

The paper presents a theoretical description of the method and a computational example showing the possibility of COA assessment in a certain example of military operation. The network model of the operation was made, the values of the parameters of this operation were estimated, and the values of indicators determining the success of the operation in a fixed time were calculated.

It is worth noting that the developed model was implemented as a computer application. This application was used in practice to evaluate the COA prepared by the participants of various staff exercises, which were mentioned in the introduction to the work.

For the aim of a presentation of the Bayesian method with MCA application, we focused on the first phase of the whole military operation. We presented the possible improvement of the introductory situation by appropriate realization recommended moves, and the analysis has disclosed the way. The same analysis can be provided in the phase 2 DELAY.

References

1 Antkiewicz R., Gąsecki A., Najgebauer A., Pierzchała D., Tarapata Z.: Computer Support For Joint Operation Planning Processes, Proceedings of the Military Communications and Information Systems Conference MCC'2008, 23-24 September, Cracow (Poland) (2008)

2 Aberdeen, D., Thiebaux, S., and Zhang, L. (2004). Decision-Theoretic Military Operations Planning, Proceedings of the 14th International Conference on Automated Planning and Scheduling, ICAPS 2004, Whistler, British Columbia, Canada. *June* 3-7: 402–411.

3 Chang, K.C., Lehner, P.E., Levis, A.H. et al. (1994). On causal influence logic. Technical report, George Mason University. *Center of Excellence for*: C31.

4 DeGregorio, E., Janssen, A., Wagenhals, W., and Messier, R. (2004). *Integrating Effects-Based and Attrition-Based Modeling, 2004 Command And Control Research And Technology Symposium The Power Of Information Age Concepts And Technologies, September 14-16.* Denmark: Copenhagen.

5 Drużdżel, M.J. (2009). Rapid Modeling and Analysis with QGENIE, Proceedings of the International Multiconference on Computer Science and Information. *Technology* 4: 157–164.

6 Dickson, K.D. (2007). Operational Design: A Methodology for Planners. *Journal of the Department of Operational Art and Campaigning Joint Advanced Warfighting School*: 23–38.

7 Falzon, L. (2006). Using Bayesian Network Analysis to Support Centre of Gravity Analysis in Military Planning. *Europ. Journ. of Operational Research* 170 (2): 629–643.

8 Haider, S. and Levis, A. (2007). Effective Course-of-Action Determination to Achieve Desired Effects. *IEEE Transactions On Systems, Man, And Cybernetics; Part A: Systems And Humans* 37 (6): 1140–1150.

9 Hoffman, M. (2005). On the Complexity of Parameter Calibration in Simulation Models. *JDMS, The Society for Modeling and Simulation International* 2 (4): 217–226.

10 Najgebauer, A., Antkiewicz, R., Pierzchała, D., and Rulka, J. (2018). Quantitative Methods of Strategic Planning Support: Defending the Front Line in Europe. In: Świątek J., Borzemski L., Wilimowska Z. (eds) Information Systems Architecture and Technology: Proceedings of 38th International Conference on Information Systems Architecture and Technology – ISAT 2017. ISAT 2017. Advances in Intelligent Systems and Computing, vol 656. Springer, Cham. In: *Print ISBN 978-3-319-67228-1.*

11 Kukacka M. Bayesian Methods in Artificial Intelligence. WDS'10 Proceedings of Contributed Papers, Part I, 25–30, 2010. ISBN 978-80-7378-139-2 © MATFYZPRESS

12 O'Connor, D. (2006). *Exact And Approximate Distributions Of Stochastic PERT Networks.* Dublin: University College.

13 Rosen, J.A. and Smith, W.L. (1996). *Influence Net Modeling with Causal Strengths: An Evolutionary Approach.* Monterey (USA), Naval Post Graduate School: Command and Control Research and Technology Symposium.

14 The Joint Doctrine & Concept Centre Ministry of Defence UK: Joint Operations Planning (2004)

15 Wagenhals, L.W., Levis, A.H., Course of action development and evaluation, in Proceedings of the 2000 Command and Control Research and Technology Symposium (2000)

16 Wagenhals L., Levis A.: Course of Action Analysis in a Cultural Landscape Using Influence Nets, Proceedings Of The 2007 IEEE Symposium On Computational Intelligence In Security And Defense Applications (CISDA'2007), ISBN 1-4244-0698-6, Honolulu (Hawaii, USA), April 1-5 (2007)

17 Zaidi A. K., Mansoor F., Papantoni-Kazakos P (2007). *Modeling with Influence Networks Using Influence Constants: A New Approach. Fairfax.* Washington (USA: George Mason University.

7

Combining Wargaming and Simulation Analysis

Mark Sisson

United States Strategic Command, Omaha, Nebraska, USA

Introduction

To quote General Anthony Zinni, "I don't think we should study things in isolation. I don't think a geographer is going to master anything, or an anthropologist is going to master anything, or a historian is going to master anything. I think it's a broad-based knowledge in all these areas, the ability to dissect a culture or an environment very carefully and know what questions to ask, although you might not be an expert in that culture, and to be able to pull it all together" (Zinni, 1998). With this advice in mind, and an attempt to study problems from a broader perspective, this chapter proposes a broad iterative approach for analysis focused on combining wargaming and simulation analysis (along with other analytical tools) to support decision-making informing real-world decisions. This ability is necessary due to the "increased complexity, ambiguity, and speed in future warfare" (TRADOC G-2, 2018). The proposed approaches are influenced by gaps in analytical processes and the contextual nature of applying analysis to real-world problems. The figure below shows the outline of the proposal.

As the figure indicates, gaps are prevalent between analytical techniques. However, some of these "gaps" (bolded in Figure 1) can be aligned. The gaps in objectives, scenarios, and assumptions, and a framework supporting portfolio development (schema) are examples of this. Although "gaps" can be aligned, different analytical tools bring different strengths and weaknesses resulting in different analytical perspectives (Gen. Zinni's "broad-based knowledge"). For example, wargames are often beneficial to explore decision-making of complex

Simulation and Wargaming, First Edition. Edited by Charles Turnitsa, Curtis Blais, and Andreas Tolk.
© 2022 John Wiley & Sons, Inc. Published 2022 by John Wiley & Sons, Inc.

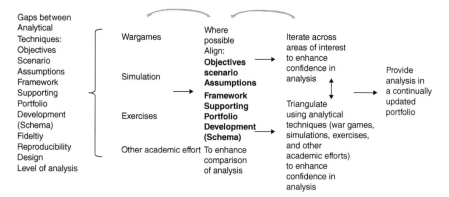

Figure 1 Potential iterative process combining analytical processes.

environments with incomplete information (Gen Zinni's "what questions to ask") but have limited repeatability. While simulations are repeatable, they are challenged by analyzing complex problems. Exercises often have high setup costs and limited repeatability.

In addition to limitations of analytical methods, any solution to a complex problem is context-dependent. In other words, actionable insights into complex problems may work under certain conditions but not others (Quigley, 2008). Since complex problems are context-dependent, a portfolio based on a framework or schema of varied analytical perspectives provides a basis for improved understanding. The schemas can provide a level of abstraction sufficient to support decision-making (Hard, Tversky, & Lang, 2006).

Current Efforts Underway

There are strategic analysis efforts underway to address these analytical problems. Perhaps the most recognizable strategic analysis effort is performed at the Office of Net Assessments located at the Pentagon (Source Watch, 2017) and called net assessments. A net assessment is more of a practice than an analytical methodology (Bracken, 2006). It is a practice, not art (like military judgment), nor is it a science (like physics) according to Mr. Bracken. A net assessment is a way of viewing the strategic environment from distinct perspectives. This perspective is driven by a focus on strategic interactions that are shaped by complex organizations.

Net assessment had its origins in the need to integrate Red and Blue strategy in a single place. According to the article, *Net Assessment: A Practical Guide* (Bracken, 2006), net assessment takes into consideration both Red and Blue actions. It produces an overall "net" assessment of a competitive situation.

Additionally, one of the greatest contributions of net assessment for strategic analysis is that it looks at longer periods. The thinking is that change is imperceptible on any given short period and can produce larger effects when viewed over longer periods. An example provided by Mr. Bracken is a view of China and whether it will attack Taiwan. In the short term this may not seem feasible, but a different picture arises if you look long term. Take for example, 30 years ago, China had almost no missiles facing Taiwan. Today the number is very significant.

A valuable insight from net assessment is that strategy is more than rivalry. Sometimes national security studies look at the issue as a one-on-one conflict, and if that conflict can be resolved then the problem is solved (Bracken, 2006). Unfortunately, in a complex environment, this is not true, which is why a portfolio of options is beneficial.

Net assessments, like the proposed methodology in this chapter, use scenarios and wargames games as part of their practice. However, net assessment does not often use complex mathematical computer models to understand a problem. Practitioners of net assessments believe that many of these mathematical models are misleading, since the uncertainties of the models are often not discussed, and their precision is not accurate in an imprecise world (Bracken, 2006). Additionally, there is a concern that assumptions are made to support modeling not the actual environment. In place of modeling complex and thinking simple, net assessment tries to model simple and think complex. Therefore, net assessments often use relatively simple models, numbers, and trends to think about what they mean. In contrast to net assessments, the proposed methodology in this chapter attempts to model broadly and iteratively within a viable framework or schema.

This different methodology may be useful since the current assessment of the environment through net assessments is not meeting all of the leadership needs. In a memo titled "Guidance," the previous Defense Secretary Ashton Carter directed the Office of Net Assessments (ONA) to focus its expertise more on current defense policy issues rather than hypothetical future threats (Franz-Stefan, 2015).

Methodology

To find opportunities, and not just challenges, the proposed methodology takes a broader and iterative approach. The methodology is driven by the objective of the analysis (see Figure 2).

With the objective in mind, an appropriate schema and alignment of "gaps" can assist in a type of meta-analysis. A framework or schema (e.g. Strengths, Weaknesses, Opportunities, Threats) is beneficial due to the different analytical

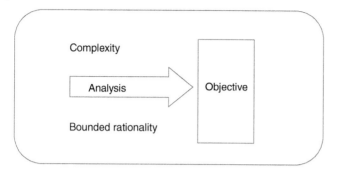

Figure 2 Objectives should drive analysis.

perspectives among different models and the contextual nature of providing actionable insights into complex problems. These frameworks or schema provide a level of abstraction that supports decision-making, but minimize the potential of getting bogged down in the noise of qualitative analysis. Furthermore, solutions to complex problems are constantly evolving and require continuous adjustment. Additionally, schemas along with the objective of the analysis can be used to clarify focus areas for analysis, and this focus needs to be consistent across the analytical effort, not "sprinkled on" at the end of the analysis, but part of the design.

Frameworks or Schemas to Support Portfolios

Analysis of complex problems should focus on providing actionable insights to support a foundation for action. This foundation is a portfolio of insights. These framed insights are scaffolds to support thinking (strategic, operational, innovative, etc.). For example, if the objective is strategic, then a Strengths, Weaknesses, Opportunities, and Threats (SWOT) framework or schema is useful. This schema provides sufficient abstraction to allow a perspective across multiple analytical efforts as well as indicating areas of additional analysis up to and including the strategic level (see Table 1 below). Another benefit is that the SWOT schema can be supported by the different analytical perspectives available within the analytical tools (e.g. wargaming gaming, simulation).

Table 1 SWOT framework – possible schema.

	Helpful	Harmful
Internal origin	Strengths	Weaknesses
External origin	Opportunities	Threats

Porter's five forces diagram

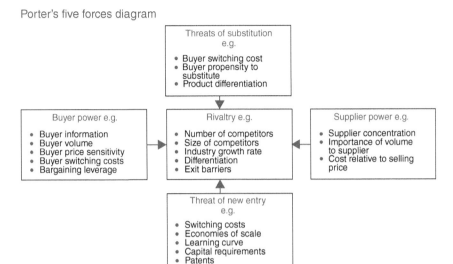

Figure 3 Porter's five forces diagram.

A summary of Porter's four corner's analysis

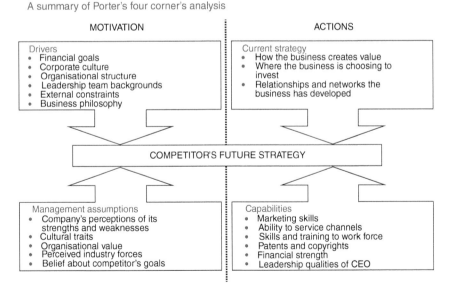

Figure 4 Porter's four corner's analysis.

Another framework or schema to support a portfolio is derived from Mr. Porter's components of competitor analysis (Porter, 1980). In this model, Drivers (future goals/strategies), Can Do/Is Doing (current strategies), Management assumptions (themselves, and their environment), and Current capabilities inform a competitor response (See Figure 4).

Where the focus is a greater understanding of factors affecting profitability, then Porter's five force analysis might be appropriate. Here the schema consists of supplier power, buyer power, competitive rivalry, a threat of substitution, and a threat of new entry (Porter, 2013) (See Figure 3).

If the objective of the analysis is at the operational or tactical level then perhaps the Center of Gravity framework is appropriate. The Center of Gravity framework is comprised of the Center of Gravity (CG), Critical Capabilities (CC), Critical Requirements (CR), and Critical Vulnerabilities (CV). CG entities are the focus of the strength of a system. CC support the center of gravities. CR are conditions, resources and means that are essential for a center of gravity to achieve its capability. CV are those critical requirements, or components thereof, that are vulnerable in a way that will contribute to a center of gravity failing (Strange and Iron, 2005).

If the objective is innovation, then a schema exploring assumptions, perspectives, and connection of ideas is useful (Seelig, 2013). This schema includes capturing insights into connecting/combining ideas, reframing problems, and challenging assumptions of the proposed way forward.

Other schemas may be appropriate depending on the objective of the analysis, these schemas may be Diplomatic Informational, Military, and Economic (DIME), Political, Economic, Social and Technological (PEST), or even Theory of the Resolution of Invention-related Tasks (TRIZ - Russian acronym for "теория решения изобретательских задач, teoriya resheniya izobretatelskikh zadatch," english translation: "theory of inventive problem solving."). Often however, the desired schemas are a combination of differing frameworks, and provide a framework for decision-making when incorporated into a portfolio.

Comparability

Actionable insights also need to be perceived as feasible and beneficial for decision-makers to act. This is where the comparison of insights across a schema is beneficial. For example, to increase confidence in wargaming insights, a common framework or schema for comparing strategic insights across war games is useful. One framework is the Strengths, Weaknesses, Opportunities, and Threats (SWOT) schema. This framework of SWOT along with design, scenario, and assumptions allows a comparison of multiple wargames (See Figure 5). The SWOT framework was selected since it provides a better understanding of the strategic

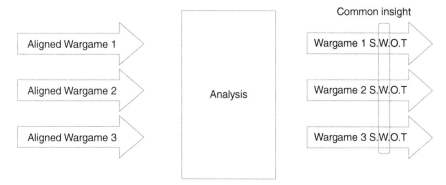

Figure 5 Comparing common features across multiple wargames improves confidence in results.

choices (MindTools, 2017); however, other frameworks would be appropriate depending on the level of analysis (e.g. Operational level gaming may consider Center of Gravity Analysis).

Using this SWOT framework across possible futures allows a more efficient comparison of insights in the context of wargames. Where patterns of strength, weakness, opportunity, and threats appear consistently across wargames, this suggests an area for further exploration. For example, say you find in the context of multiple wargames you are not able to overcome the time and distance advantages of your adversary in wargames (Threat). You may decide to shape the environment in advance of a conflict. In this case you may move forces closer to the potential conflict.

This is what the analysis of RAND wargames accomplished in 2014 and 2015 suggested. The RAND wargames focused on a possible invasion of the Baltics (Shlapak and Johnson, 2015). The games' findings are unambiguous and frightening: The findings indicate that as currently postured (Department of Defense, 2006), NATO cannot successfully defend the territory of its most exposed members to include the Baltic states. RAND utilized multiple games using a wide range of expert participants for this assessment. In the wargame, the longest period for Russian forces to reach the outskirts of the Estonian and Latvian capitals of Tallinn and Riga (Department of Defense, 2006) was approximately 60 hours. In the event of such an attack, such a rapid advance would leave NATO with a limited number of options. One option would be a costly counteroffensive, with significant escalation potential (perhaps even to the nuclear realm). Another option would be to concede at least a temporary defeat. This option would have significant implications for the future of NATO and perhaps even Europe. However, the RAND wargame does provide a method to enhance deterrence and potentially preclude a disastrous Russian attack. The potential solution would be a movement of forces forward, a force of about seven brigades, including three heavy armored brigades (Department of Defense, 2006). It was assessed that with this coalition force Russia would not be able to present NATO with a *fait accompli*. A Russian attack would instead start a prolonged and serious war between Russia and a materially far wealthier and more powerful NATO coalition. As the RAND wargames indicate a comparison of multiple wargames can improve the confidence in results.

This technique is not new; in fact, this comparison of multiple studies is currently used in qualitative research and is called meta-ethnography. Meta-ethnography is a useful method for synthesizing qualitative research and for developing models that interpret findings across multiple studies, or in our case multiple analytical efforts (Atins & Lewin, 2008). This technique is often used in health research.

Emergence

Used to support the technique of comparison is the iteration of analytical events across areas of uncertainty to promote the emergence of insights. A single analytical event is often too small a sample size to generate convincing insights from the emergent behavior of interactions. For example, the design and execution of iterative wargames may allow credible emergent insights to be confirmed, as seen with the RAND example.

One type of iteration may occur across different levels of conflict, as shown in Figure 6. These different futures can be developed through consideration of key drivers in the environment. These potential futures are areas of exploration through iterative analysis. If after strategic analysis some elements of SWOT appear consistent, these elements may be robust enough for incorporation into a portfolio.

Other areas of iteration may occur across key weaknesses as done in the RAND study. Or the iteration could occur across potential opportunities, assumptions, or other elements of a schema. Whatever area of iteration is selected, it needs to be driven by the objective of the analysis.

Triangulation

While wargames are less expensive than actual operations, even a game of limited scope requires a great deal of time and work (McHugh, Fundamentals of War Gaming, 2005). Additionally, it is not possible to capture the impact of some intangibles in a wargame to include psychological and complexity factors. As

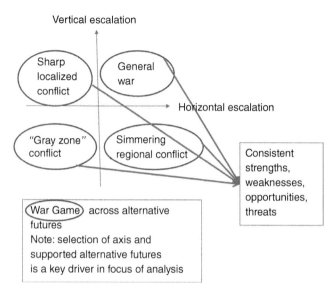

Figure 6 Iterating across alternative futures increases confidence in results.

Mr. McHugh points out, "... missing are the impressions of combat, the genuine tensions and the far-reaching responsibility." These simplifications, although necessary for the wargame, might be captured in other tools such as Subject Matter Expert (SME) interviews.

To mitigate the limitations of wargames, a broader set of tools needs to be utilized. To quote Dr. Perla, "Alone, wargames, exercises, and analysis are useful but limited tools for exploring specific elements of warfare. Woven together in a continuous cycle of research, wargames, exercises, and analysis each contribute what they do best to the complex and evolving task of understanding reality" (Harrigan, Kirchenbaum, & Dunnigan, 2016).

To improve this understanding of reality, a focus using multiple tools to provide consistent results, or "triangulate," on actionable insights is useful. An example of this cycle of research in support of triangulation is shown below. This diagram reinforces the iterative nature of the analysis, where appropriate tools are necessary to develop, explore, and assess concepts (See Figure 7).

This iterative process to develop elements of SWOT requires multiple tools, of which a partial list is provided below.

Exercises

The US Department of Defense (DOD) describes an exercise as a military maneuver or simulated wartime operation involving planning, preparation, and

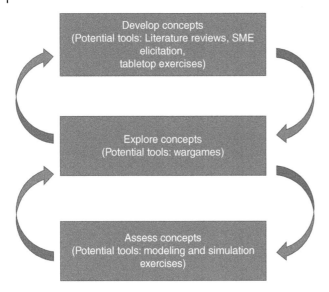

Figure 7 Potential cycle of research.

execution that is carried out for training and evaluation (DOD, 2017). As you might expect exercises are important tools in support of analysis especially when combined with other tools like wargaming.

Artificial Intelligence

Artificial intelligence has many potential applications for a holistic iterative approach to developing insights. One such tool is text analytics, which continues to improve in an ability to organize/classify unstructured data. According to IBM, approximately 80% of all information is unstructured (Monkeylearn, 2019). In the context of wargame data, a significant portion is text. Because of the messy nature of the wargaming text, analyzing, understanding, organizing, and sorting into potential schemas are challenging.

One method to overcome this challenge is using machine learning to transform text into a numerical representation in the form of a vector (Monkeylearn, 2019). One common vector is word frequency from a predefined list of words. Other tools to develop vectors are naive Bayes, support vector machines, and deep learning.

Naive Bayes has the advantage that you can get good results with limited data and computational resources (Monkeylearn, 2019). On the other hand, Support Vector Machines needs more computational resources, but can achieve more accurate results (Monkeylearn, 2019). Perhaps the most sophisticated and capable tool for text analysis is Deep learning. Deep learning is inspired by how the human brain works using Convolutional Neural Networks (CNN) and Recurrent Neural Networks (RNN).

Wargames

Peter Perla, a recognized authority on defense-related wargaming (Rothweiler, 2017) defines a wargame as "an exercise in human interaction, and the interplay of human decisions and the simulated outcomes of those decisions" (Perla, 1990). He understood that wargames, while short of the actual physical impacts of war, provide a means for learning the "dynamics of warfare." As the author of *The Art of Wargaming: A Guide for Professionals and Hobbyists*, Dr. Perla believes that to support decision-making, the design team must use a combination of tools (Rothweiler, 2017). These tools include two broad categories of wargames. One is the seminar wargame. In this type, experts are assembled and presented with the problem scenario and participate in a facilitated discussion. Take for example a seminar wargame to assess the threat of Russian occupation of Arctic territory. Key stakeholders are assembled, to include: the US government, European Command (EUCOM)/ US Northern Command (NORTHCOM), Russia, Canada, the Scandinavian countries, and NATO. This cycle is often repeated three to four times over a four-day seminar wargame with a final report drafted for the sponsor. A weakness of seminar games however is the possibility for the discussion to move toward generalities, often providing little useful information to support decision-making.

Dr. Perla believes another useful tool for strategic analysis is the matrix game (Rothweiler, 2017). This type of wargame often looks like commercial strategy games (Ruse, Panzer General, Axis & Allies, Risk, etc.). An example of this type of wargame is a Homeland Defense strategic matrix wargame. Here the players are divided into eight teams of regional/topical experts representing the United States, Mexico, Canada, the Caribbean nations, the Central American nations, and the European nations with Caribbean interests, a team representing international terrorism, and a team representing international organized crime (Rothweiler, 2017). The objective of the wargame is to explore strategies of layered defense. A strength of this type of wargaming is the analysis of alternative futures. Alternative futures allow for an exploration of strategic "what-ifs." This analysis supports plans and policies. Another strength is the experiential learning that occurs in this wargame. Weakness in these games often comes in the quality of the players – high-quality players are required. Another drawback of this type of wargame is that it requires more preparation than a seminar game (Rothweiler, 2017).

Another key contributor to wargaming literature is Francis McHugh, author of the *Fundamentals of War Gaming*. Mr. McHugh sees the value of wargames through the synthetic reality they provide (McHugh, Fundamentals of War

Gaming, 2013). Additionally, he believes wargames possess the advantages of simulation, thus providing a methodology for studying and examining almost any type of military operation from a minor skirmish to nuclear conflict. He is also a proponent of the experiential benefits of wargaming, like Dr. Perla. Mr. McHugh emphasizes the cost-effectiveness, of wargaming, "Gaming provides a means of gaining useful experience and information in advance of an actual commitment, of experimenting with forces and situations that are too remote, too costly, or too complicated to mobilize and manipulate, and of exploring and shaping the organizations and systems of the future" (McHugh, Fundamentals of War Gaming, 2005). This is critical in strategic scenarios, such as atomic warfare, where there are no clear historical examples to provide guidance.

In addition to these strengths, there are significant limitations to wargames (which can potentially be mitigated through other tools), Mr. McHugh points out. For example, while wargames are less expensive than actual operations, even a game of limited scope requires a great deal of time and work (McHugh, Fundamentals of War Gaming, 2005). Additionally, it is not possible to capture the impact of intangibles in a wargame to include psychological and complexity factors. As Mr. McHugh points out, ". . . missing are the impressions of combat, the genuine tensions and the far-reaching responsibility." These simplifications, although necessary for the wargame, must be considered in evaluating the wargame.

Along with Dr. Perla, and Mr. McHugh, other key contributors to wargaming include Mr. Herman, Mr. Frost, and Mr. Kurz. Their book *Wargaming for Leaders* (Herman, Frost, & Kurz, 2009) focuses on both military and, perhaps more importantly, business applications for wargaming. These authors view wargaming as a tool to anticipate future conflicts. A key strength is that modern wargaming is critical when dealing with incomplete information. Another strength of modern professional wargaming is that it provides a methodology to get at the things that one leader, no matter how visionary, cannot grasp on his or her own. This is done by bringing together the experts on the issue at hand and allowing them to explore the future in a low-risk environment and to find answers to questions that may not have been considered. A key component of this is exploring the adversary's weaknesses (Herman, Frost, & Kurz, 2009). A strength and a weakness (depending on context) are that wargames tell you what might happen (not necessarily what will happen).

Computer Simulation Models

A simulation model is one where the system is abstracted into a computer program. Simulation models are often useful and can be used to model complex systems without the need to make simplifying assumptions and with sufficient detail

Figure 8 "Pyramid" of models.

(Rajgopal, 2017). On the other hand, caution is necessary with simulation models because they are easy to misuse, which is one reason simulations are not used in net assessments.

Models have different levels of fidelity dependent on the required analysis. These models can build upon one another as depicted in Figure 8. High-fidelity systems/engineering models provide precise data; for example, a model of a bullet trajectory could be captured in a systems engineering model. This could be placed in an engagement model, such as BRAWLER, which simulates air-to-air combat (Defense Systems Information Analysis Center, 2107). The information from Brawler can support a Mission area model such as EADSIM, which could in turn feed a campaign level model such as STORM. As the pyramid below suggests, high-fidelity data from lower in the pyramid support those lower fidelity models above it.

Mathematical Models

Another category of models is mathematical models. In this type of model the characteristics of a system or process are represented by a set of mathematical relationships. Mathematical models can be deterministic or probabilistic. In a deterministic model, all parameters used to describe the model are assumed to be known (Rajgopal, 2017). With probabilistic models, the exact values for some of the parameters may be unknown but its distributions are sometimes capable of characterizing the desired event.

Most mathematical models are described by three main elements: decision variables, constraints, and the objective function(s). The decision variables are often used to model specific actions that are under the control of the decision-maker. An analysis of the model will seek specific values for these variables that are desirable from one or more perspectives. Linear regression is a common mathematical model.

Using exercises, and modeling and simulation to enhance confidence in strategic insights is key for "triangulation." The "triangulation" methodology can clarify specific strengths, weaknesses, opportunities, and threats.

Experimentation

In the context of military operations, experimentation is a broad category. Experimentation could be conducted to "discover, test, demonstrate, or explore future military concepts, organizations, and equipment and the interplay among them, using a combination of actual, simulated, and surrogate forces and equipment" (Academics of Sciences Education Medicine, 2019).

Building Portfolios

Taking actionable insights derived through comparison, emergence, and triangulation provides a foundation for building portfolios. Portfolios may be developed using multiple formats, a graphical depiction is selected here (Figure 9), because it is often useful to start the conversation with decision-makers. These portfolios provide decision-makers a foundation to shape and react to the environment.

Below is a proposed regional portfolio, which would have supporting analysis.

More complex portfolios, which provide additional insight may be preferable. If using the SWOT schema then the Threats, Opportunities, Weakness, and Strengths, (TOWS) Strategic Alternatives Matrix provides a portfolio with another

Figure 9   An assessment of a strategic option.

TOWS Strategic Alternatives Matrix

	External Opportunities (O) 1. 2. 3. 4.	External Threats (T) 1. 2. 3. 4.
Internal Strengths (S) 1. 2. 3. 4.	**SO** *"Maxi-Maxi" Strategy* Strategies that **use strengths** to **maximize opportunities.**	**ST** *"Maxi-Mini" Strategy* Strategies that **use strengths** to **minimize threats.**
Internal Weaknesses (W) 1. 2. 3. 4.	**WO** *"Mini-Maxi" Strategy* Strategies that **minimize weaknesses** by taking advantage of **opportunities.**	**WT** *"Minii-Mini" Strategy* Strategies that **minimize weaknesses** and **avoid threats.**

Figure 10 TOWS strategic alternatives matrix.

perspective. The TOWS Strategic Alternatives Matrix is depicted in Figure 10 (MindTools, 2017).

This matrix shows how Strengths, Weaknesses, External Opportunities, and External Threats can be fused into elements of strategy. An example of how this iterative strategy using the TOWS matrix may bring together the concepts described in this chapter.

Starting with a portfolio with risk on the y-axis, where risk is identified as the inability to achieve an objective (high risk is where analysis shows an inability to achieve an objective). The x-axis is confidence in the strategic insight (for example, a weakness shown in multiple wargames, modeling and simulation, and exercises would have high confidence). Please note variability in risk and confidence is depicted in the size of the strategic insight in the portfolio. A desirable depiction of an element of strength, weakness, opportunity, and threat would have low risk, and high confidence (see Figure 11).

Figure 11 An assessment of a strategic option.

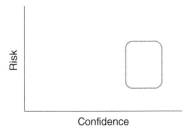

Risk

Confidence

Let us assume an objective is to defend an ally. Performing the proposed iterative strategic analysis, it is found that a strength is cyber (significant variability in risk and confidence – see Figure 12), weakness is found to be an inability to move forces sufficiently fast to stop an adversary before occupying the allies country, opportunity is found in exploiting the adversary cyber network, and threat is identified as an attack on the ally.

Using the Maxi-Maxi, a strategic threat would be to use cyber to develop capabilities against the adversary. Figure 12 provides a visual depiction of the proposed portfolio. This option shows high confidence and relatively but significant variability in risk.

The Mini-Maxi strategic thread may be to use cyber to speed up a capability to move forces forward. As depicted in Figure 12, there is high confidence in the risk (and confidence) of inability to move forces forward. The opportunity of using cyber to speed up the movement of forces in the depiction below shows high confidence in the opportunity but a significant risk in weakness.

The Mini-Maxi strategic thread may be to use cyber to slow down an adversary's attack on the ally. This portfolio option shows a great deal of variability on the strength and high risk on the threat, and therefore may not be desirable (See Figure 13).

Figure 12 Using strengths to maximize opportunities assessment.

Figure 13 Using weaknesses to take advantage of opportunities assessment.

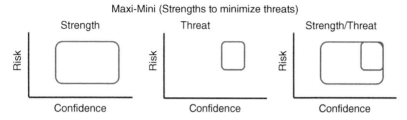

Figure 14 Using strengths to minimize threats assessment.

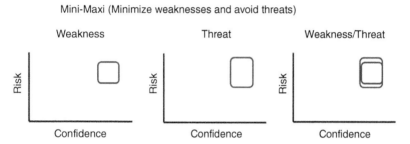

Figure 15 Using weakneses to avoid threats assessment.

The Maxi-Mini strategic thread (Figure 14) may be to use cyber to convince an adversary not to attack an ally – a proactive strategy that even if it fails may allow time for follow-on actions.

The Mini-Mini strategic thread (Figure 15) may be to move forces forward to preclude an attack by the adversary.

This example is simplistic; however, it shows how the holistic methodology providing decision-makers a set of options may be valuable. A significant value of the analysis would be in experimenting with strategic threads and then assessing how the strategic portfolios change (minimizes risk and increases confidence). In the example above, if you were to combine the strategic threads (develop cyber, and move forces together) this may provide a way of reaching the objective of defending any ally with reduced risk.

Conclusion

The proposed techniques presented above provide a set of processes to improve the credibility of insights in support of decision-making. The holistic processes include comparing insights across wargames, using iterative wargames to capture emergent insights, using multiple analytical tools to triangulate insights, and then

Table 2 Different techniques to support decision-making.

	Problem	Mitigation strategy
Comparability	No common framework to compare results across wargames	Use SWOT or some other relevant framework
Emergence	Small sample set	Iterative games
Triangulation	Wargame results are contextual	View from multiple perspectives
Organization	Wargame results are located across multiple documents	Package results by region

organizing results into portfolios to support decision-makers (see problems and mitigation strategies in Table 2) to improve strategic analysis.

The portfolios are then used to provide support for the development of strategy through a set of strategic options that have been assessed through an iterative analytical process. The analytical process can be revisited as additional information is made available. The proposed portfolios could be a series of strategic options evaluated for potential risk and benefit. The risk and confidence in strategic insights would be part of the analytical process.

In closing, the security environment has changed – this is consistent with the National Defense Strategy (DoD, 2018), which states that the increasingly complex global security environment is characterized by challenges to open international order and the re-emergence of long-term, strategic competition between nations. These challenges require techniques that can combine wargames and other analytical tools to support decision-makers.

References

Atins, S., & Lewin, S. (2008, April 16). *NCBI*. Retrieved from BMC Medical Research: https://www.ncbi.nlm.nih.gov/pmc/articles/PMC2374791/pdf/1471-2288-8-21.pdf

Bensoussan, B. and Fleisher, C. (2013). *Analysis Without Paralysis*. New Jersey: Pearson Education.

Bracken, P. (2006). Net Assessment: A Practical Guide. *Parameters*: 90–100.

Defense Systems Information Analysis Center (2017, November). *DSIAC*. In: *Retrieved from Models*. https://www.dsiac.org/resources/models.

Department of Defense (2006). *Deterrence Operations Joint Operations Concept*. In: *Retrieved from Joint Chiefs of Staff*. http://www.jcs.mil/Portals/36/Documents/Doctrine/concepts/joc_deterrence.pdf?ver=2017-12-28-162015-337.

DOD. (2017, August). *DTIC*. Retrieved from DOD Dictionary: http://www.dtic.mil/doctrine/new_pubs/dictionary.pdf

DoD (2018). *National Defense Strategy.* In: *Retrieved from dod.defense.gov.* https://dod. defense.gov/Portals/1/Documents/pubs/2018-National-Defense-Strategy-Summary.pdf.

Franz-Stefan, G. (2015, June 12). *The Future of Net Assessments at the Pentagon.* Retrieved from The Diplomat: https://thediplomat.com/2015/06/the-future-of-net-assessment-at-the-pentagon/

Hard, B., Tversky, B., and Lang, D. (2006). Making sense of abstract events: Building event schemas. *Memory and Cognition:* 1221–1235.

Harrigan, P., Kirchenbaum, M., and Dunnigan, J. (2016). *Zones of Control: Perspectives on Wargaming.* MIT Press.

Herman, M., Frost, M., and Kurz, R. (2009). *Wargaming for Leaders.* Booz Allen Hamilton: McGraw-Hill.

McHugh, F. (2005). *Fundamentals of War Gaming.* Newport: US NWC Press.

McHugh, F. (2013). *Fundamentals of War Gaming.* New York: Skyhorse Publishing.

MindTools (2017, September 28). *SWOT Analysis.* In: *Retrieved from MindTools.* https://www.mindtools.com/pages/article/newTMC_05.htm.

Monkeylearn (2019). *Text Classification.* In: *Retrieved from Monkeylearn.* https:// monkeylearn.com/text-classification/.

National Academics of Sciences Education Medicine (2019). *National Academies Press: OpenBook.* In: *Retrieved from National Academies Press: OpenBook.* https:// www.nap.edu/read/11125/chapter/4#30.

Perla, P. (1990). *The art of wargaming: a guide for professionals and hobbyists.* Annapolis: Naval Institute Press.

Porter (2013, June). *Porter's Five Forces of Competitive Position Analysis.* In: *Retrieved from CGMA.* https://www.cgma.org/resources/tools/essential-tools/porters-five-forces.html.

Porter, M. (1980). *Competitive Strategy: Techniques for Analyzing Industries and Competitors.* New York: The Free Press.

Quigley, K. (2008). *Responding to Crises in the Modern Infrastructure: Policy Lessons from Y2K.* Hampshire: Palgrave Macmillian.

Rajgopal, J. (2017, November). *University of Pittsburgh.* Retrieved from PRINCIPLES AND APPLICATIONS OF OPERATIONS RESEARCH: http://www.pitt. edu/~jrclass/or/or-intro.html

Rothweiler, K. (2017, March 29). *Wargaming for Strategic Planning.* Retrieved from Strategy Bridge: https://thestrategybridge.org/the-bridge/2017/3/29/ wargaming-for-strategic-planning

Seelig, T. (2013). *InGenius: a Crash Course on Creativity.* New York: Harper One.

Shlapak, D., & Johnson, M. (2015). *Reinforcing Deterrence on NATO's Eastern Flank.* Retrieved from RAND: file:///C:/Users/sisso/Downloads/RAND_RR1253.pdf

Source Watch. (2017, November 9). *The Center for Media and Democracy.* Retrieved from Office of Net Assessment: https://www.sourcewatch.org/index.php/ Office_of_Net_Assessment

Strange, J. and Iron, R. (2005). *Understanding Centers of Gravity and Critical Vulnerabilities.* Retrieved from Air University http://www.au.af.mil/au/awc/awcgate/usmc/cog2.pdf.

G-2, T.R.A.D.O.C. (2018). *The Red Team Handbook.* In: *Retrieved from US Army.* https://usacac.army.mil/sites/default/files/documents/ufmcs/The_Red_Team_Handbook.pdf.

Zinni, A. (1998). *"Non-Traditional Military Missions: Their Nature, and the Need for Cultural Awareness & Flexible Thinking" in Capital "W".* In: *War.* Quantico: Marine Corp University.

8

The Use of M&S and Wargaming to Address Wicked Problems

Phillip E. Pournelle

Commander, United States Navy, Retired, Fairfax, Virginia

The United States Department of Defense (DoD) faces significant challenges as it confronts competitors in a rapidly changing environment.[1] The rapid proliferation of new capabilities in the hands of competitors and opponents who employ them in new and innovative ways has created wicked problems that require innovative solutions.[2] Modeling and Simulation (M&S) tools cannot effectively provide solutions to these wicked problems on their own and in many respects the validity of answers from these tools should be viewed with suspicion.[3] The proper use of an iterative cycle of research employing multiple tools will likely deliver the best results. The cycle of research employing wargames, operations analysis (including M&S), and exercises will provide the best answers to wicked problems. Each of these approaches can strengthen the other and avoid certain weaknesses. However, each of these tools requires the employment of best practices to avoid certain pathologies that can weaken the overall effort.[4]

The reason why M&S will not be effective in most cases is the problem will be dominated by the decisions humans will be making, particularly under conditions of uncertainty. Simulation simply has not solved the problem of

---

1 Summary of the 2018 National Defense Strategy of the United States.
2 The Assessments and Recommendations of the National Defense Strategy Commission, 13 November 2018. https://www.usip.org/publications/2018/11/providing-common-defense
3 GAO, Defense Strategy: Revised Analytic Approach Needed to Support Force Structure Decision-Making, 14 March 2019. https://www.gao.gov/products/GAO-19-385
4 Perla, Peter, *The Art of Wargaming*, Naval Institute Press, 1990.

Simulation and Wargaming, First Edition. Edited by Charles Turnitsa, Curtis Blais, and Andreas Tolk.
© 2022 John Wiley & Sons, Inc. Published 2022 by John Wiley & Sons, Inc.

realistically representing human decision-making under varying conditions. Further, if the problem includes new innovations then it is highly likely the interactions between elements of the competing sides will not be fully understood. Worse, the range of decisions (or decision space) of the competing actors will not be fully understood. While computers are very fast, they are currently only capable of taking actions that the programmer has written for them. Therefore, the use of M&S, a tool focused on systems in interaction, would be premature. The decision space of the humans will need to be explored first using wargames, a tool focused on human decision-making under conditions of uncertainty.

In this chapter, we will examine the best practices for the employment of two elements of the cycle of research for the analysis of wicked problems, particularly for defense-related issues. We will begin with the motivation for conducting the analysis and the applicability of this process particularly to wicked problems. We will discuss the need for proper framing of the problem and how wargaming can assist in this effort, particularly in the development of the phenomenology of the competition. We will then explore how M&S can best be employed to support the analysis. We will identify some of the pathologies resident in wargaming and M&S and how the combination of tools employing best practices can avoid their pitfalls.

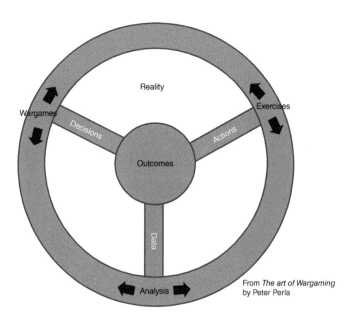

Figure 8.1 Peter Perla's Cycle of Research (Perla, Chapter 9).

Why Are We Doing This?

Most operations analysis whether for business or government is devoted to the support of the development or execution of a strategy toward an objective. To fully support that strategy, it is important for the analyst to understand what a strategy is and what are its supporting elements. Richard Rumelt provides an excellent summary of what makes a good strategy:

> "A good strategy is, in the end, a hypothesis about what will work ... A good strategy has at a minimum, three essential components:
>
> - a *diagnosis* of the situation,
> - the choice of an *overall guiding policy*,
> - and the design of *coherent action...*
>
> In general, strategic leverage arises from a mixture of anticipation, insight into what is most pivotal or critical in a situation, and making a concentrated application of effort... The most critical anticipations are about the behavior of others, especially rivals."[5]

To be successful, a Leader must examine a problem and tell his subordinates how to act in order to meet an objective. His commands will be based on his understanding of the problem and how his actions will bring about the desired state of affairs. The challenge, highlighted in the last sentence, is the leadership of the competition is attempting to take actions to bring about his desired goals, or at least frustrate or pervert his competition's efforts. This creates the wicked problem. Multiple sides have an objective and design actions to bring them about but must adjust their actions according to the actions and presumed actions of the opposing sides. Good strategy, in anticipation of other's actions, will make modifications to the plans before actual implementation. When implemented in the field the plan will change.[6] Not only does a plan never survive contact with the enemy, but plans or orders are seldom carried out precisely as ordered, another element of the human dimension.

These are the classic elements of a wicked problem. Any implementation of a solution to the problem will inevitably change the problem, particularly because the opposing leadership will make adjustments accordingly as the competition develops.

Therefore, any effective analysis to support a commander must as part of a diagnosis of the problem take into account the perspective of the opposing side(s). The analysis then becomes a dialectic, a collision between the competing

---

5 Richard P. Rumelt, *Good Strategy/Bad Strategy*, 2011
6 "No battle plan ever survives contact with the enemy." Attributed to Helmuth von Moltke the Elder in the 1800s.

hypotheses, synthesized to make a prediction of what the outcome will be and/or a recommendation on what to do.

Further complications in the analysis arise when the challenge includes the assessment of new capabilities, actions, doctrines, etc. that have not been assessed before. While the understanding of the effectiveness of artillery may be very well understood, the introduction of directed energy weapons, unmanned systems, new sensors, and other technologies may not be well understood. Further their full implications probably have not been fully explored. More than likely these factors are the very thing the analysis has been called upon to assess.[7]

However, the commander and the analyst must start from somewhere and that is usually a scenario. We suggest employing problem-framing techniques[8] including the use of structured analytic techniques[9] in the development of the scenario. The elements of a scenario include the actors, their objectives, their capabilities, the environment, and other factors. All of these of course are critical elements of a wargame.[10]

Before we discuss wargaming for the purposes of innovation we need to define the term for the purposes of this chapter. The Department of Defense defines wargames as follows:

Wargames are representations of conflict or competition in a synthetic environment, in which people make decisions and respond to the consequences of those decisions.[11] The same document goes on to state Wargaming is most effective when it contains the following elements:

- People making decisions under uncertainty.
- A fair competitive environment.
- The game should have no rules or procedures designed to tilt the playing field toward one side or another.
- Adjudication.
- Consequences of actions taken.
- Iterative.

This definition is mostly for the purposes of planning and analysis but matches that of most analytic wargames. However, there are multiple types and styles of wargames designed according to their purpose.

---

7 Dmitry (Dima) Adamsky, *The Culture of Military Innovation*, Stanford Security Studies; 1 edition (January; 1 edition (January 27, 2010).

8 Heijden et al. *The Sixth Sense: Accelerating Organizational Learning with Scenarios*, Wiley; 1 edition (August 16, 2002)

9 Lipmanowicz & McCandless, *The Surprising Power of Liberating Structures*, Liberating Structures Press; 1 edition (October 28, 2014)

10 Perla, Peter, *The Art of Wargaming*, Naval Institute Press, 1990.

11 Joint Publication 5-0: Joint Planning, 16 June 2017. It goes on to identify best practices for DoD wargaming.

Figure 8.2 describes the general category of games according to their primary purpose in a two-by-three matrix. Games can generally be categorized by the intentions of their design: to create knowledge, convey knowledge, or to entertain. Most games are designed to entertain the general public. Most games members of the military are familiar with are intended to convey knowledge, particularly for training. Military members have often participated in games designed to assess their ability to accomplish tasks. For our purposes in this chapter we are talking mostly about games designed to create knowledge. While all games attempt to include some element of all three (create, convey, entertain), the sponsor or consumer will have a primary purpose and it can be dangerous to take a game designed for one purpose and make claims about its suitability for another, particularly misconstruing "lessons learned" from training and education games as evidence of future environments. Games can also be categorized by the nature of the problem, whether it is structured or unstructured. In our case, because we are wrestling with innovation and new warfare domains, we are usually confronted with an unstructured problem and attempting to move it to a more structured form.

Framing the Problem

For business and defense analysts wrestling with a wicked problem, the objective is to attempt to add structure to the problem, where appropriate. Once the analyst has a structured problem, or at least an element of the problem can be structured, M&S can provide powerful insights, particularly when examining systems in competition.

	Creating knowledge	Conveying knowledge	Entertainment
Unstructured problem	Discovery games	Education games	Role playing games
Structured problem	Analytic games	Training games	Commercial Kriegsspiel (e.g. risk)

Figure 8.2 General categories of wargames (Pournelle, "Designing Wargames for the Analytic Purpose" MORS Phalanx, June 2017. Derived from *Dr. Jon Compton's "Analytical Wargaming" (Compton 2014) and Elizabeth Bartels presentation, "Gaming: Learning at Play"* published in *OR/MS Today, August 2014.* It is also noted due to the competitive nature of the commercial wargame market, these games have been the source of a lot of innovation in the mechanics of wargame design).

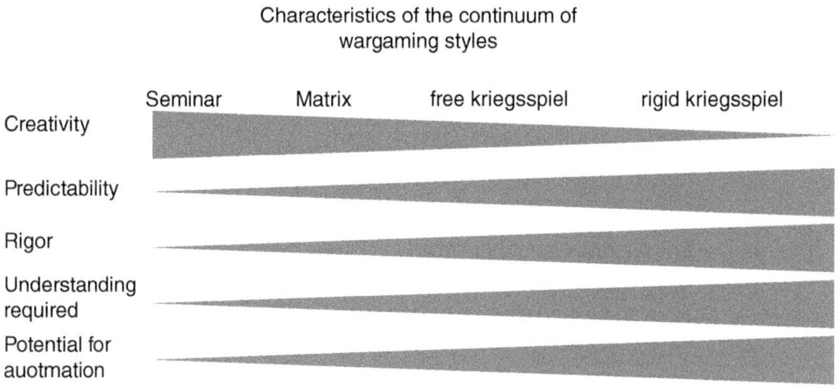

Characteristics of the continuum of
wargaming styles

Figure 8.3 The continuum of wargaming (Pournelle, "Designing Wargames for the Analytic Purpose" Phalanx, MORS, June 2017. Derived from Dr. Peter Perla's Spectrum of Analysis in the Art of Wargaming).

The main challenge in adding structure to the problem is gaining an understanding of the interactions of the competing sides. As our understanding of the interactions improves and the phenomenology describing these interactions develops, we can develop and employ more rigorous models to capture and adjudicate these interactions. This evolution of our understanding in the cycle of research and incorporation of it into the rules is the Continuum of Wargaming.

The Continuum of Wargaming is a concept of this evolution from uncertain interactions into well-developed rules. The completion of this continuum is necessary before tools like M&S can be employed. If an analyst attempts to develop an M&S before working down this continuum, he risks creating an invalid model. For an M&S to be valid requires two components: 1) the interactions between the elements/agents/sides are fully understood and 2) the decision space of the elements/agents/sides are fully known. To avoid a likely invalid model in M&S, the analyst must move down the continuum to the point where a rigid kreigspeil[12] with rules that capture the range of actions the opposing sides may take and satisfactorily adjudicates these interactions combined with an understanding of the range of actions (and why) the competing sides may take.[13]

The boundaries between the styles of games are not clear-cut and, in some games, different styles may be employed for different problems in the same game. As our understanding of the interactions and associated phenomenology improves and rigorous understanding develops, the analysts and game designer move down

---

12 kriegsspiel is German for war chess or wargame.
13 At a minimum the analyst should capture in his report what action the M&S does not capture, those actions and interactions which are not accounted for.

the continuum until a rigid kriegsspiel has been developed. After observing many games, the analyst will become armed with a rigorous understanding of the interactions of the opposing sides and associated rules to adjudicate them, the analyst is now prepared to develop a valid M&S.[14] The following is a summary description of the styles of games in the continuum:

Generally, seminar games are conducted when addressing new phenomenon or capabilities and it is difficult if not impossible to adjudicate the interactions of the competing sides. In this case the goal of the seminar game is to identify the range of actions of the competing sides and a rudimentary exploration of the areas of interaction. The range of actions and interactions becomes the basis of the analysis (or exercises) in the next step in the cycle of research.

The role of a referee in the seminar game is mostly to facilitate the extraction of concepts and capabilities of the topic from the players. The referee is not the expert on the topic, though some understanding is necessary. The players are assumed to be the experts on the topic and the game is designed to enable them to explore the range of actions they might take. The referee must then work with the players to attempt to map out the interactions between actions taken by the players and make some initial assessments as to what the results might be. Adjudication may be extremely vague, but at a minimum the goal is to highlight where actions by the two sides will be in direct conflict. For example, in a cyber operations-themed wargame if one side attempts to infiltrate the other's communications network and exfiltrate the common operating picture (COP), and the other creates a false COP as a honeypot, this would be an obvious interaction that bears future analysis. Post-game analysis would explore how the two sides would develop the means to take their respective actions; what indicators each side might receive due to the employment of these techniques; and the probabilities of success, failure, or effective deception.

Generally, matrix game[15] are conducted when the game examines relatively new areas and uses a gathering of experts in the field to use their expertise in the adjudication of the interactions. Participants act as both players and contribute to adjudication. When contributing to adjudication, the players step back from their role in the game and contribute based on their expertise. The referee is not an expert on the topic. The referee's goal is to first elicit from a team or player their

---

14 Game designers may wish to employ automated tools, even M&S, to assist in adjudication of some elements of a developing game. However, they must be careful to balance issues of precision, speed, and transparency when doing so. There is a danger of alienating the players from the game if they view the tools as a black box and are unable to understand the relationship between their choices in actions and outcomes. We will take this up later.

15 John curry, *matrix game for Modern Wargaming Developments in Professional and Educational Wargames Innovations in Wargaming Volume 2*, lulu.com, 14August 2014.

action, what they think they will accomplish, and why it will succeed. Then each of the participants is encouraged to contribute to the discussion about what will likely occur.

The role of a referee in a matrix game is to come to a consensus about what the likely outcome will be. In some (many) cases this will not be certain, and a range of outcomes may be illustrated, in which case dice or some other randomization tool will be employed to determine **an** outcome. Like in planning, this decision of an outcome becomes an assumption to keep the game moving. The results of the adjudication then become the new state in the game, which other players will use as the basis of their actions, each of which is subject to the same adjudication process.[16]

Post-game analysis would employ the rich discussion by the experts about the potential actions and interactions and the reasoning behind them. Analysts then research the issues highlighted in the game and refine them. This refinement creates new rules to follow for future games. Participants (the invited experts) are encouraged to challenge and refine these new rules in future iterations of the game. Once the experts are satisfied by the rules, they can be used to (hopefully) accelerate the future games and devote their efforts to new issues in the conflict.

For example, the use of hypersonic weapons may not be fully developed, and many informed opinions are held by experts in the field. A game exploring these innovative weapons should probably include technologists who understand their abilities and planners and operators who will be forced to employ them or defend against them. As the engagements and interactions between associated technologies are examined by the participants, post-game analysis would be devoted to these engagements and new rules would result. For example, what is the single shot probability of kill (SSPK) of an S-500 air defense system versus a scramjet missile?

Generally, a Free kriegsspiel is a game where the referee (or White Cell) is the expert on the issues in the game. The referee must still draw out from the player or team their actions, why they are doing it, and why they think it will succeed. The referee then uses his or her expertise (often assisted by rules developed for the issues) to adjudicate the interactions. In a Free Kreigspeil game devoted to analysis and exploration, the objective is to understand the players' actions and the reasons behind them. The objective is an exploration of their decision space rather than the interactions themselves. Hopefully, with the expertise developed for the game(s), players can rapidly explore the range of actions and their consequences. Post-game analysis is devoted to the actions of the teams and why they did them.

---

16 Just like in planning, the analyst will go back and re-examine these assumptions in follow-on analysis and games.

Research should be devoted to understand if the range of actions the players took are considered realistic, probable, or possible.[17] As in any game, care must be given to the post-game survey, hotwash, or other data collection techniques to capture areas where the players may have wanted to act, but the lack of rules dissuaded such actions; these are areas for additional analysis and/or the need to return to a less rigorous game style.

For example, our understanding of advanced weapon systems like Anti-Ship Cruise Missiles (ASCMs) is highly developed. The incorporation of a few experts with some thumb rules in the White Cell would enable adjudication of the employment of these weapons and their interaction with specific air defense systems. This adjudication may be supported by a computer or other automation system, which the referees may adjust based on player inputs. However, if the goal of the effort was to encourage innovation and a player came up with a new use for the ASCM (or a counter measure, etc.) then the referee should either encourage the novel approach and use a Matrix style discussion for adjudication and/or place the concept on the list of issues to analyze and explore following the game.

Generally, a Rigid Kriegspiel is a game where the role of the referee is simply to enforce adjudication based on rules that have been written down. The referee should still be interested in why the players are taking an action, but the probable outcomes and risks should be known by the players. These are often displayed in the form of combat resolution tables (CRTs), probabilities, or other tools. If the rules are written down and not too voluminous, players should be able to rapidly explore the range of action and consequences of the actions. As in any game, care must be given to the post-game survey, hotwash, or other data collection techniques to capture areas where the players may have wanted to act but felt there were not rules to enable such actions; these are areas for additional analysis and/ or the need to return to a less rigorous game style.

For example, in a naval wargame conducted in the interwar period, detailed rigid rules were available for naval gunfire. With an understanding of the rate or fire, weight of shell, and other factors the results of a gun duel could be determined using a combat resolution table (CRT) or complex formula. Today these rules could be automated using a computer or other automated device.

An automated or computer-based wargame is a Rigid Kriegspiel. All the rules about interactions are incorporated into the code and execution is done by the machine in accordance with the prewritten instructions. The players' actions and communications interchange remain flexible, but the actual changes in the environment are executed in accordance with the rule set.

---

17 For actions considered possible but not realistic (like flying airplanes into buildings) see Talib, *The Black Swan: The Impact of the Highly Improbable*, Random House, 2007.

Games may employ these different styles in a single game. Gunfire may be addressed by rigid rules expressed in a CRT. Meanwhile, torpedoes and ASCMs may be adjudicated by experts making adjustments in a computer system. Space systems may require adjudication by a range of experts who are also players in the game. Cyber operations may require capture in a seminar manner due to challenges in understanding or classification.

However, it is not until all the interactions of a game have been captured in a set of rigid rules and the decision space of the competing sides is understood is a scenario ready for incorporation in an M&S. To jump directly from a seminar level of understanding into an M&S is fraught with peril, for it means the modeler is proposing a model of both the interactions and the decision space of the actors without it having been explored and understood by the experts in the field(s). This would be a premature incorporation of a simulation into the modeler's representation of reality. At a minimum the modeler must caveat what has been done in the input, and therefore the output of his model.

Recoding a computerized game to incorporate a new concept or capability can be an expensive and time-consuming proposition, particularly if there are validation requirements. Therefore, it is crucial for an analyst to make certain all of the relevant issues are captured in the less rigid gaming techniques then design a manual rigid kriegspiel before embarking on coding or modifying existing code to the analytic purposes.[18]

M&S Support to Wargames

M&S can be a very powerful tool in the design and support of a wargame. The key to a successful wargame is to achieve a proper balance between realism and playability. If the players (or sponsors) believe the game is not realistic or accurate, they will not fully engage or withdraw (or the sponsor may cancel the game or discard its findings). If in the pursuit of accuracy, the game is too complicated or the adjudication takes too much time, the players will become bored and disengage or withdraw. If the game designer is able to achieve the proper balance, the players will enjoy a willing sense of disbelief and happily engage. An extremely well-designed game will enable the players to enter a state of flow where they fully take on their roles in the imaginary space and lose themselves within the game.[19]

So, the game designer faces a serious challenge of how to address these two often competing goals and the solution is to do the analysis before the game begins.

---

18 Remarks of Paul Vebber, former game designer at matrix game (the corporation not the style of games).

19 Mihaly Csikszentmihalyi, *Flow: The Psychology of Optimal Experience*, Harper & Row, 1990. Entering the state of flow is often referred to as being in "the zone."

While automated in-stride adjudication has been the holy grail in wargaming for many decades, it rarely has been successful except in circumstances where a rigid kriegsspiel is instantiated in a human-in-the-loop simulation (often with uneven results). Instead, game designers have often been very successful when they conduct detailed analysis of the interactions before the game is played and CRTs are built. These CRTs are in essence a model of the subordinate levels of the conflict. For example, a CRT for air-to-air warfare is a model for that type of conflict in a joint operational level game. When aircraft from the opposing sides clash, the CRT takes in to account the factors involved in the fight and determines the outcome.

Using M&S in advance the analyst can generate a CRT that captures the important factors. In our example we have two flights of opposing aircraft that will engage each other. The game designer anticipating this kind of clash will occur employs an authoritative M&S tool for air-to-air combat. Using this tool, he conducts a series of simulation runs and records the outcomes in each. For each batch of runs the analyst systematically makes changes to the inputs. For example, the number of aircraft on each side may be adjusted, the weapons load out may be adjusted, and the risk the pilots are willing to take (e.g. fire from long range or wait and engage at a closer range). Additional adjustments may be how much fuel the two sides have, which often is based on range from the launch point or an aerial refueling tanker; another, how much surveillance support they have, which often is based on the range from the nearest friendly airborne early warning aircraft, or the amount of electronic support and jamming the opposing sides have. Using the detailed M&S, the game designer or analyst can examine the input and output of these systematic changes to see what the impact is. Using regression analysis and Latin Hypercube sampling, the analyst can isolate what factor contributed in what manner to the outcome of the engagement and build the CRT accordingly.[20] The key is for the analyst and game designer to understand what factors significantly contributed to the outcome and avoid excessive detail in the construction of the CRT.

Pathologies and How to Avoid Them

M&S and wargaming have certain pathologies an analyst must take care to avoid. The wargaming community has devoted considerable effort to identifying and addressing pathologies that can arise in games.[21] We will address a few of the

---

20 Iman, R.L.; Helton, J.C.; Campbell, J.E. (1981). "An approach to sensitivity analysis of computer models, Part 1. Introduction, input variable selection and preliminary variable assessment". Journal of Quality Technology. 13 (3): 174–183. doi: 10.1080/00224065.1981.11978748. Iman, R.L.; Davenport, J.M.; Zeigler, D.K. (1980). Latin hypercube sampling (program user's guide). OSTI 5571631.
21 See Weuve, et al, Wargame Pathologies, Center for Naval Analysis, September 2004 https://www.cna.org/CNA_files/PDF/D0010866.A1.pdf.See also:

pathologies that can arise when using M&S and wargaming and suggest ways using the two techniques together can reduce the likelihood of such pathologies corrupting the results.

Many of the pathologies of operations research (OR) arise out of its nature:

> Operations Research is defined as "a scientific method of providing [decision makers] with a quantitative basis for decisions." Here, the key words are scientific and quantitative... although the mathematics used in analysis is objective, the choices of which models to use and which parameters are most important, the assumptions that underlie the analysis, and sometimes even the method of solving the mathematical problem can be subjective... In making the translation, analysis must simplify and often discard much that is not reproducible or readily predictable – including at times, human behavior.[22]

All models are abstracts of reality giving rise to the important proverb:

> Essentially all models are wrong, but some are useful.[23]

All models cannot address all potential factors, lest they become incomprehensible. Further excessively detailed M&S have been criticized for being untraceable.[24] Therefore, a pathology will arise when analysis using these tools claims excessive precision in the answers, getting things exactly wrong.

The first most frequent pathology regarding precision is assuming the model will provide an accurate and valid point solution. While services like to employ a precise number in their public relations campaign (e.g. "355 ship Navy"), a good analyst knows the results of their models provide insights into adequacy of forces and at minimum acknowledges there is an error band or confidence interval regarding the final answer. Further the most valuable insights from any model are gained when comparing excursions to baselines, seeing how the inputs cause changes to the outputs, particularly in regard to risk in stochastic models.

The second most frequent pathology regarding precision is the length of time it takes to build the simulation, populate it with objects and actors, instantiate the scenario, etc. Some large simulations can take years to construct, troubleshoot, validate, and operate. This is especially true for models with excessive details or

---

22 Perla, Peter, *The Art of Wargaming*, Naval Institute Press, 1990. Citing Morse and Kimball, *Methods of Operations Research*.

23 Attributed to George E.P. Box along with the following: "Just as the ability to devise simple but evocative models is the signature of the great scientist so overelaboration and overparameterization is often the mark of mediocrity."

24 Among them Dr. Christine Fox of APL, former Acting Deputy Secretary of Defense. See her comments here: https://www.youtube.com/watch?v=tR18jMftSuk&t=3871s at time 1:01:09.

uneven levels of specificity.[25] By the time the model is ready to perform its function, the conditions set by the sponsor have changed. Essentially the answer is no longer of value to the customer. Further, if the model takes too long to generate enough runs to gain sufficient data and confidence intervals, then it is not possible to provide effective comparisons of inputs to outputs to gain meaningful insights to the customer.

Models are a very effective tool for the examination of systems in competition. The challenge is when the system being modeled is very dependent on the perspectives and action of human beings. To be tractable, campaign analysis and other methods of OR must employ certain Measurements of Effectiveness (MOE) and Measurements of Performance (MOP), but these may not lend themselves or may be undermined by human behavior. Many of these human factors are of a qualitative nature (morale, fatigue, operator quality, will to fight, etc.). Some factors such as doctrine can be used as a stand-in for what opposing sides will do, but some countries are notorious about not following their doctrine.[26]

The most damaging pathology of the use of M&S, like any other method of OR, is using it for the wrong task or crowing out other techniques, leading to myopia. Classical OR and stochastic M&S most often use the law of large numbers in an effort to generate a confidence interval on a particular measurement describing a goal. Given a set of assumptions and a range of rational inputs, the likelihood of certain events is determined. The challenge is, military operations are most likely in response to surprise or other unanticipated events (e.g. the attack on Pearl Harbor, the Battle of Midway, the attack on the United States on 11 September 2001, etc.) or tasked to prepare for future events. Given the nature of most classical OR and M&S in particular, they are not suitable tools for hypothesizing impactful seemingly random events (like a "Black Swan")[27] because there is no data to draw upon. However, techniques like M&S might be effective in analyzing these possible future histories after they have been identified by tools such as wargaming.[28] Further leaders can be trained to prepare for such random events and systems analyzed for their robustness in response to them through scenario learning (i.e. wargaming)[29]

---

25 For example, the Joint Warfare System (JWARS).

26 "A serious problem in planning against American doctrine is that the Americans do not read their manuals." An apocryphal observation from an unknown Soviet author of Cold War vintage.

27 Taleb, Nassim Nicholas (2010) [2007]. *The Black Swan: the impact of the highly improbable* (2nd ed.). London: Penguin. ISBN 978-0-14103459-1.

28 Perla, 1990.

29 van der Heijden, et. Al, *The Sixth Sense: Accelerating Organizational Learning with Scenarios*, Wiley; 1 edition (August 16, 2002). Taleb, Nassim Nicholas (7 April 2009), *Ten Principles for a Black Swan Robust World*, http://www.fooledbyrandomness.com/tenprinciples.pdf.

One pathology that occurs in M&S is premature coding on a new (or newly identified or in vogue) phenomenon. The modeler seeing the importance of the issue, runs off to incorporate the phenomenon into the model, adding a fudge factor and then advertising to the clients they have addressed the issue. The pathology is the phenomenon was not addressed into the design of the original model and the fudge factors are both artificial and do not address the issue itself. For example, when Net Centric Warfare came into vogue, the designers of Joint Integrated Contingency Model (JICM) decided to install fudge factors into the model that adjusted the combat effectiveness of combat forces (particularly large ground force units) according to a scoring of the how the modeler thought the various units had "information dominance." The challenge is properly modeling the (not new, but in vogue) phenomenon of Network Centric or Information Warfare. Models (either M&S or in wargames) should address how a unit perceives the battlefield which in turn adjusts the unit's behavior and performance. Applying fudge factors based on faulty regression analysis devoid of an understanding of the underlying factors are artificial and can create certain pathologies.[30] It took time for the Center for Army Analysis (CAA) to identify the prominent factors using subordinate models and make adjustments to the methods in JICM, not just artificially adjust the combat coefficients in the existing model. Further CAA now compares their results to other models that more explicitly examine the perceptions of the agents in the models (e.g. STORM+) to understand the impact of this phenomenon in addition to the use of wargaming before instantiating the scenario in the models.

Wargames suffer from their own set of pathologies and we will address some of the most common ones. Most of these can be addressed by employing the best practices identified in JP 5-0.

The most important and common pathology of wargames arises out of the execution of events that claim to be wargames but are not. Some of these wargame-like events can be valuable, many are a waste of time and talent. Rehearsal of Concepts (RoC drill), Map Exercise (MapEx), and other events can be valuable in assisting a team to visualize how their actions, synchronized together, are intended to accomplish a task. They are not wargames because there is no opposition creating friction, nor a clash of opposing theories of victory (a dialectic), yet they are still valuable if those limits are understood and honored. The pathology arises when the limits are not understood, and excessive claims are made regarding the results.

A Bunch of Guys and Gals Sitting Around a Table (BOGGSAT) is just that, an unstructured freewheeling discussion about the general courses of action people might take but no actual decision is made, nor are the consequences of their

---

30 Brian McCue, *U Boats in the Bay of Biscay: An Essay in Operations Analysis*, Xlibris, Corp. November 2008.

actions examined. BOGGSATs may not actually include opposing sides. BOGGSATs and brainstorming sessions have been proven to be ineffective for the task of analysis.[31] It is unfortunate that there are a lot of BOGGSATs being conducted by people who claim to be professional wargame designers. The best way to avoid this is to make sure there is an appropriate adjudication process and consequences for actions taken by the players. If this is not possible due to a lack of understanding of a key phenomenon in the game (as is often the case in seminar games), the game report should at a minimum document the interactions of the competing sides to identify areas for future analysis.

Pathologies can also arise when wargame designers employ the same type and style of game regardless of the issues, the questions being posed, or the state of understanding of the phenomenon in the game. This is akin to using the same analytic tool for all issues, such as shoehorning all problems into being an optimization problem or a Markov chain. The application of the wrong type or style of games may lead to false conclusions or failure to identify the right questions for future analysis. Failure to revise the game design and move down the continuum of gaming will likely result in duplicative efforts without advancing understanding of the problem, such as a constant repeating of seminar games with vague and inconclusive statements in the game report. The best way to avoid this is a deliberate effort to use the game(s) to identify issue for further analysis using OR tools including M&S and move down the continuum.

Pathologies can arise when there is insufficient rigor in the game design, execution, and post-game analysis. If the game designer does not conduct sufficient research into the topic of the game or selecting the best methods in the design appropriate to the topic, the game will be flawed, and several potential pathologies may arise. If there is insufficient research into the topic, the models used in the game will not match reality. If insufficient research is given to the design of the game, it may be difficult for the players to play it and the staff to run it, or the data collection will fail to provide meaningful results. Games that are too detailed may drive the players to the wrong level of decisions (e.g. making tactical-level decision in a strategic-level game) or take too long to execute. The solution to this problem is comprehensive research into the topic in preparation and game design including the use of M&S in building CRTs and other supporting material.

---

31 Downes-Martin, et al. "Validity and Utility of Wargaming", Report of Working Group 2 of the MORS Wargaming Special Meeting, October 2017. https://paxsims.files.wordpress.com/2017/12/mors-wargaming-meeting-2017-working-group-2-final-report-20171208.pdf See also "The illusion of group productivity: A reduction of failures explanation", Barnard Nijstad, Wolfgang Stroebe & Hein Lodesijkx, European Journal of Social Psychology, 36, 3148 (2006). For a general overview of how brainstorming has long been debunked see "Groupthink: The brainstorming myth", Jonah Lehrer, The New Yorker, January 20, 2012 online at http://www.newyorker.com/magazine/2012/01/30/groupthink (last accessed 11/18/2017) and references contain therein.

One of the most dangerous pathologies of wargames is negative learning, large groups of people learning the wrong lessons. Wargames create a powerful collective narrative in the minds of the participants.[32] If those narratives are based on false pretenses and assumptions, then the game has conducted negative learning.[33] For example, if the results of a game were strongly influenced by the ability of a team to conduct aerial refueling of its aircraft and the flying tankers were able to provide excessive fuel at an excessive range (defying laws of physics), then the participants in the game are likely to draw the faulty conclusion about how effective their air armada was in the execution of a war plan, and not understand that it is fatally flawed. In fact, such an oversight may preclude follow-on analysis in this area because it did not arise in the game. To avoid this, have experts in the field use a comprehensive planning process and provide support using analytic tools before the game is played.

The pathology of negative learning or false conclusions may arise when the participants are not experts in the field. Gathering a group of people without regard to needed expertise can cause the game to draw false conclusions, particularly when the adjudication process uses the contribution of the players to make a decision (e.g. Matrix style games). Even when the players are not part of the adjudication process, their lack of expertise can undermine the validity of any conclusions about what the

Similarly, games can suffer when players are just thrown together. If a game is intended to draw a conclusion about a strategy, doctrine, or other issue then a false conclusion may arise from a game where players on teams have only met for the first time at the game. The game designer should not expect such teams to work effectively together for the first time. When a strategy, doctrine, or command structure fails, it will be difficult, if not impossible, to determine if the cause was due to flaws in the doctrine or the lack of cohesion in the team trying to implement it.

Related to a lack of cohesion in teams is the lack of planning by those teams. Without building coordinated plans, team actions will be haphazard and again, the game designer will not be able to determine the cause of failures in games. We suggest that for military challenges, teams should employ the appropriate doctrinal planning process to the problem faced by their side prior to initiation of moves within a game.[34]

One of the most frequent pathologies of analytic games arises from poor construction of a Red team. Red is the usual designation for the opposition teams in games run by Americans. (Interestingly, China's own forces are referred to as Red while opposition forces are Blue.) Based on the purpose of a game, the role and

---

32 Perla 1990.
33 At a minimum, games should "Do no Harm." See Beall, "Defense Innovation Through Wargaming," MORS Phalanx, September 2015 page 58.
34 United States Joint Staff, *Joint Publication 5-0 Joint Planning*, 16 June 2017.

range of actions of Red may be adjusted. In basic training games, Red performs certain actions to stimulate Blue so the evaluation team can assess the response. In more advanced training games, Red performs in accordance with known doctrine. In entertainment games, Red should play to win. In games structured to create knowledge, Red should operate somewhere between known doctrine and playing to win. The further out in the future the game is designed to examine, the more flexibility Red should have toward playing to win.[35]

Pathologies can arise and invalid conclusions can be drawn from games where Red is overly restricted. If the goal of a game is to assess effective strategies, technologies, or other elements about a future conflict with a competitor, then Red should have increasing flexibility in its actions against Blue. As the time frame of the conflict is projected into the future, the range of action available to Red should increase. This is due to the fact the United States may not have a complete knowledge of Red doctrine as it exists today and far less an understanding of how that doctrine may evolve.

The same pathology may arise when the Red team does not have the same resources as Blue or others. This is not to say Red must have the same or equivalent order of battle as Blue, etc. Red should have the same number of players and appropriate expertise to match that of Blue. If the goal of the game is a dialectic examination of truth to create or discover knowledge, then both sides of the dialectic must have the same ability to examine the decision space.

Another common pathology is drawing strong conclusions from a single game. Games are a clash between competing ideas, a dialectic. As discussed earlier, there are a number of reasons why a team may not have performed well, to which we can add their familiarity with the rules of the game. It often takes several games for teams to absorb the game mechanics and develop into an effective team. Therefore, analysts and game designers should be careful about drawing conclusions based on a single game, multiple games may be required before conclusions can be drawn with sufficient confidence.

Combining Wargaming and M&S

Wargames and M&S are very different disciplines but can be complementary. A good analyst will employ the appropriate tool to a wicked problem in an iterative process to gain insights. Each of these tools can then be employed to support the other.[36] Here we will provide a general guide on to how to employ these tools in combination to develop a valid M&S and gain appropriate insights into a wicked problem.

---

35 See the Cafferey Triangle in Caffrey's *On Wargaming*, Naval Institute Press, 2019.
36 Perla, Peter, *The Art of Wargaming*, Naval Institute Press, 1990. Chapter 9.

As discussed earlier, the first use of M&S is in the development of CRTs or their equivalent to address interactions in a wargame. In turn the wargame should produce areas of interest for future analysis that may require M&S. These will then produce new CRTs for future games, etc. As the game rules develop, elements will move down the continuum of gaming until more and more areas can be adjudicated less by expertise (matrix game and Free Kriegspiel) and more by a written set of rules (Rigid Kriegspiel).

Different elements of a game may employ different adjudication techniques. For example, during the period between World Wars I and II, the Naval War College took this approach. Gunfire was thought to be well understood (based on gun exercises) and handled in a rigid kriegspiel manner employing elaborate tables using range, rate of fire, weight of shell, and other factors for resolution. Aviation, in its nascent stage, was handled in a Matrix-like manner using expertise from pilots participating in the games and related exercises. Early electronic warfare was not well understood and handled using seminar techniques. As more rigid rules were developed, they enabled analytic energy to be devoted to less-understood areas. In the same manner, modern games can revise the rules as games are iterated and informed by analysis and exercises or experiments.

As a body of games are completed, the associated reports should highlight the decision space of the competitors in the games. Successful decisions can be captured in the form of doctrine. Players in a game series can develop these, particularly as guides for how subordinate units should respond to particular situations. This developing doctrine then informs the decision tables modelers use to express behavior of units within an M&S.

Eventually most of interactions between competing units can be described by a fixed set of rules. Further, enough play between teams will have explored all of the relevant potential moves they may take and the conditions when they would make them. This is when M&S can be employed, and the tool can produce valid results. There will be boundary conditions. Some interactions may never be captured with rigid rules. The analyst will have to determine how best to capture these factors with parametric or other conditional analysis and note them. For example, it may be determined some form of deception cannot be fully analyzed in which case the M&S or other analysis will be examined in both the cases, success and failure of the deception. Alternatively, there may be actions the competing sides could take (e.g. escalation to use of nuclear weapons) that the analyst or sponsor determines is outside the boundary conditions of the overall study. These should be noted accordingly in the final report.

This process of iterating games with the intention of refining the rules can assist the analysts in avoiding the pathologies of both M&S and wargaming (or any analytic technique). The games inform the decision-making required to build a valid model. M&S informs players and the analyst of the consequences of actions taken.

Good games use the data gathered in analysis to inform players about the probable outcomes of their choices and the analysts use the game to prioritize where the next round of analysis should go. Game and analysis should be grounded in the truth of exercises and experiments. For best effect, the games informing this iterative process should ensure the teams have players with the right expertise and employ a comprehensive planning process to build a coherent plan.

While this whole process may seem onerous, it is crucial to ensure valid analysis is conducted, particularly in the modern era where innovation and new warfare areas have begun to change our understanding of the competition. Those who gain in this understanding, will likely win the competition.

Part IV

Wargaming and Concept Developing and Testing

9

Simulation Support to Wargaming for Tactical Operations Planning

Karsten Brathen, Rikke Amilde Seehuus, and Ole Martin Mevassvik

FFI – Norwegian Defence Research Establishment, Kjeller, Norway

Introduction

Wargaming has a long history and is applied to a broad range of problems at different levels of command (Perla, 2012; Caffrey Jr., 2019). This chapter addresses wargaming as applied in planning of operations as outlined in planning and decision-making processes. More specifically, it covers how simulations can support wargaming for development, analysis, and comparison of courses of actions (COAs) at the land tactical level (brigade). To support this type of wargaming with simulation an understanding of the purpose of the wargame and how wargaming is performed is needed. The simulation support needs to be tailored and although they share a number of characteristics, simulation support to planning is certainly different from simulation support for training. Wargaming as part of the operational planning process at the land tactical level is characterized by time constraints and limited resources. Compared to manual processes, introducing simulation support could offer the opportunities to explore and analyze larger numbers of possible COAs and reduce the time needed for wargaming. Such support must however not come with an additional burden with respect to processes and resources. To be accepted by the planning staff, the simulation support needs to be integrated within the operational systems' environment, be easy to use, and require a minimum support for setup and running.

The chapter describes the characteristics of wargaming in operations planning, the opportunities and benefits simulation support could potentially provide and how these characteristics transfer into simulation user needs and requirements, to

Simulation and Wargaming, First Edition. Edited by Charles Turnitsa, Curtis Blais, and Andreas Tolk.

bring added value to the wargaming. Some of these requirements challenge the state of the art in military modeling and simulations (M&S), although significant progress within important M&S technologies for tactical wargaming has been made in recent years. Examples of such technologies are command and control (C2) and simulation interoperation technologies, M&S of C2 and battle command, and service-oriented simulation architectures and infrastructures. FFI has established a proof-of-concept demonstrator, the SWAP (Simulation-supported Wargaming for Analysis of Plans) system, as a tool for technology assessment and user requirement capture and evaluation. SWAP is a federated system of C2-planning systems and simulations components together with supporting infrastructure. The chapter describes important technology building blocks for simulation support to tactical COA wargaming. How these are employed in SWAP and experiences provided by experiments with SWAP is outlined.

The chapter informs operational planning personnel about the potential for simulation to support their wargaming and outlines an example of how such a system may look like. For simulationists, it outlines the major technology building blocks that are needed for a simulation support capability to COA wargaming and points to areas within military M&S where further research and development are needed.

Operational Planning and Wargaming

There is not one universally accepted definition of wargaming. The North Atlantic Treaty Organization (NATO) defines wargaming as "*A simulation, by whatever means, of a military operation involving two or more opposing forces, using, rules, data, and procedures designed to depict an actual or assumed real life situation.*" The essential element of a wargame is however human decision-making in a competitive and dynamic environment, where the decisions drive how the wargame evolves and where results are adjudicated in a structured way. The purpose of all types of wargames is to put players in a realistic enough environment where decisions can be examined and explored, and shared understanding between players can be created. Wargaming may be applied from the strategic to the tactical levels of military activities. For the higher levels, the problems are typically more ill-defined, whereas at the lower levels, the problems are more well defined and constrained, implying a qualitative to quantitative scale of adjudication from higher to lower levels. The aim of wargaming can either be to improve decision-making in real situations or educate and train people to make better decisions in simulated situations, i.e. for discovery or for learning. Wargaming for tactical operations planning belongs to the discovery type of wargaming.

A brigade operation involves a large number of assets and personnel to perform all the different intertwined and complex activities to succeed in battle. Planning of brigade operations is thus a major undertaking requiring a highly trained planning staff. To perform such complex planning, a defined analytical process is needed. The process should give normative procedures for analyzing a mission, developing, analyzing, and comparing COAs, selecting a COA, and producing a plan and an order. Such a process is expressed in the US Army Military Decision Making Process (MDMP) (US Army, 2005) and the Norwegian Army Planning and Decision Making Process (PDMP) (Norwegian Army, 2015). Both have the same methodological and logical basis, as shared with most military analytical staff processes, but also, not surprisingly, have some differences. For the sake of the purpose of this chapter, the stages or steps of MDMP and PDMP need not to be distinguished. One difference is, however, how many COAs are wargamed, but the wargaming itself and the aim of the wargame are the same. In the PDMP, *hasty wargames*, where only key personnel are involved in a review of important events, are performed to develop, analyze, and compare COAs and also to select a COA. The selected COA is analyzed and refined into a plan by a *prepared wargame*, where all specialists and staff branches are involved so that all aspects of the COA can be covered. The MDMP prescribes wargaming of the alternative COAs. In Banner (1997), the two different approaches are discussed.

Figure 1 shows the steps of the MDMP, including their main input and output. We will not outline all the steps and stages but focus on the parts of the process where wargaming is applied and the elements that are important for considering supporting simulation. These parts are COA development/analysis and COA comparison and these two parts should constitute a major part of the time available to perform the whole planning process. The main starting point for development of own COAs is the commander's intent, stating briefly and clearly what own force must do and the conditions the force must meet to succeed with respect to the adversary, terrain, and the desired end state (US Army, 2017). The commander's intent drives the planning process. Development of the initial draft of a possible own COA is a highly creative process that we do not go into here.

A COA states how a mission is to be accomplished. If time permits, several substantial different COAs are developed against the adversary's most likely and most dangerous COAs. A COA is expressed with a sketch, see Figure 2, the task organization and textual statements. A COA sketch is made up of standard military symbols and tactical graphics on a map of the area of operation. A COA sketch together with the text express the five Ws (Who, What, When, Where, Why) for the subordinate units. For further development, analysis (and comparison) of COAs hasty wargaming is employed in PDMP.

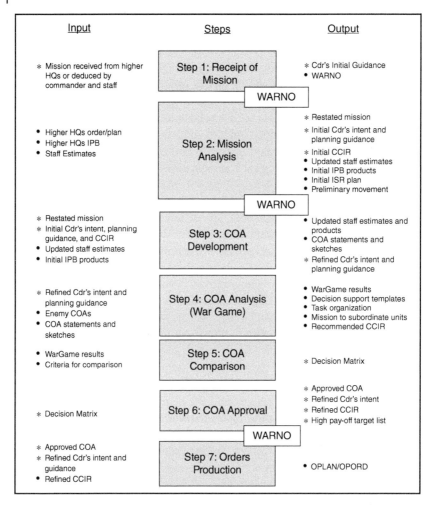

Figure 1 The Military Decision Making Process (MDMP). From (US Army, 2005).

From the comparison of alternative COAs, a COA is selected to be developed further into a detailed plan. The main tool is prepared wargaming, where one of the important outputs is the complete synchronization matrix, which is a type of a Gantt chart, see Figure 3. The columns represent time periods, the rows the functional units. The cells include statements about the tasks and actions to be performed in a five Ws format, including coordination and dependencies. The wargaming may focus on different parts of the operations, such as the start of the operation, transitions between phases, and special battle activities. The

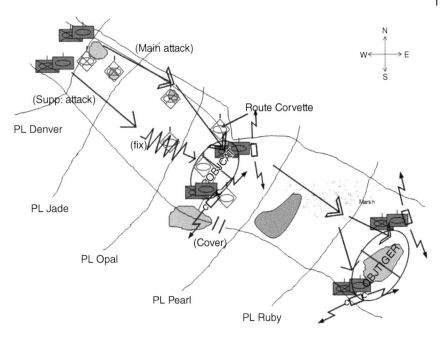

Figure 2 A course of action (COA) sketch.

wargaming is typically performed in an *action–reaction–counteraction* fashion with as many of the staff and specialists as possible present.

The wargaming helps the commander and the staff to visualize the flow of battle in order to discover and analyze weaknesses and possibilities. COAs are examined and populated with details to develop the plan. It entails an iterative step-by-step review of own actions and the opponents' reactions. As mentioned, it is a critical step in the planning and should be allocated more time than any other step of the MDMP/PDMP. The wargaming is however a huge cognitive challenge and is limited by the knowledge and experience of those who take part. Wargaming is also time-consuming, which often limits how many own and opponents COAs that can be examined or to what extent the review can be done. Superior and subordinate commands may also conduct planning as a collaborative effort, leveraging the knowledge and manpower at several commands simultaneously. This, however, implies distributed interactions that further challenge a common and shared understanding.

Wargaming is inherently a human activity, and the product is not only a COA and a plan, but the wargaming process itself is of utmost importance to build a shared understanding and cohesion within the team of the commander and the staff. The wargaming also helps developing an understanding of the factors that

Unit	Phase 1 from PL DENVER to PL AUSTIN	Phase 2 from PL AUSTIN to PL PEARL		Phase 3 from PL PEARL to PL RUBY	Phase 4 from PL RUBY to TIGER		
1–221 Tank Sqn	**Main attack** along axis corvette	**Seize** objective area CAT	**Secure** objective area CAT	**Main attack** along axis corvette	**Main attack** along axis corvette	**Seize** objective area TIGER_NORTH	**Secure** objective area TIGER_NORTH
1–223 Mech Cpy	**Main attack** along axis corvette	**Seize** objective area CAT	**Secure** objective area CAT	**Main attack** along axis corvette	**Main attack** along axis corvette	**Seize** objective area TIGER_NORTH	**Secure** objective area TIGER_NORTH
1–222 Tank Sqn	**Main attack** along axis corvette	**Seize** objective area CAT	**Secure** objective area CAT	**Follow and support** along route route_TIGER until phase line PL_RUBY	**Supporting attack** from phase line PL_RUBY along route_ TIGER_SOUTH	**Seize** objective area TIGER_SOUTH	**Secure** objective area TIGER_SOUTH
1–224 Mech Cpy	**Main attack** along axis corvette	**Seize** objective area CAT	**Secure** objective area CAT	**Follow and support** along route_TIGER until phase line PL_RUBY	**Supporting attack** from phase line PL_RUBY along route_ TIGER_SOUTH	**Seize** objective area TIGER_SOUTH	**Secure** objective area TIGER_SOUTH

Figure 3 A synchronization matrix.

will influence the events that can actually happen, likely counter to an envisioned COA. One could say that the purpose of the wargaming is, paradoxically, to prepare for actions when the selected COA fails. By having considered and explored different options, the commander and the staff are primed to handle unexpected outcomes by effectively adjusting their plan.

What are the Benefits of Simulation Support to COA Wargaming?

In general, wargaming at the tactical level is more suitable for simulation support than at the higher levels, since the issues under scrutiny are more objective and quantitative and thus more amendable to be expressed in models for simulations (Lawson III, 2016). Simulation support to tactical planning wargaming gives a more objective adjudication compared to manual judgments and subject matter expert opinions. Simulation output does not only include combat results, but as importantly the results of other tasks, such as deployments and maneuvers. A brigade operation is a huge undertaking with a large number of entities that are to perform a large number of different tasks in a concerted manner. Simulations facilitate time–space considerations that are almost impossible to do manually. Thus, looking at the simulation support as a book keeper for all the details, including time–space, fuel consumptions, logistics, terrain, weather, etc. is as important as the evaluation of outcomes of the combat situations of the operations. Evaluation of combat situations is of course much harder. The evaluation can be based on entity level or aggregated levels models, but regardless, the adjudication is based on complex computations out of scope for human reckoning.

Manual wargaming restricts the number of COAs that can be considered due to the time needed for wargaming each COA. Simulation support shortens the time needed to explore a larger number of options or allows more thoroughly analysis of one option within the same amount of time. Alternatively, one may wargame the same number of COAs within a considerably shorter amount of time. This is achieved by a simulation running much faster than real time (i.e. scaled real time with a factor much greater than one) and of course the adjudication process is much faster in a computer simulation than when performed manually.

Simulation-supported wargaming comes with an inherent capability to automatically record data as the wargame unfolds. This logging capability facilitates the capture of the data needed for subsequent briefings and content of the plan and the order. For example, if a plan or an order is amended with a video animation of the recorded simulation of the selected COA, it is fair to assume that the plan is more easily and correctly comprehended than a plan expressed only in formatted text and static graphics.

Principles of Technology Support to Wargaming for Operations Planning

There are numerous ways that technology can support wargaming (Reddie, 2019), and there have been several initiatives and research efforts on computer-based support and automation to military decision-making (Rasch et al. 2003, Kott et al. 2005, Surdu and Kittka 2008, CERDEC Public Affairs, 2015). Some of them are related to tools to support different parts of the process for wargaming of COAs. Tools to generate COA sketches are discussed in Forbus et al. (2001) and Cohen et al. (2015). A tool, Course of Action Development and Evaluation Tool (CADET), is outlined in Rasch, Kott, and Forbus (2003) and Kott et al. (2005) that generates a detailed synchronization matrix based on a COA sketch and COA statements. Simulation of COAs is outlined in Kimeche and de Champs (2004) and Surdu and Kittka (2008). Our own work on support to wargaming of COAs, and specifically simulation of COAs (SWAP), is elaborated below.

In this section, we discuss the general characteristics and principles by which simulation support to wargaming of COAs should have and be designed by. Despite the past research and technology development efforts, currently, the operational planning process is basically a manual process only supported by general computerized tools such as text, presentation and map overlay editing tools. Typically, such tools lack the ability to record the semantics of the data provided. Simulations need semantically defined input data. To avoid the additional burden of preparing the simulation input data, the C2 and planning tools for defining and expressing COAs should also capture the semantics, so that preparation of the simulation data for wargaming is blended into the COA development process and the tools that the planning staff is using. In this way, preparation of the simulation input, the COA sketch, own task organization, and adversary order of battle and a preliminary synchronization matrix, does not add additional efforts and the need for additional simulation support staff.

Given that the wargame process is instrumental in itself, the simulation should emphasis support and not automation. Simulation support of COA wargaming falls within the broader class of decision support systems (DSSs). A DSS can be classified as passive, active, or cooperative, depending on how the user interacts with it. In Hannay, Brathen, and Hyndøy (2015), it is argued that a DSS generating an "optimal" COA or suggesting a COA is not the goal (an active DSS), but rather the simulation support should facilitate insight and consequences of a COA as the wargame is executed (cooperative DSS). What wargamers need is simulation support that allows them to effortlessly explore COAs, including actors in the area of interest and the many possible outcomes, as they supervise and interact with the simulation execution. What they do not need is a "black box" that tells them the figures of the outcome from a certain

COA. Thus, simulations exploring a space of possible outcomes, to perhaps generate an "optimal" COA are not the focus. What is needed is cooperative decision support favoring simulation support that shows consequences of a COA in a transparent manner. This translates into a case-based approach to the simulation support for COA wargaming, as opposed to a statistical based approach. A case-based approach provides one possible outcome for a selected case (i.e. a COA) (Hannay, Brathen & Hyndoy, 2015). Transparency and representativeness are therefore important, in which support to the cognitive understanding of the COA is a key.

There is, however, merit to combine case-based simulations with a datafarming approach for discovery of COAs that subsequently potentially could be a candidate for wargaming (Schubert et al., 2017; Huber and Kallfass, 2015). In datafarming a large number of simulations are performed to be able to explore the solution space by varying model parameters based on an experimental design. However, for the wargaming itself a specific COA is examined and a case-based simulation is favored.

Combat simulations can offer much more detailed results compared to seminar and tabletop wargames (Turnitsa, 2016). However, it can come with the price of more complex usage and technical setup and running, which needs additional simulation and technical support staff. Simulation to support planning must need fewer resources than what is acceptable for instance for simulation-based training. Capturing semantic data as an inherent part of the MDMP/PDMP as described above is one contribution to achieve this. Other measures include both the underlying models of the simulation and the technical infrastructure of the simulation.

One way to reduce the simulation support staff is to include models, not only of the force structure, but also of the command structure. A COA is normally expressed at the immediate lower level of command and today simulation support staff needs to transform these tasks into lower level tasking for subordinate units and then manually enter the more detailed sets of instructions into the simulations. If models of the command structure and behavior are included, the simulation could be able to receive a COA as it is expressed and by itself decompose and task combat entities. Enabling technologies for modeling a command structure are multi-agent systems and artificial intelligence (AI) methods. M&S of the C2 and battle command organization decision-making, including doctrine and tactics, is demanding, and AI- and agent-based approaches are essential. Models of C2 and battle command, in addition to behavior models of the combat entities, make it possible to employ simulations with a small simulation support staff.

Concerning the technical infrastructure, there are specifically two measures that are important, namely, the interface between the C2 systems and the simulation systems and more generally the technical architecture of the combined C2 and simulation federation. Standards for the interface between C2 systems and simulations, denoted C2SIM, have been developed by research groups in the

Science and Technology Organization (STO) of NATO and product development groups of Simulation Interoperability Standards Organization (SISO) over the last decade.

The technical architecture should leverage Service-Oriented Architectures (SOA), combined with the distributed simulation architecture High Level Architecture (HLA), resulting in a hybrid SOA–HLA architecture (Hannay, Brathen and Mevassvik, 2017). Employing these architectural styles by micro services and web services eases the burden of deployment of the simulation and C2 software, and minimizes the software needed on the client side. The user access to the simulations can be provided through a familiar web browser.

The simulation support for COA wargaming must be versatile and easy to use, so the wargamers can handle it more or less by themselves. For example, defining COAs used by the simulation should be similar to sketching or drawing them on a paper map. On the technical side, the simulation support must not introduce yet another system for the commander and the staff. The simulation needs to be blended with the operational C2 and planning systems. The user access to the simulation should be as a service via the C2 system, giving the impression that the two are one and the same. Providing the wargaming simulation support as a service also facilitates a distributed MDMP/PSMP.

The simulation support should enable to wargame the same evolving COA several times or multiple COAs in the same amount of time it would take to conduct a manual wargame. Generally, the simulation needs to be run much faster than real time, but it should also be user controlled to allow for closer analysis of specific parts of the COA. By running simulations faster than real time and employing artificial intelligence methods to model and simulate the command structure, the need for wargaming support staff can be reduced considerably.

In general simulation support to wargaming for operation planning must provide much more flexibility, specifically with respect to management of time, than what is found in simulations for training. Also, the ability to easily adjust the simulation during the course of the wargame to support the refinement of the COA is of major importance. Although live, virtual, and constructive simulations are employed for training support, constructive simulations, also denoted computer-generated forces/semiautomated forces, are the only simulation type that should be employed for support to COA wargaming.

Enabling Technologies

Based on the principles for simulation support to COA wargaming as outlined in the preceding section, this section describes the enabling technologies. The next section will outline how these technologies have been leveraged in the SWAP system.

The enabling technologies can be divided into two categories. One is concerned with models, both models for the simulation and the models of the data that is exchanged between the C2 and simulation system (i.e. the COA itself). The other is concerned with the technical implementation of the simulation and the integration of C2 systems and simulation systems.

Models

The far most important part of simulation support for COA wargaming is the models. Without proper models, the support can be easy to use, and require minimum efforts to set up and run, but the support is worthless.

For simulating land maneuver warfare realistic movement of combat entities and their utilization of the terrain is essential. Modeling of this kinetic part of land maneuver warfare is reasonably well understood and appropriate models are available in commercial off-the-shelves computer-generated forces (CGF) tools. The major challenge is modeling the behavior of both combat entities and C2 and battle command, which require models that include representation of human decision-making. Thus, models of the movement, observation, firing, and communication are available in typical CGF tools, but they are in most cases not able to simulate, let say a combined arms attack at the company level or above. Thus, to reach the needed level of autonomous simulation, the entities have to able to interpret and carry out higher level tasks, such as "hasty attack" as expressed in a COA. The behavioral part is best modeled by an agent-based modeling approach, where the concept is to represent the overall structure for the behavior of all the entities (C2 and combat) by the interactions of intelligent agents, meaning that they are able to react to their environment, pursue goals, and communicate with other agents (Wooldridge, 2002). The result is a heterogeneous multi-agent system consisting of a hierarchy of command agents and a set of combat agents (leaf nodes in the hierarchy) that execute the tasks. The command agents are able to receive the tasks prescribed by the COA and decompose the tasks into tasks for their subordinates. They receive reports as the simulation unfolds and redirect subordinate agents. The lowest level command agents then must be able to task the combat agents that actually execute the combat activities. Modeling the behavior of the agents themselves can be best approached by leveraging AI methods and architectures such as Belief-Desire-Intention (BDI), Behavior Trees, Context-based Reasoning (CxBR), and Soar. For the executing combat entities, such as main battle tanks and fighting vehicles, a more traditional sensor–controller–actuator architecture is widely used.

The main challenge for simulation of the behavior of land combat entities, apart from modeling the behavior itself, is to model how it is affected by terrain and weather. The past, current, and forecasted weather influences the movability of the combat entities, which again influence their behavior. Thus, online weather services are needed. The terrain itself, even without considering weather, influences the behavior greatly. It is important to take into account the terrain for proper movement simulation and for simulations of tactical dispositions. This calls for terrain analysis functionality as part of the simulation. Terrain analysis is already performed during the battlefield preparation part of the planning process, but additional functionality is needed for simulation. Such functionality includes calculations of e.g. routes, view sheds, and vantage points, such as observation and attack areas and positions. Some of these terrain features are independent of mission and COA, some are associated to a specific mission and task, and some are specific to a selected COA.

In order to simulate a COA it needs to be digital and executable. This means that the COA is represented in such a way that it not only makes sense to a human but can be interpreted by a computer. The Battle Management Language (BML) introduced by Carey *et. al* (2006) attempts to bridge the C2 and simulation communities by digitizing information related to C2. They define the BML as "the unambiguous language used to command and control forces and equipment conducting military operations and to provide for situational awareness and a shared, common operational picture." A BML facilitates the exchange of orders, requests, and reports between C2 systems and simulation systems, but applies to robotic systems as well. The Coalition Battle Management (C-BML) standard (SISO, 2014) is capable of representing the 5Ws used in land operations tasking in a formal way. C-BML can be used in concert with the C2 Lexical Grammar providing additional semantics for the tasking of entities (Schade et al., 2010). The BML concept was originally targeted toward the land domain but has been applied for simulation of air operations (Brook, 2011) and tasking of robotic systems (Davis et al., 2006; Remmersmann et al. 2013, Seehuus et al. 2019). A theoretical study found that the concepts of C-BML apply to maritime operations as well (Savasan et al., 2013).

The C-BML (SISO, 2014) is based on the Joint Consultation, Command, Control Information Exchange Data Model (JC3IEDM), developed by the Multilateral Interoperability Programme (MIP, 2007), in order to facilitate interoperation of different C2 systems. For the purpose of wargaming and COA simulation, C-BML is capable of capturing tasks, the assignment of tasks to units, controlling features that apply to the COA (e.g. phase lines and boundary lines) and temporal relationships between tasks (i.e. synchronization of tasks). The Military Scenario Definition Language (MSDL) (SISO, 2008) is normally used to represent the task organization, but this is supported by C-BML as well.

The NATO STO has since 2004 contributed to the technical verification and operational validation of C-BML (Tolk et al., 2004). It was recommended that MSDL and C-BML should be harmonized and merged into one standard as the current SISO C-BML and MSDL standards overlap and are based on different data models. A new standard, C2SIM, is currently developed creating a common core data model that covers tasking, reporting, and initialization. The different applications of C2SIM, for instance land operations or robotic systems, are supported through extensions (e.g. the Land Operations Extension – LOX). For additional semantics and future capabilities, e.g. reasoning, the C2SIM core is represented as an ontology. This next generation C2SIM standard have been tested during exercises (Pullen et al., 2018) as important steps toward verifying the technical approach and the operational validation of the standards. Reaching a NATO C2SIM Standardization Agreement is the end goal.

The C2SIM family of standards does not specify how a COA or tasking of units should be performed between simulated units, e.g. within an HLA federation. The NATO Education and Training Network Federation Object Model (NETN FOM) supports tasking and reporting through two different modules (NATO, 2018): the HCBML – a standard way of exchanging C-BML using HLA, and the Low-Level BML (LLBML). The LLBML, originally developed by Alstad et al. (2013), provides a standard way for command agents to task executing agents.

System Implementation

Lessons identified from experimentation and demonstrations with simulation-supported COA wargaming with the Norwegian Army (Bruvoll et al., 2016) indicate that a useful operational capability must be easy to use and require few operators and additional technical personnel. Thus, the COA simulation capability must be able to utilize the existing defense information infrastructure.

For defense information systems in general, service orientation is pertinent and SOA enabling employment of cloud technologies is now the state-of-the-practice architectural approach. Functionality must be loosely coupled in terms of data, processing, space and time, so that it can be used in different contexts. Also for simulation systems, SOA is gaining popularity and is denoted as Modeling and Simulation as a Service (MSaaS). The goal is that setting up simulations should be rapid and easy, enabled by reusability by standardization of common functionality and composability (so far only from an information technology perspective) through loose coupling and standardized service descriptions. The MSaaS framework is based on a cloud infrastructure to facilitate simulation access "on demand, anywhere." Containerization can be applied for run-time composition and execution in cloud-based environments, and has been investigated and applied in the context of MSaaS (van den Berg et al., 2017, van den Berg et al., 2017b).

The development in MSaaS facilitates interoperation of C2 systems and simulation systems and reduces the need to introduce additional separate simulation support systems to wargaming for tactical operations planning. MSaaS also reduces the amount of resources and time needed for setup and execution management of the simulation system. Both of these aspects contribute to the reduction in the need for technical support staff. An example of MSaaS employed for simulation support to COA wargaming can be found in Asprusten and Hannay (2018).

Although MSaaS has gained a lot of popularity, it does not mean that the traditional architecture for distributed simulations, High Level Architecture (HLA), becomes legacy. HLA can be viewed as a SOA operating on highly specialized rules optimized for simulation purposes and under restrictions not found in the general purpose SOA. To cater for this, a hybrid architecture framework, which allows specialized architectures, such as HLA, to reside intact in a larger SOA, has been proposed (Hannay, Brathen & Mevassvik, 2017) from which the NATO MSaaS reference architecture is derived (Hannay and van der Berg, 2017).

There are a number of important considerations that have to be made for the successful interoperation of C2 systems and simulation systems. This however does not only apply to the basic architecture. Future C2 systems should be designed specifically to support the interaction with simulation systems. One of the most important lessons identified from experimentation is related to time management. C2 systems typically do not include time management and run in real time, whereas simulations may produce reports that are time stamped into the future, and are not limited to real-time execution. Other factors to considerations are that simulation systems may provide both ground truth and perceived truth, while C2 systems normally only support the latter. In Gautreau et al. (2014) an overlay to the Distributed Simulation Engineering and Execution Process (DSEEP) is proposed to capture these differences, but also, a C2SIM reference architecture is needed.

SWAP

The SWAP system is a proof-of-concept demonstrator of a DSS for COA wargaming that incorporates functionality to assist planning in the land domain. The initial objective for SWAP was to create a tool that could simulate a COA in order to reveal its strengths and weaknesses (Hyndoy et al., 2014). During the development of SWAP, it was also realized that a digital tool for creating COAs can be beneficial in itself, especially if it incorporates analytical functionality to assist COA development. Over the last six to seven years several versions of the SWAP system have been established. This section outlines the main capability and functionality of SWAP to indicate the main building blocks of simulation support for wargaming of COAs.

The SWAP system is being developed in close cooperation with the user community in the Norwegian Army, and in February 2019, 52 cadets from the

Norwegian Military Academy tested SWAP. With SWAP, users can develop digital, executable COAs and simulate them without support by dedicated simulation operators. SWAP demonstrates the concept of a cooperative DSS that is available from any computer through a web-based user interface. SWAP offers sufficient possibilities for stakeholders to be able assess the concept and suggest further development.

The SWAP system facilitates digital planning by displaying forces on a digital map and providing functionality for tasking units and creating tactical graphics and control measures. The aim is a user interface to demonstrate that making digital semantic COAs can be just as easy and fast as drawing on paper.

Drawing a COA on a digital map, as opposed to on a paper map, opens new possibilities for collaboration between users at different physical locations and integration with C2 and battle management systems. In the current version of the SWAP system, the order of battle, expressed in MSDL, is imported from a C2 system. A connection to a specific C2 system, the Norwegian Command and Control Information System (NORCCIS), has been demonstrated with an earlier version of SWAP (Bruvoll et al., 2015). Also, the final COA can be imported directly into a C2 system, thus eliminating the need to enter the data manually into the C2 system after the COA wargaming is completed.

SWAP takes the concept of digital COAs one step further, into *executable* digital COAs. The COA created by the user interface is translated into a machine-interpretable, unambiguous language, making it possible to simulate the COA without additional simulation operators in line with the idea behind the work on the C-BML and C2SIM.

SWAP supports some essential tasks to perform brigade operations. SWAP supports two types of movements of combat entities, one for as fast as possible movements and one for cautious advancement. The user can define target areas and command maneuver units to seize or support by fire those areas. Similarly, engineering units can be ordered to breach an obstacle sketched as an area. Units can be synchronized by using phase lines. Figure 4 illustrates a COA sketch in SWAP.

SWAP provides functionality for automatic terrain analysis. Currently, it incorporates two terrain analysis tools, one for tactical route planning (Bruvoll, 2014; Tolt et al., 2017) and one for identifying vantage points. The terrain analysis functionality can be employed both as analytical tools and as services for the behavior models of the simulation.

The route planner in the SWAP system finds the best route given a set of prioritized criteria. Criteria are accessibility, cover (from direct fire), concealment, and threat. Priorities are predetermined for the different tasks, but by adapting the priorities based on the type of vehicles and the task, more realistic movement should be possible without the user having to provide more details. Routes can be adjusted by adding via-points. The route planner can also be used as an analytical

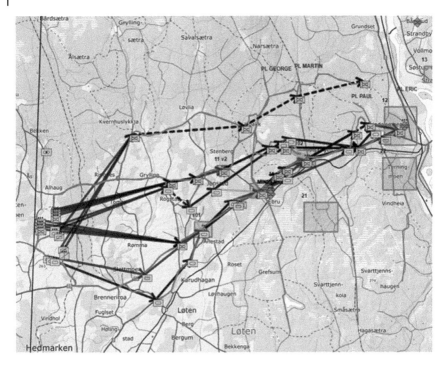

Figure 4 COA sketched in SWAP.

tool by giving the user the possibility to manipulate priorities. This requires a high-resolution terrain model. In addition to finding the best route with the given constraints, the route planner estimates travel time for a given route for a single battle tank, taking soil type and terrain inclination into consideration.

The vantage point calculation tool finds positions that have line of sight to a selected target area and at the same time are close to cover from direct fire from the same area. The user can use this tool to find good positions for, e.g. observation, attack, or the support of an attack.

When creating a COA at a given level of detail has been developed, it can be simulated. The unit movements are presented to the user and status information such as health, fuel and ammunition supplies is updated as the COA unfolds. As discussed earlier, such a simulation can be used to discover weak elements in the plan, such as synchronization issues and show potential consequences of decisions. But most importantly, it makes it possible to compare different COAs.

The simulation system consists of a C2 simulation part and a combat simulation part. The combat simulation part is based on VT MAK VR-Forces, an off-the-shelf simulation framework for CGFs. The multi-agent system for the C2 simulation interprets, decomposes, and executes a digital COA expressed with C-BML

(Løvlid et al., 2018; Løvlid et al., 2013). The multi-agent system (i) interprets higher level tasks, (ii) synchronizes the execution of tasks, and (iii) decomposes tasks at higher echelons to tasks for lower echelons. Interpreting higher level tasks means translating tasks such as "seize" and "support by fire" into a set of lower level tasks such as "move to location" and "set rules of engagement." Synchronizing the execution of tasks involves making certain units wait for each other at phase lines. Synchronization is also performed automatically when units are tasked to support or attack the same area. It is possible to task units at all levels of the task organization, i.e. one can task a platoon, a company or a battalion. The multi-agent system sends tasks from the lowest echelon to the combat simulation, and receives situation reports back from the combat simulation. The situation reports are displayed to the user together with unit statuses.

SWAP is built upon the concept of MSaaS (Hannay and van den Berg, 2017; Cayirci, 2013; van den Berg et al., 2017), meaning that the SWAP system consists of several loosely coupled back-end services and front-end services (applications) that can be reused individually for different purposes. The route planning service, for instance, is used by the SWAP front-end as well as the simulation system (back-end services). The services are implemented as Docker containers (minimal virtual machines), which enables rapid deployment of multiple instances of SWAP without having to install and configure each instance (Asprusten and Hannay, 2018; van den Berg, Siegel, and Cramp, 2016).

SWAP is a distributed system. For example, the SWAP front-end is a web application that can run in any web browser. It loads maps from a web map server and connects to the route planning service and the vantage point service. It connects on demand to an available simulation federation running in a cloud environment. An enterprise service bus implements message broker services to handle both data exchange internal to the SWAP system and with external systems.

SWAP builds on open international standards to federate the different parts. As mentioned, the task organization and initial positions are expressed in MSDL (SISO, 2008). The user interface application translates the user input into C-BML (SISO, 2014). The digital COA is interpreted by the multi-agent system, which decomposes tasks into low level tasks that most simulation systems for CGFs can interpret. The multi-agent system communicates with the combat simulation over HLA (IEEE, 2010), using the RPR-FOM (SISO, 2015) with the addition of the NETN-module to LLBML (NATO, 2018).

SWAP Experiment

Asking upfront what users want from new technology is often not the right approach for user involvement. New technology presents new possibilities for performing work processes differently that may not be obvious to users before the technology is experienced (Rogers 2003). In February 2019, an experiment was conducted, where 52 cadets from the Norwegian Military Academy used the SWAP system to develop and evaluate COAs. The purpose of the experiment was to collect insights into the requirements and potential for simulation-supported COA wargaming.

In the experiment, traditional manual planning was used as a reference to evaluate planning using SWAP. The cadets were divided into 17 groups of approximately three people. Two days before the experiment, the cadets were introduced to the brigade-level tactical problem to be solved, and on the day of the experiment, they were told which one of three battalions to plan for. After practical preparations, the cadets had two hours to produce a minimal decision briefing with several alternative COAs and one recommended. The experiment was designed as a crossover study, where each group planned both with and without the use of SWAP, half of the groups using the SWAP system first and the other half using traditional means first. After each planning session, the cadets completed a questionnaire.

The functionality in SWAP was presented to the cadets two days prior to the experiment, and the cadets were given 30 minutes at the day of the experiment to become familiar with SWAP. Several researchers and technical staff were available for questions during the experiment. When using SWAP, cadets were told that trying out and providing feedback on SWAP was the main objective, possibly at the expense of delivering a satisfying decision brief.

The cadets were able to take advantage of the functionality in SWAP when preparing their decision briefings. The cadets used the route planner and vantage points tools, and the descriptions of the COAs in the decision briefings showed how cadets had used results from the simulations when comparing their COAs and discussing possible losses and logistic issues. Generally, the feedback from the cadets was positive, and they seemed able to envision how a system like this could work in a real setting.

Despite the very short time available for training, the cadets were able to use all the functionality fairly easily. The cadets were able to develop COAs. However, tasks for indirect fire and limitation of tasks for synchronization were missed. Another limitation was that the use of phase lines was the only means to synchronize tasks and units. This resulted in many additional phase lines. They also called for the possibility to adjust the task organization as the COA was developed.

Generally, the terrain analysis services were well received, and the cadets could see the potential of having such tools to help analyze the terrain, especially when the terrain is unfamiliar. They suggested additional tools, such as a distance tool. The cadets also asked for more control over the route planning functionality by how the aspects of time, accessibility, concealment, cover, and threat were prioritized.

In order to fully exploit the possibilities with a DSS like the SWAP system, it must be possible to run the simulations much faster than real time. In the current version, the simulation ran at about seven times real time, which turned out to be too slow.

When observing how the cadets used the tool, it was confirmed that the most sensible way to use the tool is not to simulate a whole COA in one go, but to do so in steps or phases, following the action–reaction–counteraction approach. It should be possible to sketch the first tasks, simulate them, and use the results to decide how to proceed in a tight iterative loop of development and refinement.

Conclusion and Way Forward

Planning of military operations is a major undertaking and involves considering a large number of factors. Simulation support to operation planning has the potential to ease the cognitive burden that is put on the planning staff as well as making it possible to shorten the time needed for the planning and explore a larger number of COA options. Simulation is able to more objectively take into considerations a wider range of factors than human players. However, to be accepted in the time-constrained and resource-limited environment of tactical operation planning, simulation support must not come with an extra burden of processes and resources.

The main enabling technology building blocks to introduce simulation support to COA wargaming are now available. This chapter has outlined the most important of them like C2 and battle command M&S, C2SIM, and MSaaS. To be accepted, taking the cooperative DDS approach is considered to important.

The major challenges ahead for operational use are further development of the models for tactical land operations and user acceptance. And we have come to a stage of technology readiness level that allows us to explore how simulation support to COA wargaming will impact the planning process itself. Also, it should be explored whether the clear distinction between planning and execution of operations can to be challenged, by bringing simulation support to COA wargaming to replanning, and decision-making during execution of operations.

References

Alstad A., Løvlid R. A., Bruvoll S., Nielsen M. N. (2013). *Autonomous battalion simulation for training and planning integrated with a command and control information system* (FFI-report 2013/01547*).* FFI.

Alstad A., Mevassvik O. M., Nielsen M. N., Løvlid R.A., Henderson H. C., Jansen R. E. J., de Reus N. M. (2013). Low-level battle management language. *Proceedings of 2013 Spring Simulation Interoperability Workshop.* Orlando, FL.

Asprusten M. L., Hannay J. E. (2018). Simulation-supported wargaming using M&S as a service (MSaaS). *Proceedings of NATO Modelling and Simulation Symposium 2018*, 14-1–14-11.

Banner G. T. (1997). Decision making – A better way, *Military* Review, *LXXVII(5)*, 63–68.

Brook, A. (2011). UK experiences of using coalition battle management language. *Proceedings of the 2011 IEEE/ACM 15th International Symposium on Distributed Simulation and Real Time Applications, USA,* 160–167.

Bruvoll S. (2014). *Situation dependent path planning for computer generated forces*, (FFI-report 2014/01222). FFI.

Bruvoll S., Hannay J. E., Svendsen G. K., Asprusten M. L., Fauske K. M., Kvernelv V. B., Løvlid R. A., Hyndøy J. I. (2015). Simulation-supported wargaming for analysis of plans, *Proceedings of NATO Modelling and Simulation Group Symposium 2015*, 12-1–12-16.

Bruvoll S., Svendsen G., Asprusten M. L., Kvernelv V. B., Løvlid R. A., Hannay J. E., Fauske F. M. (2016). *Simulation-supported wargaming for analysis of plans (SWAP)*, (FFI-report 16/00524). FFI.

Burland B. R., Hyndøy J. I., Ruth J. L. (2014). Incorporating C2-simulation interoperability services into an operational command post, *Proceedings of the 19th International Command and Control Research and Technology Conference*, Arlington, VA.

Caffrey Jr. M. B. (2019). *On wargaming. How wargames have shaped history and how they may shape the future*. Naval War College Press. Newport, RI.

Carey S., Kleiner M., Hieb M. R., Brown R. (2001). Standardizing battle management language–A vital move towards the army transformation. *Proceedings of IEEE Fall Simulation Interoperability Workshop*, Orlando, FL.

Cayirci E. (2013). Modeling and simulation as a cloud service: A survey. *Proceedings of 2013 IEEE Winter Simulation Conference*, IEEE, 389–400.

CERDEC Public Affairs (2015). Commander's virtual staff. Army applies computer automation to operational decision making, *Army Technology Magazine*. 3(3). 16–17.

Cohen P. R., Kaiser E. C., Buchanan M. C., Lind S., Corrigan M.J., Wesson R. M. (2015). Sketch-Thru-Plan: A multimodal interface for command and control. *CACM*. 58(4). 56–65.

Davis, D., Blais, C., Brutzman, D. (2006). Autonomous vehicle command language for simulated and real robotic forces. *Proceedings of 2016 Fall Simulation Interoperability Workshop*, Orlando, FL.

de Reus, N. M., de Krom, P. P. J., Mevassvik, O. M., Alstad, A., Sletten, G. (2008). BML-enabling national C2 systems for coupling to simulation. *Proceedings of 2008 Spring Simulation Interoperability Workshop*, Providence, RI, 628–638.

Forbus K. D., Ferguson R. W., Usher J. M. (2001). Towards a computational model of sketching, *Proceedings of 6th International Conference on Intelligent User Interfaces*, ACM Press, 77–83.

Gautreau B., Heffner K., Khimeche L., Mevassvik O. M., and de Reus N. A. (2014). A proposed engineering process and prototype toolset for developing C2-to-simulation interoperability solutions, *Proceedings of 19th International Command and Control Research and Technology Symposium*, Alexandria, VA.

Hannay J. E., Brathen K., Hyndøy J. I. (2015). On how simulations can support adaptive thinking in operations planning. *Proceedings of NATO Modelling and Simulation Group Symposium 2015*, 18-1–18-14.

Hannay J. E., van den Berg T. W. (2017). The NATO MSG-136 reference architecture for M&S as a service. *Proceedings of NATO Modelling and Simulation Group Symposium 2017*. 3-1–3-18.

Hannay J. E., Brathen K., Mevassvik O. M. (2017). A hybrid architecture framework for simulation in a service-oriented environment, *Systems Engineering*, 20(3), 235–256.

Headquarters Department of the Army (2005). *Army planning and orders production*. FM 5-0. Washington, DC.

Headquarters Department of the Army (2017). *Operations*. FM 3-0. Washington, DC.

Huber D., Kallfass D. (2015). Applying data farming for military operation planning in NATO MSG-124 using the interoperation of two simulations of different resolution, *Proceedings of the 2015 Winter Simulation Conference*. 2535–2546.

Hyndoy J. I., Mevassvik O. M., Brathen K. (2014). Simulation in support of course of action development in operations. *Proceedings of Interservice/Industry Training, Simulation and Education Conference 2014*, Orlando, FL. 1844–1854.

IEEE (2010). *Standard for modeling and simulation (M&S) high level architecture (HLA)*. (IEEE Standard 1516-2010). IEEE.

Kimeche L., de Champs P. (2004). M&S decision support for course of action analysis, APLET. *Proceedings of NATO Modelling and Simulation Group Symposium 2004*. 4-1–4-21.

Kott A., Budd R., Ground L., Rebbapragada L., Langston J. (2005). Building a tool for battle planning: Challenges, tradeoffs, and experimental findings, *Applied Intelligence*, 23, 165–189.

Lawson III J. (2016). M&S support to wargaming, *CSIAC Journal of Cyber Security & Information Systems*, 4(3). 28–35.

Løvlid R. A., Alstad A., Mevassvik O. M., de Reus N., Henderson H., van der Vecht B., Luik T. (2013). Two approaches to developing a multi-agent system for battle command simulation. *Proceedings of the 2013 Winter Simulation Conference*. 1491–1502.

Løvlid R. A., Bruvoll S., Brathen K., Gonzalez A. (2018). Modeling the behavior of a hierarchy of command agents with context-based reasoning, *Journal of Defense Modeling and Simulation: Application, Methodology, Technology*. 15(4). 369–381.

Multilateral Interoperability Programme (2007). *The Joint C3 Information Exchange Data Model (JC3IEDM)*. Edition 3.1a.

NATO (2018). *STANREC 4800-AMSP-04 NATO Education and Training Network Federation Agreement and FOM Design*, 2018.

NATO Collaboration Support Office (2015). *MSG-085 standardization for command and control – simulation interoperability: Final report*. Paris, France.

Norwegian Army (2015). *Staff handbook for the Army. Plan and decision making process* (In Norwegian).

Perla P. (2012). *The art of wargaming. A guide for professionals and hobbyists*. J. Curry, Ed. www.wargaming.com, 2012.

Pullen J. M., Khimeche L., Galvin K. (2018). C2SIM in CWIX: Distributed development and testing for multinational interoperability, *Proceedings of NATO Modelling and Simulation Symposium 2018*. 13-1–13-13.

Rasch R., Kott A., Forbus K. D. (2003). Incorporating AI into military decision making: An experiment, *IEEE Intelligent Systems*, 18(4). 18–26.

Reddie A. W., Goldblum B. L., Lakkaraju K., Reinhardt J., Nacht M., Epifanovskaya L. (2018). Next-generation wargames. Technology enables new research designs, and more data, *Science, 362(6421)*. 1362–1364

Remmersmann T., Tiderko A., Schade U. (2013). Interacting with multi-robot systems using BML. *Proceedings of 18th International Command and Control Research and Technology Symposium.*

Rogers E. M. (2003). *Diffusion of innovations*. (Fifth ed.). Free Press.

Savasan H., Caglayan A., Hildiz F., Schade U., Haarmann B., Mevassvik O. M., Heffner K. (2013). Towards a maritime domain extension to coalition battle management language: Initial findings and way forward. *Proceedings of IEEE Spring 2013 Simulation Interoperability Workshop*, San Diego, CA.

Schade U., Hieb M. R., Frey M., Rein, K. (2010). *Command and control lexical grammar (C2LG) specification*. ITF/2010/02. Wachtberg, Germany. Fraunhofer-Institute for Communication, Information Processing and Ergonomics.

Schubert J., Seichter S., Zimmermann A., Huber D., Kallfass K., Svendsen G. K. (2017). Data farming decision support for operation planning. *Proceedings of 11th NATO Systems Analysis and Studies Panel Operations Research and Analysis Conference*. 7.2-1–7.2-20.

Seehuus R. A., Rise Ø. R., Hannay J. E., Wold R., Matlary P. (2019). Simulation support for military planning. *Proceedings of Interservice/Industry Training, Simulation, Education Conference 2019*. Orlando, FL. paper 19212, 1–13.

Seehuus R. A., Mathiassen K., Ruud E. L., Simonsen A. S., Hermansen F. (2019). Battle management language for robotic systems: Experiences from applications on an UGV and an USV. In J. Mazal (Ed.), *Modelling and Simulation for Autonomous Systems. MESAS 2018.* (pp. 302–320). Lecture Notes in Computer Science, vol 11472. Springer.

SISO (2008). *Standard for Military Scenario Management Language (MSDL)*, SISO-STD-007-2008.

SISO (2014). *Standard for Coalition Battle Management Language (C-BML) Phase 1*, version1.0. SISO-STD-011-2014.

SISO (2015). *Standard for Real-Time Platform Reference Federation Object Model (RPR FOM) version 2.0.* SISO-STD-001.1-2015.

Surdu J. R., Kittka K. (2008a). The deep green concept. *Proceedings of Spring Simulation Multiconference (SpringSim '08)*, Society for Computer Simulation International, 623–631.

Surdu J. R., Kittka K. (2008b). Deep green: Commander's tool for COA's concept, *6th International Conference on Computing, Communication and Control Technologies, CCCT 2008*, Orlando, FL.

Tolk A., Galvin K., Hieb M. R., and Khimeche L. (2004). Coalition battle management language. *Proceedings of 2004 Fall Simulation Interoperability Workshop*, Orlando, FL. Paper 04F-SIW-103.

Tolt G., Hedström J., Bruvoll S., Asprusten M. (2017). Multi-aspect planning for enhanced ground combat simulation, *IEEE Symposium Series on Computational Intelligence (SSCI)*. 2885–2892.

Turnitsa C. (2016). Adjudication in wargaming for discovery, *CSIAC Journal of Cyber Security & Information Systems.* 4(3). 28–35.

Wooldridge M. (2002). *An introduction to multiagent systems.* Chichester, UK: John Wiley & Sons Inc.

van den Berg T. W., Huiskamp W., Siegfried R., Lloyd J., Grom A., Phillips R. (2017). Modelling and simulation as a service: Rapid deployment of interoperable and credible simulation environments – an overview of NATO MSG-136. *Proceedings of 2017 Fall Simulation Innovation Workshop.* Orlando, FL.

van den Berg T. W., Siegel B., Cramp A. (2017). Containerization of high level architecture-based simulations: A case study, *Journal of Defense Modeling and Simulation: Applications, Methodology, Technology, 14(2)*, 115–138.

10

Simulation-Based Cyber Wargaming

Ambrose Kam

Lockheed Martin, Moorestown, NJ, USA

Motivation and Overview

Cybersecurity threats are growing exponentially. The "barrier to entry" for individuals interested in carrying out cyberattacks to institutions, and even nation-states, is lower than ever before. Yet, cyberattacks can have devastating impacts-from losing someone's savings or credentials to causing social and economic chaos through disruption of the national services. Better cybersecurity approaches are needed to defend mission-critical systems and network infrastructure; unfortunately, traditional cybersecurity solutions like anti-virus and network firewall are reactive in nature. We need to better understand the cyber threats and develop cyber solutions. This drives the need for simulation-based wargames and tabletop exercises, so that defenders can better anticipate the adversary's attack vectors and employ the appropriate strategy in a realistic manner. This is especially relevant as our military systems are expected to support multi-domain operations (MDO) in a highly contested environment. Modeling and simulation (M&S) techniques should be used to explore different what-if attack scenarios, so cyber protection team (CPT) can be trained in a heuristic manner. Human-in-the-loop virtual simulation could be leveraged to assess cyber operators real-time decision-making abilities and their overall skill level. With a simulation-based cyber wargaming environment, we can now apply our technical know-hows to train the next generator cyber warfighters so that they can better anticipate the adversary's tactics, techniques, and processes (TTP) in a very dynamic and complex operating environment.

Simulation and Wargaming, First Edition. Edited by Charles Turnitsa, Curtis Blais, and Andreas Tolk.
© 2022 John Wiley & Sons, Inc. Published 2022 by John Wiley & Sons, Inc.

Introduction

Modeling and simulation (M&S) and Operations Analysis (OA) have long been considered a staple in the systems engineering process. As a matter of fact, M&S technologies have been used for decades by many organizations within the Department of Defense (DoD).The first computerized simulation was an "air defense simulation"[1] created in 1948 by the Army Operations Research Office (ORO) at Johns Hopkins University.ORO also created the first Monte Carlo-based digital computer simulation – Computerized Monte Carlo Simulation (CARMONETTE) in 1953 to represent company or battalion level tank/anti-tank engagements.[2] Today's M&S tools have the capability to accurately represent complex systems and their interactions. The impacts are far-reaching. From systems definition to development to testing to manufacturing to deployment, M&S plays a significant role throughout the system engineering process.

Perhaps, it should not be a surprise that M&S technologies are being applied in the training arena as well. With the ever-rising cost for performing live testing, M&S has emerged to becoming a complementary alternative. Besides costs, the availability of the systems-under-development is also a major contributing factor. In order to train the operator or test the operating procedures, it is often necessary to leverage M&S to support these activities before the systems are ready for deployment. For example, a prototype of a new system could be represented digitally in a realistic M&S environment, and operators can perform scripted or unscripted tasks in the scenario. This way Concept of Operations (CONOPS) can be refined and system (or system-of-systems) performance can be predicted.

Besides virtual prototyping and CONOPS analysis, M&S offers a number of advantages over some other forms of training such as classroom lectures, system drills or even field exercises. One of the most important is cost. Modern simulators are high-fidelity devices, which are able to accurately replicate all aspects of an operation. It follows that it is more cost-effective to provide training to teach a pilot how to engage enemy aircraft with various missiles in a flight simulator, than it is to let him loose in an actual aircraft with a full load of missiles, after classroom training. Using simulators in training also reduces the maintenance cost of the real equipment. Cost can also be saved in terms of the networked simulators being able to link together simulators, which may be thousands of miles apart.

---

1 The "Air Defense Simulation" was used to study North American air defense capabilities and naval anti-aircraft guided missile systems. This simulation ran on one of the first Univac computers.
2 Congressional Modeling and Simulation Caucus, Modeling & Simulation Technological Advancements, March 2011.

Hence, joint exercises can then take place between forces, which do not need to leave their premises.[3]

The second advantage with M&S is the ability for the instructor to control and monitor the student. Not only does M&S yields cost savings, but it can also increase training transfer effectiveness and shorten the training cycle. The student will be able to achieve expected level of competence before actually having hands-on the real thing. This gives the student and instructor the level of confidence that the student will be able to do it right from the very first time. A validation exercise done by the South African Army showed that students going through the gunnery training simulator achieved 30–40% quicker reaction time and 14% better first hit results than those who did not use the simulator.[4]

The third aspect of using M&S is its ability to cope with environmental limitations. Databases and scenarios can be created which replicate non-local environments, which might be encountered on future deployment. Units and individuals can train at their home base without the need to travel to other areas located many miles away. This also allows units to undertake mission rehearsal exercises whereby in war, actual combat missions can be pre-fought.[5]

Lastly, safety is another M&S benefit. By improving the levels of training before allowing individuals to use actual equipment can only improve safety. This safety factor applies to the individual undergoing training, those around him on the exercise and finally, the safety of the actual equipment.[6]

Indeed, M&S training has been used even in fighting and cultural interactions. For example, M&S simulators like Virtual BattleSpace 2[7] or First Person Cultural Trainer (FPCT)[8] have been given to US troops in their home base before heading to Iraq and Afghanistan. With the complexity of M&S technologies and the visualization capabilities, the demand for live testing can be better managed. This is the reason why M&S has and will continue to play an important role for both government and civilian user communities providing critical support to the security of our nation.

---

3 CPT Boh Choun Kiat.(April-June 1999). Advancement in simulation technology and its military application.*Journal of the Singapore Armed Forces*, V25 N2.www.mindef.gov.sg/safti/pointer/back/journals/1999/Vol25_2/5.htm.
4 Helmoed-Romer Heitman. (January 1997).*Core Element of SA Army Simulation.* Jane's Defence Weekly, p. 9.
5 CPT Boh Choun Kiat.(April-June 1999). Advancement in simulation technology and its military application. *Journal of the Singapore Armed Forces*, V25 N2. www.mindef.gov.sg/safti/pointer/back/journals/1999/Vol25_2/5.htm.
6 CPT Boh Choun Kiat. (April-June 1999). Advancement in simulation technology and its military application. *Journal of the Singapore Armed Forces*, V25 N2. www.mindef.gov.sg/safti/pointer/back/journals/1999/Vol25_2/5.htm.
7 http://www.bisimulations.com/products/vbs2
8 http://forgefx.blogspot.com/2010/02/iraq-afghanistan-3d-training-simulation.html

There are, in general, three types of M&S: live, virtual, and constructive (LVC). Constructive simulation is one where both the operators and systems are modeled in a scenario; virtual simulation is one where only the systems are modeled but human operators are controlling them; live simulation is one where both operators and systems are real. Each of the 3 LVC simulation (along with the operations analysis techniques) has benefited different phases of the product/system lifecycle. For example, constructive simulation might be better suited for concept of operations (CONOPS) definition and top-level system requirements capture. Virtual simulation could be leveraged to support tactics, techniques and processes (TTP) improvements and live exercises are mostly for training and wargaming.

Over the past couple of decades, there have been several key technological advancements that greatly expanded the power and complexity of what M&S can achieve – allowing M&S capabilities to be leveraged to support a full spectrum of demonstrations and analyses activities including requirements definition (and derivation), mission effectiveness study, performance analysis, analysis of alternatives (AoA), training, and system design & test. These areas are incidentally very relevant to cyber. One of the major shifts in DoD is that cybersecurity is no longer a "bolt on" product. Acquisition processes like DFARs are dictating that cyber must be fully integrated into the program since the inception phase. Cybersecurity is a new warfare area in the Navy, meaning it has the same importance as tactical features such as weapons, radar, and command and control. The DoD now treats a cyber attack as the same as a conventional attack via weapons. Either method can lead to mission kill, crippling an asset, and potential loss of life. This shift has dramatically changed the calculus on developing every system. And by employing a simulation-based wargaming environment to include cyber as part of the multi-domain operations would go a long way in establishing a new norm, and break the current silos different services are working together.

Recently, M&S and OA have been applied to the cybersecurity domain area in the same manner as with sensor, weapons, command & control and logistics & sustainment. In many ways, cyber is a multi-faceted problem that poses a different set of challenges. Cyber is inherently very dynamic and complex as the attack surfaces span from hardware to software to end-user and to policy. Unlike traditional weaponry, cyber cuts across multiple mission areas and can be used to support both offensive and defensive operations. For these reasons, cyber techniques need to be examined and assessed their effectiveness before deployment. Since the adversary's network is not available for actual testing for the obvious reasons, simulations would be the next best thing to answer the What-If questions. These can be done through LVC M&S. With LVC, cyber effects can be examined at different levels serving different purposes.

Cyber Simulation

Modeling and simulation (M&S) have long been considered a critical element within systems engineering. Up until recently, M&S has not been applied to the cybersecurity domain area. One big area of concern is threat modeling. This is an inherent problem because of the complexity of the attack vectors; in addition to the typical seven layers in Open System Interconnect (OSI), one has to consider operators and user account passwords, so this become a much more dynamic problems than communications and networking. Perhaps, one way of solving this is to leverage government validated vulnerability databases.

The National Vulnerability Database (NVD)[9] and Common Attack Pattern Enumeration and Classification (CAPEC)[10] are valuable resources on the Internet to understand cyber threats and their impacts; they are commonly referenced by network professionals, software developers and cyber analysts over the course of the information system development process. The M&S framework in this paper highlighted the need to parse the NVD and CAPEC so that validated cyber threats can be ingested in a simulation environment. One of the main challenges in modeling and simulation (M&S) development lies in verification and validation (V&V) of the simulation itself. Without sufficient V&V rigor, the simulation results and the analytical results produced will not be perceived as credible. Rather than creating cyber vulnerabilities and developing malware from scratch, it is far more efficient and accurate to bring in cyber effects from these recognized data resources, and use them in cyber threat assessment analysis.

DoD recently issued the Joint Publication 3–12 which provides guidelines for military cyber operations. One of the important parts of JP 3–12 is how the military organizes different cyber effect categories. The commonly known "5D effects" (as identified in the JP3–12) include destroy, disrupt, deny, degrade, and deceive. The Figure 10.1 shows examples of the "5D" cyber effects and the consequences of these effects for the oil/gas industry.

There are many cyber simulation tools both in the DoD and commercial sectors. One that takes advantage of the JP3–12 5D cyber effects, CAPEC and NVD is called Cyber Attack Network Simulation (CANS). A cyber simulation should have a highly-flexible solution that can be used for both offensive and defensive cyber operation assessment. It has been extended to support a number of cyber use-cases, including – but not limited to – cyber threat analysis (with or without an external combat simulator), wargaming, cyber mission planning, operator

---

9 https://web.nvd.nist.gov/view/vuln/search
10 https://capec.mitre.org

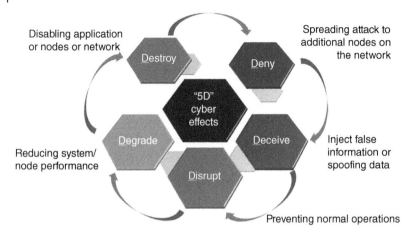

Figure 10.1 Joint publication (JP) 3–12: "5 D"cyber effects.

training, and tabletop exercises. By incorporating the exploits from NVD and CAPEC, it provides a new level of realism that the cyber mission forces are looking for. Systems engineers can use a cyber simulation to conduct architectural risk assessments and identify gaps in security coverage that require controls or lockdowns, and it may even be used to assist penetration testers and to train network/ system administrators and other cyber defenders. Figure 10.2 below shows CANS framework.

To evolve a constructive based cyber simulation to support wargaming, there has to be changes to the internal software architecture. A cyber wargaming engine should use a client–server based architecture, in which the simulation engine is decoupled from the user interfaces. This provides the ability to use multiple in attack launchers (to simulate multiple attackers) and network visualizers, as well as different roles (e.g. Red/White/Blue Cells).This is a critical step in supporting

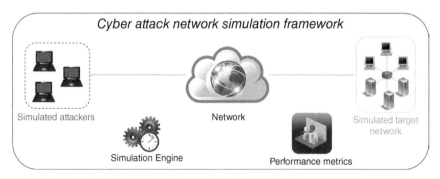

Figure 10.2 Cyber Attack Network Simulation (CANS) architecture.

the wargaming use case where it typically involves multiple teams (and multiple players on the same team).With this feature, each cyber operator can cooperate or act independent to execute different 5D effects to attack a given system(s).On the defensive side, the opposing color team can try to defend their network or system by executing defensive solutions; common techniques like IP whitelisting, firewall configurations, regular software patching, vulnerability scans, host, or network-level intrusion detection system (IDS) can be executed to mitigate the effects of the cyberattacks.

Gameplay begins when the Game Server has been initialized and all participants have joined the game. During the game, participant activities will be to fulfill their objectives or to simulate other human activities on the network. As participants attempt to fulfill their objectives the result of the attempts will be scored. Blue team participant play begins when the Game Server has been initialized and all participants have joined the game. Blue team participant gameplay includes detection of system anomalies, investigation of system anomalies, choosing courses of action in response to detection of system anomalies, detecting system vulnerabilities, courses of action to remedy system vulnerabilities, and simulation of normal user activities. Once a participant decides that there is an anomaly they will decide what course of action they will perform, if any, from the list of available actions that are available to them. Courses of action will be initial determined during game creation but can also be given to them during gameplay. Courses of actions can be applied to modules, network devices, software, and links. Courses of action can also be in the form of alerting another team or team member or alerting a team or team member of a completely different team. Courses of action will be scored if there is a course of action to system anomaly relationship exists in the ontology. The participant may have several or no courses of actions available to remedy an anomaly. Some courses of action may be more correct than others resulting in a higher score. A course of action that is completely incorrect will result in a score of zero if the anomaly it is attempting to remedy is related to an objective.

Red team participant gameplay may include these things from the Miter's ATT&CK framework with respect to the Blue team's network:

- Initial Access
- Execution
- Persistence
- Privilege Escalation
- Defense Evasion
- Credential Access
- Discovery
- Lateral Movement
- Collection

- Command and Control
- Exfiltration
- Impact

Red Cell participant gameplay will also include the completion its defined objectives. The types of Blue Cell detections and courses of action/remedies should be directly related to the different parts of this framework. The red team participant will have the opportunity to perform some type of software/exploit execution. The types of execution tactics, techniques, and procedures (TTPs) available will be defined during game creation or added to the red team participant's attack catalog during gameplay. During gameplay red team participant exploits may have the ability to persist. This of course is directly related to the specific persistence techniques being used by the red team participant to exploit a specific vulnerability. The type of execution TTPs available will be defined during game creation or added to the red team participant's attack catalog during gameplay. During gameplay the red team participant may have privilege escalation exploits allowing them to escalate privileges on a given system. The type of privilege escalation TTPs available will be defined during game creation or added to the Red Cell participant's attack catalog during gameplay. During gameplay the red team participant may have credential access exploits allowing them to attain user credentials. The type of credential access TTPs available will be defined during game creation or added to the Red Cell participant's attack catalog during gameplay. During gameplay the red team participant may have discovery exploits allowing them perform reconnaissance or intel mission about different parts of the system. The type of credential access TTPs available will be defined during game creation or added to the red team participant's attack catalog during gameplay. During gameplay certain actions taken by the participant will result in the completion, partial completion, or the missing of objectives. The relationship of participant actions to objectives will be determined during game creation but may also happen during gameplay. The completion of objectives may be used for scoring, to end the game or determine the winner.

Being a discrete event simulation tool, CANS was designed to support Monte Carlo-based analysis; this approach is more scalable at assessing different attack patterns in a simulated than testing them in a real environment. Users can execute what-if scenarios both from the defensive as well as offensive standpoint. Cyber is inherently very dynamic and complex as the attack surfaces span from hardware to software to end-user and to policy; and CANS can provide insights into the vulnerability of a system given a cyber defensive strategy. To elevate the results at the mission analysis level, CANS can interface with a higher-level simulation tool like AFRL's Advanced Framework for Simulation, Integration, and Modeling (AFSIM) over DIS.

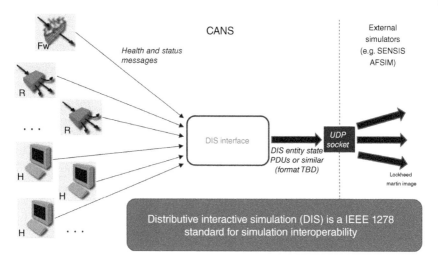

Figure 10.3 Distributed Interactive Simulation (DIS) event sequence.

There are many standards that govern the information technology realm. Likewise, a number of simulation specific standards have emerged over the past decades. Standards provide significant benefit toward achieving interoperability among disparate simulation systems and integration of simulation into many defense-related processes.[11] Regardless of the type of simulation or the application domain within which the simulation is applied, standards aim to increase the chance of successful integration. Like many other engineering or science disciplines, standards provide proven, widely accepted frameworks within which designers and developers can achieve a higher level of interoperability than if they were to create simulation applications without any guidance whatsoever.[12] The Institute of Electrical and Electronic Engineers (IEEE) currently manages these two standards: Distributed Interactive Simulation (DIS) as IEEE 1278 and the high level Architecture (HLA) as IEEE 1516.[13]

Since cyber is not officially part of DIS v6.0 (it will be in v7.0), the team has developed custom packet data unit (PDU) messages to carry the cyber effects from CANS into AFSIM. Specific DIS PDU enumeration of the cyber effects are captured in the table 10.1 below. Figure 10.3 shows the DIS interface flow diagram.

11 The interested reader is encouraged to visit the Simulation Interoperability and Standards Organization (SISO) website for more detailed information (SISO can be found atwww. sisostds.org).

12 Jonathan Searleand Brennan, John. "Simulation Components," http://ftp.rta.nato.int/ public//PubFullText/RTO/EN/RTO-EN-MSG-067///EN-MSG-067-06.pdf, pp. 6–8.

13 JonathanSearle and Brennan, John. "Simulation Components," http://ftp.rta.nato.int/ public//PubFullText/RTO/EN/RTO-EN-MSG-067///EN-MSG-067-06.pdf, pp. 6–8.

Mission Analysis Tool

The Air Force Research Lab (AFRL) in Wright-Patterson Air Force Base owns the Advanced Framework for Simulation, Integration, and Modeling (AFSIM); AFSIM is a government-approved software simulation framework for use in constructing mission and engagement-level analytic simulations for the Operations Analysis community. The primary use of AFSIM applications is the assessment of new and advanced weapon system concepts, and the methods of their employment. The most recent AFSIM 2.3 release contains significant improvements in usability by providing analysts with an expanding integrated scenario development and results visualization capability through the AFSIM-Wizard application. In addition, the first Alpha release of an AFSIM-based virtual experimentation and war-gaming capability (called AFSIM-Warlock) is also being delivered with AFSIM 2.3. Figure 10.4 below shows the multi-color team environment in Warlock.

Since AFSIM is also IEEE 1278 DIS compatible, the team experimented running AFSIM and CANS together in the same simulation environment. The advantages are twofold. First, CANS can provide detail level cyber effects that AFSIM previously cannot. The intention of AFSIM was never meant to be a mission analysis tool that models IP or embedded cyberattacks. AFSIM natively can provide cyber effects at the system or platform level, but it does not model any CVEs (from NVD, for example) or CAPEC's attack patterns. CANS, on the other hand, was

Figure 10.4   AFSIM Warlock operator-in-the-loop custom interface (sample).

Table 10.1 Generic cyber effects in CANS/AFSIM.

Enumeration	System impacted	Effect	Additional data	Associated CANS effect	Implemented in AFSIM	Implemented in CANS	Tested
0	Sensor	Turn on (all)	N/A	Repair	☐	☐	☐
1	Sensor	Turn off (all)	N/A	Deny	☐	☐	☐
2	Comm	Turn on (all)	N/A	Repair	☐	☐	☐
3	Comm	Turn off (all)	N/A	Deny	☐	☐	☐
4	Weapon	Turn on (all)	N/A	Repair	☐	☐	☐
5	Weapon	Turn off (all)	N/A	Deny	☐	☐	☐
6	Track Mgr	Turn on (all)	N/A	Repair	☐	☐	☐
7	Track Mgr	Turn off (all)	N/A	Deny	☐	☐	☐
8	Platform	Turn on all components	N/A	Repair	☐	☐	☐
9	Platform	Turn off all components	N/A	Deny	☐	☐	☐
10	Sensor	Spoof (all tracks)	N/A	Action spoofing	☐	☐	☐
11	Weapon	Firing delay increase	Number of additional seconds	Packet flooding	☐	☐	☐
12	Track Mgr	Purge all tracks	N/A	Deleted data	☐	☐	☐

never meant to show cyber effects beyond the system and platform level. The whole notion of cyber impacts at the overall mission level (e.g. number of leakers, red/blue attrition, etc.) is foreign to CANS. Hence, this marriage of CANS/AFSIM is mutually beneficial. The team has tested the following cyber effects running over DIS. When the CANS cyber effects have been received by AFSIM, it will implement those effects accordingly. This approach is much better than just arbitrarily shutting down systems or breaking communication links in a given vignette.

With this DIS approach along with the CANS and AFSIM operator interfaces described in the earlier sections, Lockheed Martin has a cyber wargaming engine that can work either in an integrated multi-domain operations (MDO) environment or as a cyber-only exercise. This concept has been unveiled to various wargaming and operator training communities, and the feedback has been overwhelmingly positive.

The definition of wargaming should also be expanded to accommodate for a new breed of cyber exercises involving operator-in-the-loop performing their mission in a virtual environment. There are several reasons for this thrust. First, as opposed to guns, tanks and airplanes, cyber warfare is entirely fought within the information system paradigm involving the classic Open System Interconnect (OSI) architecture. Hence, those who are "cyber warriors," are already familiar with the virtual computer environment. Second, these cyber warriors are typically young officers who grew up in the video game culture. They are crafty in the multi-player, multi-sided gaming environment. Hence, the traditional instructor-led classroom style may not work well with them. To captivate their interest, cyber wargaming should be performed in an interactive manner, and in an environment these young cyber warriors are already familiar with. The future of cyber wargaming should still involve traditional setup of a traditional wargame (i.e. red/blue/white cells, mission objectives, rules of engagements, etc.), and need to be executed in a virtual environment powered real-time M&S engines. Similar to the classic game of "capture the flag," the participants from both red/blue sides will exercise their cyber knowledge while morphing their offensive/defensive strategies. It is through this type of wargaming exercises, observers can discover new cyber attack vectors and identify system/operationally vulnerability in a system-of-system environment. In the past, DoD has executed seminar type of cyber wargaming with resounding success. The next logical extension is to fight the cyber and electronic warfare in a safe digital environment using M&S tools. A can be used to extract data and after-action report (AAR) analysis.

In a cyber wargaming environment, cyber defenders would choose to perform activities such as hardening the environment, monitoring detection systems, or responding to any evidence of a breach. Each side would have a catalog of activities to choose from, each with precomputed (or straightforwardly derivable) impacts and results that would result after selection, though these pre-ordained results would not be made known to the players in advance. As each side performs their activities, the game would consider the effects of those actions and update the system state maintained by the wargaming platform, whether through a white team judgment, a simulation component, an actual action performed on the target system, or some combination. Throughout the activity, the test director and white team would monitor the game's progress, adding new stimuli and tuning the system as required by the chain of events.

Wargames

Wargame is an overloaded term with different meanings at different organizations. Dr. Peter Perla in his 1990 seminal book *The Art of Wargaming* stated that a wargame is "a warfare model or simulation whose operations does not involve the activities of actual military forces, and whose sequence of events affects and is, in turn, affected by the decisions made by players representing the opposing sides." Some would like to adhere to this definition while others have similar definitions. In general, a wargame is not a technique for producing rigorous, quantitative or logical dissection of a problem or for defining precise measures of effectiveness by which to compare alternative solutions.[14] A wargame affords an opportunity to study a warfare problem through the human decisions and interactions of its participants as they endeavor to defeat their opponent.

Wargame designers can leverage Department of Defense (DoD) scenarios or made-up scenarios to provide a warfighter relevant and relatable environment for the exploration of cyber warfare in the context of a broader conflict. Players from the Red and Blue Cells would typically operate at the operational/tactical level of warfare and the information they were provided was appropriate for decisions at that level. Operational level considerations are necessary to understand the aggregated effects of cyber operations.

The White Cell provides control and oversight of game execution. While Cell members were indoctrinated with the objectives of the game, game rules, and details of each move. They were responsible to ensure the game addressed the stated objectives. Scenario designers should identify a Major Scenario Event List (MSEL); and the White Cell has the responsibility to determine if and when injects are required – should the Red/Blue teams deviate too much from the key events in MSEL list. The intended purpose of the injects was to (i) steer the red or the blue team back to the original intent of this Cyber Wargame, and (ii) to force

---

14 Curry, John and Perla, Peter. (2011). Peter Perla's *The Art of Wargaming*.

the Red/Blue team to react to unexpected event sequence (such as weather, unexpected adversary asset maneuver/deployment, or unexpected usage of a system, etc.).The Red Cell represents the US adversary in the wargame; in some literature, Red Cell is the opposing force or (OPFOR).The Blue Cell represent the US friendly side. In the context of the wargame, each side's objective is to defeat the opposing team.

During the wargame, the scenario should have several moves. One team could be attacking the other side; while in the next move, the other team could launch counter-attacks. Each move should be planned to last for two to three hours. At the beginning of each move, White Cell provides a scenario context briefing (including the objectives, location of assets, rules of engagement [ROE], etc.).The Red and Blue Cells then divided into break-out sessions and plan their respective courses of action (COA).At the end of the move, the Red and Blue Cells should be given time to brief their plans in plenary session which included all participants. Following the move, all participants completed a survey to provide feedback on the move and their assessments of key actions their cell took.

The concept of wargaming can be applied to cyber so that it can be better understood from the offensive and defensive standpoint. Cyber wargaming is a tool that is useful to organizations for assessing current and future capabilities, planning, examining possible scenarios, and training staff. While a cyber wargaming exercise may not help defend against emerging techniques and technologies directly, the use of the process model to support "what if" scenarios for forward-leaning views of changing technologies and potential attack vectors can be useful. It could then be coupled with a red team exercise to validate results. The use of cyber intelligence sources and analytics of actual events (like the Computer Emergency Response Team [CERT] report of the Ukraine power grid cyber incident in 2015) to re-assess the effectiveness of products and processes in mitigating attacks serves as an augmentation to the continual defend-and-adapt process of cyber operations.

Cybersecurity has always been a threat for all modern military platforms. Attack vectors can come from different sources and forms. It can be as low as the embedded chip to as high as the network or user account password. There are also secondary and tertiary vectors like heating/ventilation & air conditioning (HVAC) system for the sailors' sleeping quarters, industry control system (ICS) for power generation or cooling system of a turbine or radar. These systems once compromised can directly or indirectly affect the sensor/weapon system performance. This is the reason why DoD is taking cyber threats very seriously. For example, the acquisition of the Office of Undersecretary of Defense (OSD) community has implemented the DoD 8510.01.The goal is to have cybersecurity baked into any given acquisition program. The diagram in Figure 10.5 below shows a digitized battlefield where each participant is performing his/her mission based on the

Figure 10.5 Digitizing the wargaming battlefield through simulation.

platform/system he/she is in control of. The advantages of this approach over the traditional sand top box or card-based wargame is repeatability, adjudication process and flexibility of the scenarios. In traditional wargaming environment, it is always difficult to go back and re-examine the decision-making process. This is particularly important to after action review (AAR).Adjudication can also be better facilitated with the simulation driven environment. Traditionally, the adjudication process is being handled by human subject matter expert (SMES) and his/her dice roll. This might cause dispute because one side (say, Red or Blue) might not agree with the results of the dice roll. This causes a lot of disruptions and affects the flow of the wargame. Simulation-based wargame can at least minimize this type of disagreement. The digitized battlefield might provide a better way to support scenario injects during gameplay. Scenarios and vignettes can be scripted (and tested) ahead of time to ensure strategic, tactical and technical feasibility.

To facilitate this digitized simulation-driven wargaming engine, the Lockheed Martin team developed a minimum viable prototype (MVP) that consists of a detail cyber simulation tool feeding cyber effects to a mission-level simulation tool. The prototype wargaming framework allows participates to play out the scenario in real time. For example, a human operator could fly one of the red aircraft launching cruise missiles at the Blue targets (e.g. JMOC or JROC). Another operator could be representing the tactical action officer (TAO) on the Blue surface ship, and he/she could make target/weapon pairing assignment. When the target persecution process is done, he could fire an interceptor to engage the incoming target. Another operator could be a Blue commanding officer (CO) on the JROC, and he/she could be overseeing the Find, Fixed, Target, Track, Engage, Assess (F2T2EA) kill chain process. In this set up, all the Red/Blue entities or platforms are represented in a force level simulation of

AFSIM. With the new client/server setup in CANS, human operators can inter-act with each other (say, Red attacker trying to infiltrate the network protected by the Blue cyber protection team) and their interactions could affect how the entities will perform in AFSIM. For example, JROC or JMOC might not be mak-ing the optimum target/weapon assignments because the tracks spoofed, or that the TAO in the Blue ship cannot issue the engagement order in time due to a degraded data link (see Figure 10.6 below).

Multi-Domain Operations (MDO) has been a challenge in the DoD communi-ties. Having a simulation-based environment would help because decision makers and stakeholders could explore new ways to better support MDO. From the plan-ning to execution, MDO requires cooperation between different service domain areas (including, air, space, surface, subsurface, cyber, land, and electronic war-fare).For example, is it preferable to drop bombs on an adversary's building that hosts both command & control function and network gears? Doing so might also knock out the adversary's network equipment that the friendly cyber team is counting on as entry point. At the current time, each domain area is still being planned separately and integrated before execution for de-confliction. In other words, the air planning cell will do target and weapon pairing independent from the cyber team. The deconfliction would come from the end which will require re-planning if the adversary's network gears are deemed to be critical for their mission. The ideal approach is to develop plans cooperatively with inputs from each domain area.

National Cyber Range (NCR) is a facility that provides a secure and isolated environment to perform training, testing, and evaluation of realistic cybersecurity activities.

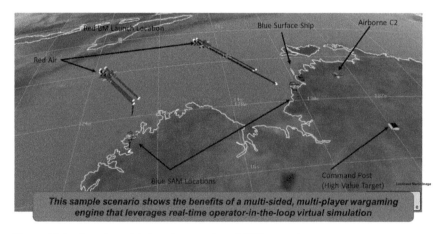

This sample scenario shows the benefits of a multi-sided, multi-player wargaming engine that leverages real-time operator-in-the-loop virtual simulation

Figure 10.6 Sample multi-domain operations (MDO) scenario.

Commercial Wargames

The commercial world is catching on to the concept and benefits of cyber wargames. Perhaps, this should not be a total surprise given the fact that enterprises are just as likely (if not more so) to be targets of cyberattacks. Ever since the first publicized cyber attack event of TJMax, in 1989, retailers and consumer product sector realized they could be victims to not just fraud and data exfiltration. The banking industry and financial institutions are facing similar challenges. Even hospitals are not immune where cases of ransomware of health records are well-documented. This prompted Chief Information Security Officers (CISOs) to explore ways to enhance cyber resiliency of their infrastructure and information systems. Training, penetration testing, red teaming, and continuous monitoring are all necessary elements of cybersecurity at the enterprise level. For example, penetration testing or red teaming has been used for some time both to provide an assessment of deployed technology and as a learning activity; training continues to hone operators' technical skills and end-user cyber awareness while continuous monitoring is critical for cyber visibility. But increasingly, enterprise CISOs are turning to cyber wargames for additional levels of realism. Findings from cyber wargaming can be used to improve event response, platform and application development, selection and integration of defensive technologies, and deployment compliance to reduce risk.

While other training venues exist, such as computer-based training (CBT) or instructor-led classroom-based seminars, none of these options offer the same sort of simulations of real world experiences that happen with a cyber wargaming event. In an actual cyber attack on an organization, considerable stress is placed upon the employees and their managers who have to make real-time decisions to not only minimize the initial cyber effects but also to recover quickly so that normal business operations could continue. It is only in a cyber wargaming setting where many of the same pressures and rapidly occurring events that compete for people's attention and focus could be replicated. In a well-orchestrated cyber wargaming event, participants will feel like they are in a real-world situation. This presents the opportunity for learning how well staff and management will perform under fire, while simultaneously assessing the strengths and weaknesses of the cyber infrastructure in the organization. Ultimately, cyber wargaming helps employees improve their skills or identify areas of needed improvement.

Cyber wargame scenario can be based on real-world actual scenarios; using a simulation environment to model systems, user behavior and applications would enhance the value of wargaming because simulations can reflect the variability of actual cyber events. Utilizing the attack vectors, adversarial TTPs, and targeted technology captured in the STIX/Trusted Automated eXchange of Indicator

Information (TAXII) ontology could produce a set of replayable testing capabilities.

Although cyber wargaming offers improved business operations by driving down cyber risks, there are still some limitations. Unknown attack vectors, unidentified vulnerabilities, and unanticipated reactions will potentially still develop. In addition, operational TTPs can change over time and affect, either negatively or positively, the ability of an institution to adapt and react. These drive a requirement that cyber wargaming scenarios be updated and adjusted to adapt, and that a program of continuous review of cyber incidents is used to inform defenses and identify decreases in mitigation effectiveness.

Commercial sector CISOs captured the same principles when these wargames are being run. There are also commercial-off-the-shelf tools. Consulting companies are getting involved often providing subject matter expertise in designing and conducting these wargame scenarios. Similar to DoD wargames, these commercial counterparts are focusing real or hypothetical scenarios. The objective is to raise awareness and sharpen the skills of the employees. For example, data breach could not only provide a good exercise to test the CPT operators to detect anomaly by analyzing user or network behaviors. At the network level, it can also test the responsiveness of the data centers and infrastructure resiliency. Recent cyber conferences or wargaming forums, several CISOs have shared success stories in their respective enterprises, and discussed how cyber wargaming has transformed the way their cyber teams mitigate cyber attack risks.

Incorporating a computational scoring system into the cyber wargaming process should also be part of ongoing high priority developmental efforts. This should provide the outcome of the applied effects, the impact to each individual participant's operations, and a measure of the outcome of selected COAs, Metrics produced by (or derived from) the simulation environment are a good measure of the ability of the participant organizations to adapt to the attack event while maintaining ongoing availability of their critical operations.

Metrics from these cyber wargaming events are typically relative measures of the results within the testing scope; most often, they are limited to a single organization or small group engaged in coordinated wargaming exercises. Improving this activity through scenarios that more strongly integrate business activities and technology operations in relation to a realistic and repeatable simulation of sophisticated adversarial cyber events. Scenarios based on actual business functions linked to organizationally deployed technologies can provide a realistic simulation of the impacts of cyber events on business operations. Scenarios that integrate elements from both current tabletop "what-if" exercises and technology-based red team exercises draw an association between the cyber defense and business process effects to examine a realistic view of an event outcome.

Develop cyber risk metrics to measure success of achieving pre-defined objectives is a critical task. This provides several benefits, including identifying and scoping areas needing improvement on the part of the participants, determining the effectiveness of technology in the defense presented, and identifying opportunities for improving future composite exercises.

Wargaming for cyber has been used by the DoD, DHS, technology providers, and private interests as a means of providing learning, testing processes and technology, and identifying gaps in achieving good security controls. Wargaming, both tabletop and red teaming, has proven to be a useful process for helping to advance the effectiveness of security and control in many operational settings. Tabletop exercises have included cross-team communication, incident response, intelligence gathering and distribution, and management response scenarios. These have effectively served as a means of identifying gaps in processes and technology, and as learning exercises for administrative actions. Red teaming, penetration testing, and capture-the-flag exercises have included both offensive, defensive, and adversarial wargaming techniques in scenario and non-scenario based events. These have also provided effective assessments of vulnerabilities in technology and process.

Along with benefits, there are limitations to cyber wargaming. For example, executing wargaming is always going to be different than the real world. Being in a highly controlled environment with specific rules of engagement will admittedly inhibit the ability to perfectly mimic real-life events, where actual actions (and reactions), situations, and attitudes may be quite different; secondly, since a scripted scenario with MSEL list needs to be followed, cyber wargaming might lack the ability to fully assess participants' spontaneity, preparedness and reaction to real-life events and surprises and thus not providing a true test of a crisis, emergency, security or system's capabilities. Nonetheless, industry experts can agree on the value cyber wargaming. Simulation-based cyber wargaming offers additional advantages than other types of non-simulation based cyber wargaming. Board game-based cyber wargame, for example, allows two or more players to assume the roles of attacker and defender. Players representing attackers attempt to move through the steps of the cyber kill chain to reach the "Action on Objectives" goal, while the defending players attempt to stop their progress through various response actions. Game-play takes place on a game board with cards, dice, game pieces, paper, pens, and play money. The cards present the actions that attackers and defenders can take, with a balance of the role of technology and tradecraft.

Future Work

To protect systems from cyber threats, it is helpful to develop simulation models to study how malware may penetrate and spread through a system of software applications and networked components. Cyber test ranges are generally useful in studying the detailed malware mechanisms by which cyber threats could gain a foothold and propagate. However, once understood on a small system, a given malware threat, being merely a deterministic software routine, operates the same way every time on a given platform and software environment. Therefore, once characterized, it is no longer necessary to continue modeling or performing the detailed operations when investigating effects on larger scale systems. Instead, only the effects of an exploit need to be modeled, and whether a given platform would be susceptible.

Some threats may not be purely deterministic. For example, the time required for a given threat to succeed on a given platform may be a random function. In these cases, the abstract model can faithfully capture the behavior by describing and replicating the stochastic statistical behavior as captured from test-range data.

Cyber vulnerabilities present growing threats to complex networked systems and network centric operations. Several approaches have been employed to study and understand cyber effects. A prevailing method utilizes cyber test ranges consisting of physical computers and virtual machines on which actual malware is exercised on quarantined networks. Such test-ranges provide crucial information about the details of malware mechanisms, but are expensive to operate and scale upward while offering limited availability. All of these are calling for better and faster modeling & simulation (M&S) techniques to model/study existing and emerging threats. M&S can also combine the effects of "traditional" threats (missiles, aircraft, tanks, and submarines) in a realistic operational scenario so that the decision makers can have an appreciation of the threat space across multiple spectrums (physical and virtual).Cyber protection teams (CPT) and cyber subject matter experts (SME) would consider using simulation tool as a complementary product to the test range activities.

While it is true that human responses are needed to fight cyber threats in the near future, automation and eventually autonomous system could be applied in orchestrating cyber defensive responses. With the advent of the artificial intelligence and machine learning (ML), it is conceivable that these technologies should be tested in a cyber wargaming environment. This is driven by two factors. One is the shortage of qualified cyber professionals. Human teams usually consist of individuals who are trained and experienced in live penetration of computer systems, including infrastructure hacking, web-application hacking, and specialties in compromising and persisting in Windows, Linux, or other platforms. Such cyber professionals are certainly in high demand. Pulling them away from their regular duties will always be a challenge. The other factor at play is the availability and improved capability of AI/ML. Machines have proven they can defeat human master players in Go and chess. With high performance computing (HPC) and advanced ML algorithms, simulating adversary activities in terms of specific capabilities and motivations of the adversary (or class of adversaries) is no longer a pipedream. Through optimization, ML can begin modeling complex nature of cyber events and attack patterns.

Summary

Modeling and simulation activities are invaluable in developing complex systems, such as a new weapon systems or energy management system. Ultimately, customers benefit from better overall cost, schedule, and technical performance on programs as a result of M&S technologies. It has been widely documented that M&S with analysis provides end users with the ability to better understand "what if" studies and answering the "so what" question involving an array of system components, and their interactions provide cost savings, cycle time reductions, and lifecycle cost savings. As systems become more complex, so does the M&S efforts needed to support them. For this reason, M&S will continue to play a role in future efforts.[15]

One of the key organizational objectives for cyber wargaming is to support assessment of current capabilities. Planned and controlled wargames present a unique opportunity to identify the strengths and weaknesses of an organization's network and systems architecture, its defensive technologies, and its processes and procedures. By subjecting the current defensive regime to cyberattacks in a simulated environment, valuable lessons can be learned with minimal risk.

While traditional types of cyber wargaming (e.g. tabletop and red teaming) are well established, creating a middle-tier simulation-based cyber wargaming combines the two extremes promises to yield additional benefits to the advantage of the entire sector. By creating a simulation-based wargaming exercise that borrows from both realms, additional insights can be gleaned regarding technology capabilities or gaps and inherent cyber risks in engineering and architectural choices, without the extreme cost and effort of conducting detailed technical experiments.

The importance of cyber wargaming and operator training especially in the context of multi-domain operations (MDO),simulation tools designed for wargaming and training. This effort involves advancing the existing Cyber simulation engine in support of a real-time operator-in-the-loop wargame. Cyber models will be enhanced to represent existing and emerging threats in a realistic operational scenario. Part of the effort is to leverage existing simulation environment to model possible cyber techniques used in recent conflicts (like Georgia or Ukraine) – based on open sources. AFSIM was used to give a larger mission analysis context. Through DIS, CANS, and AFSIM are being federated in the same simulation environment. The objective is to leverage detail cyber attack modeling in CANS and feed the 5D effects into AFSIM. By taking advantage of AFSIM's Warlock operator interface, operators can try to perform their missions with cyberattacks

---

15 Thomas Fargnoli. (2008). *Common Modeling & Simulation Framework Services*. Lockheed Martin MS2 White Paper.

happening to their systems or infrastructure. This can be done without compromising real tactical systems with malware. Through the wargaming exercises, DoD can further refine Cyber Techniques, Tactics, and Procedures (TTPs) in a realistic operational scenario.

Virtual simulation exercises with real operator-in-the-loop capabilities have been around for a couple of decades. However, simulation-based cyber operator training in a multi-domain operations (MDO) context does not yet exist; currently, cyber operators are being trained in isolation from other warfighters. As a result, it is difficult for cyber operators (both offensive and defensive) to understand how their courses of action impact other domain areas and vice versa. As in the real-world, cyber operations need to be planned and coordinated with subject matter expects from other mission areas. The proposed idea of using cyber wargaming is to provide a more realistic MDO environment for cyber mission operators; cyber wargaming identifies and can be used to validate innovative cyber defense technologies that have the potential to improve cybersecurity and reduce risk from cyber sources. Additionally, this framework can be easily extended to support wargaming. With a simulation-driven engine, wargame designers do not need to rely on heavily scripted scenarios with limited repeatability value. The simulation tools can produce metrics which could be invaluable to after action reviews (AAR).

One approach is to leverage the Cyber Attack Network Simulation (CANS) and run it in conjunction with AFSIM over Distributed Interactive Simulation (DIS) communications. The goal is to have detailed cyber effects being ingested into the mission level simulation environment of AFSIM. With Warlock, domain area operators (e.g. tactical action officer of a surface ship, 5th Gen fighters, radar operators, etc.) can now take control of some of the platforms/systems in the MDO scenario running in AFSIM; CANS, by itself, also supports multi-team/multi-player features for cyber operations. With CANS and AFSIM running together, cyber tactics, techniques, and processes (TTPs) can be explored. The novelty of this concept is that AI and Machine Learning techniques can be easily added to augment the value proposition of this simulation-based cyber operator training and wargaming environment. Since attack vectors are conceived by AI through different permutations and combinations of existing and emerging threats, new scenarios can be composed together well beyond the imaginations of the human scenario development team. Just like in the real world, none of the scenarios in this wargaming environment encountered by the Blue cyber team will be scripted and rehearsed ahead of time. The more prepared the cyber operators are, the better they can recognize potential vulnerabilities, and the quicker they can enhance mission success. The motto "Train like we fight" should be applied to cyber as well as traditional domains, given the complexity nature of multi-domain operations (MDO).

References

1 Applebaum, A., et al.(2016). Intelligent, automated red team emulation.*Proceedings of the Annual Computer Security Applications Conference*(December 2016).

2 Bodeau, D. and Graubart, R. (2013). *Characterizing Effects on the Cyber Adversary: A Vocabulary for Analysis and Assessment*. MTR 13432, PR 13–4173, The MITRE Corporation.https://www.mitre.org/sites/default/files/publications/characterizing-effects-cyberadversary-13-4173.pdf

3 Bodeau, D. et al. (2018). *Cyber Threat Modeling: Survey, Assessment, and Representative Framework*. HSSEDI, The Mitre Corporation.

4 Committee on Technology for Future Naval Forces (1997). *Becoming 21st Century Force, Modeling and Simulation*, vol. 9. U.S. Naval Studies Board National Research Council.

5 Department of Defense Instruction (DoDI) 8500.2.*Information Assurance (IA) Implementation*. Secretary of the Navy Instruction (SECNAVINST) 5239.3B.*Department of the Navy (DoN) Information Assurance (IA) Policy*.

6 Security Technical Implementation Guide–Application Security and Development, Defense Information Systems Agency (DISA).(23 January 2014).

7 CJCSM 6510.02. (5 November 2013).Information Assurance Vulnerability Management (IAVM) Program.

8 Deloitte (2014). An introduction to cyber war games. *CIO Journal* http://deloitte.wsj.com/cio/2014/09/22/an-introduction-to-cyber-war-games.

9 DoDI O-85301. (8 January 2001). Computer Network Defense (CND).

10 DoD 5200.1-R. (January 1997) Information Security Program.

11 DoDI 8500.01.(14 March 2014). Cyber Security.

12 DoDI 8500.01E.(23 April 2007). Information Assurance (IA).

13 DoDI 8500.2. (6 February 2003). Information Assurance (IA) Implementation.

14 DoDI Risk Management Framework (RMF) for DoD Information Technology (IT) 8510.01.(12 March2014).

15 Federal Information Security Management Act (FISMA) of 2002.

16 Fox, D., McCollum, C., Arnoth, E., and Mak, D. (2018). Cyber wargaming:framework for enhancing cyber wargaming with realistic business context. *MITRE*.

17 Johnson, A. (2019). *CANS: Cyber Wargaming Pre-Requirements*. Lockheed Martin Internal Document, Lockheed Martin.

18 NISTSpecial Publication (SP) 800-53 Revision 4;, 2013

19 Perla, P., *Art of Wargaming: A Guide for Professionals and Hobbyists*, 1990

20 Perla, P., et al.(2002). Wargame-Creation Skills and the Wargame Construction Kit.https://www.cna.org/cna_files/pdf/D0007042.A3.pdf

21 The MITRE Corporation. (2009). One Step Ahead: MITRE's Simulation Experiments Address Irregular Warfare,https://www.mitre.org/publications/projectstories/one-step-ahead-mitres-simulation-experiments-address-irregular-warfare

22 The MITRECorporation. (2015). Adversarial Tactics, Techniques, and Common Knowledge (ATT&CK). https://attack.mitre.org/wiki/Main_Page

23 Strom, B. et al. (2017). *Finding Cyber Threats with ATT&CKTM-Based Analytics,"* MTR 170202, PR 16-3713. The MITRE Corporation.

24 UcedaVelez, T. and Morana, M.M. (2015). *Risk Centric Threat Modeling: Process for Attack Simulation and Threat Analysis.* Wiley.

25 Williams, G. and Kam, A. (2016). *LM Cyber Effects Wargame.* Lockheed Martin Internal Document, Lockheed Martin.

11

Using Computer-Generated Virtual Realities, Operations Research, and Board Games for Conflict Simulations

Armin Fügenschuh[1], Sönke Marahrens[2], Leonie Marguerite Johannsmann[3], Sandra Matuszewski[4], Daniel Müllenstedt[5], and Johannes Schmidt[1]

[1] Brandenburg Technical University Cottbus, Senftenberg, Germany
[2] German Institute for Defense and Strategic Studies, Hamburg, Germany
[3] German Air Force, Wunstorf, Germany
[4] German Air Force, Kalkar, Germany
[5] German Air Force Command, Cologne, Germany

Introduction

In the years 2014–2017, one of the authors (A. Fügenschuh) organized four trimester-long interdisciplinary courses at the Helmut Schmidt University/ University of the Federal Armed Forces Hamburg on Simulation and Wargaming. One used "classical" conflict simulation board games (commercial board games) [6], three were based on electronic simulations: First we used the game "Command: Modern Air/Naval Operations (C:MA/NO)" [4] and then two times the professional simulation environment "Virtual Battlespace 3 (VBS3)" [3, 7, 8]. The suggestion for these courses came from the Defense Planning Office of the Bundeswehr in Ottobrunn (S. Marahrens). Computer-aided analysis tools with different resolutions had already been in use there for various questions. The new development of such a system from scratch, being tailored to the special requirements of the Office, would have required a high financial outlay and a corresponding lead time. On the other hand, there were programs on the market that could be immediately downloaded and installed for comparatively little money

Simulation and Wargaming, First Edition. Edited by Charles Turnitsa, Curtis Blais, and Andreas Tolk.

(less than 100 Euros). An example of this is the program Command: Modern Air/ Naval Operations (Version v1.07, Build 678) (short: C:MA/NO) of the company Matrix Games, which can be attributed to the field of serious gaming.

The use of off-the-shelf wargames was also analyzed in [2]. C:MA/NO focuses on operations with larger entities, mostly ships and aircrafts, but it can, with limitations, also be used for the modeling of the behavior of individual soldiers. The software Virtual Battlespace 3 (VBS3, version 3.8.1) of the manufacturer Bohemia Interactive (BI) in contrast focuses on the perspective of individual soldiers, and shows the simulated world through their eyes in a way known from so-called first-person ego-shooters. The software is used by multiple armed forces including the German Bundeswehr as a training tool. In both systems, some or all of the other entities can be controlled by the software, and their behavior can be attributed as "artificially intelligent," in the sense that it follows rules that are at first unknown to the player (unless he devotes his time to study it with the purpose of extracting these rules) and it also takes randomly chosen decisions (based on computer-generated pseudorandom numbers). When it comes to decisions on the usage of his own troops, the user only relies on his experience that he gathered from many runs of the simulation environment, or put differently, he is left alone.

Consider the particular case of a UAV mission planning problem, which can be simulated in both C:MA/NO and VBS3. It is desired to plan the flight trajectories for an inhomogeneous fleet of unmanned aerial vehicles (UAVs) visiting all or at least a largest possible subset from a list of desired targets. When selected, each target must be traversed within a certain maximal distance and within a certain time interval. The UAVs differ with respect to their sensor properties, speeds, and operating ranges. If the targets are surrounded by radar surveillance, then the UAVs' trajectories should be chosen to avoid these forbidden areas. The fuel consumption rates during cruise (which are depending on the actual speed and the altitude level), climb, and descend are crucial and thus need to be considered as well. When solving this problem manually, the user of a simulation software is similarly overwhelmed by its complexity as a human UAV operator in real life. The simulation in C:MA/NO or VBS3 only allows the experimentation of different orderings for the targets, and from the outcome the user has to deduce if the mission goals were accomplished or not. At this point, artificial intelligence methods can support the user. When formalizing the UAV mission planning problem (using an algebraic language to represent the operational and technical constraints), it comes into the reach of exhaustive search methods that are able to scan the entire solution space and return a proven global optimal solution. While some users may enthusiastically use such support, others have objections to rely on such artificial intelligence (AI) methods. For the latter, we developed a simplified version of the planning problem as a board game, which can be used in the training of UAV mission planners and UAV operators. Initially, they have to plan the mission manually all by themselves, and then the help of computers and AI can be introduced as a second step. This fosters to gain confidence, even without the demand of having fully understood the underlying mathematical methods.

Public Software (C:MA/NO)

In 2015, a group of seven students analyzed the software Command: Modern Air/Naval Operations (C:MA/NO) [4]. C:MA/NO is a commercial game that has a long developing history (starting as "Harpune" in the 1990s on a Commodore Amiga), albeit a still active and loyal player community. A professional version is also available (at a much higher price), which allows to interact with the program core routines via APIs, which is necessary for automated runs and large computational studies. For our seminar and tests, the gaming version was handed out to the course participants. The students were asked to design their own scenarios, based on conflicts they recently observed by international media coverage. Data needed to describe these conflicts and to model them within the C:MA/NO environment had to come from public sources.

The C:MA/NO main screen shows a map of the conflict theater like Google Earth View. The user can either decide to play a given mission, or may design his own missions. In the latter case, a certain geographical theater must be chosen, in which the ground, air, and naval facilities and units have to be inserted from a given database. This database contains pre-modeled objects from several armed forces worldwide and their materials from the past (starting around 1950) to modern days, and also to use future equipment not yet introduced in the respective forces. As the title announces, the focus is primarily on naval and aerial equipment. The underlying values and data sets for the equipment demonstrate a high level of detail implemented by the developers, programmers, and data collectors. For example, aircrafts use a different amount of fuel depending on their flight altitude and velocity. Despite having a detailed database, it was analyzed by the students if these data sets are connected with a corresponding physical model that is operating on an adequate level of a similar accuracy. Because of its high degree of details C:MA/NO can be considered as a serious game, a term that refers to a certain type of computer games being close to a realistic simulation. The richness of detail does not necessarily refer to the graphical representation (which in C:MA/NO can be described as rather abstract or primitive in terms of today's standards), but to the mapping of the interacting objects in the simulated environment and with the player. Any number of conflicting parties can stand side by side or against each other, where one of it is usually controlled by the player. An AI system inside

C:MA/NO assumes the role of the other parties. Furthermore, if necessary or wished, the AI can assist the player in controlling his own units. Now, is C:MA/NO just a richly detailed (serious) game, or is it also a way of simulating conflicts? This was the basic question posed to our group of students. The participants found quite different answers. Based on their individual observations of selected scenarios provided and supplied by the C:MA/NO manufacturer, they first engaged on the control of the system attempting to challenge the simulation engine. In a second round, they created their own scenarios and experiments. Some of these were designed to understand and compare physical and technical aspects of a single system or component of a system within C:MA/NO. Other participants created complex scenarios in which a large number of own and opposing units faced each other in order to test the control and command capabilities in C:MA/NO.

C:MA/NO turned out to be as persuasive as the program title suggests, especially in air and sea maneuvers. Many physical and technical phenomena are captured in the simulation and are reproduced correctly, albeit with simplified models. The AI to control one's own or opposing entities makes most of the time comprehensible decisions. However, when taking a deeper look, each student was also able to detect weaknesses in the provided models in the course of his/her work, so that C:MA/NO should not be used uncritically and unchecked for real-world problems. C:MA/NO faces its biggest problems when being "abused" for the simulation of land operations. Here other tools, such as VBS3, are a better choice. Considering the short training time and the predominantly correct presentation and behavior of the systems, C:MA/NO can be recommended, if a moderate simulation benefit, e.g. grasping a military problem, is to be achieved with little expenditure of money, time, and personnel. For more advanced, refined analysis (which will require more resources), and for land operations, one has to choose other simulation environments.

The scenario designed by C. Vosseberg ([4], Chapter 4) was motivated by public reports from 2014 on the approaches of foreign (Russian) bomber squads near the NATO airspace in Norway. In order to set up a scenario in C:MA/NO that matches to the real-world situation, a chain of nine Norwegian radar station were placed along the coastal line, together with two air force bases where F-16 interceptors are based (BLUFOR). Their geocoordinates are available on Wikipedia.

The Russian Air Force (OPFOR) uses TU-95 long-range bombers and Il-78 refueling aircrafts. When an aircraft is running low on fuel, C:MA/NO automatically assigns it to the nearest airport, so that it would not crash. Hence in order to prevent OPFOR from flying over NATO airspace when low on fuel, a no-fly zone covering the NATO–European land masses had to be installed. The TU-95 were sent on an automatic mission of flying along the Norwegian coast from North to South. When running out of fuel, they now simply crash. Hence with some trial and error, Il-78 were started to meet the TU-95 at the right locations for refueling operations.

When OPFOR is detected by one of the radar systems, the BLUFOR player has to launch interceptors to follow the OPFOR bombers in short distances, adjusting their flying altitude and speed accordingly. When their fuel is largely consumed, they will automatically return to their home bases; it is possible to turn off this feature, so a more advanced player needs to decide himself on the latest point in time for a safe return.

By playing this scenario, the player–operator experiences that flight altitude and speed have a major impact on the range of his interceptors. The database offers different fuel consumption values for several discrete altitude and throttle bands. Flying at the lowest possible speed in the highest altitude consumes the least amount of fuel, whereas flying fast and low has the opposite effect. The underlying physics are properly embedded within C:MA/NO's fuel computation engine. Transferring the knowledge about these effects can also happen during a theoretical lecture on flight physics. The practical consequences on a given mission, however, are best experienced during a game playing process. When one group of interceptors returns, the BLUFOR player must have ready the next interceptors, so that a gap-free surveillance can be guaranteed.

To include some variance in the gameplay, it is possible to randomly select the altitude of OPFOR's bombers. Since the simulated radar rays to not follow earth's curvature, the detection of an incoming aircraft is expected to be later when flying at lower levels. This is properly modeled in C:MA/NO. In turns out that on very low levels it is possible to use gaps in the Norwegian radar chain, allowing to enter the country undiscovered. C:MA/NO allows to plan such fuel optimal missions for OPFOR, where a fuel-saving high altitude is chosen far away from the coastline (and the radar stations), and a lower altitude is selected when approaching the mainland.

User-Tailored Software (VBS3)

The next group of students received VBS3 for examination. BI claims on its website: "Virtual Battlespace 3 (VBS3) is a flexible simulation training solution for scenario training, mission rehearsal, and more." The predecessor VBS2 was already in use at various points in the Federal Armed Forces (Bundeswehr), but especially for training purposes. It has also been used by many other armies around the world, such as the British and the US forces. In training, AI is not needed because all living entities in the simulation are controlled by real human players connected via a computer network. The physical and technical characteristics of the systems and the interaction of the players (and their avatars) with the environment are an important factor in terms of realism and thus the usability of such simulations. VBS3 provides a graphical representation of the virtual environment in the current state of computer technology and therefore requires modern hardware equipment with fast graphic cards to achieve a sufficient frame rate. It complements C:MA/NO in the sense that it particularly focuses on land operations and assumes the perspective of a single soldier acting alone or as part of a group (company size or below).

L. Pätzold ([7], Chapter 3) analyzed the capabilities of VBS3 with respect of planning a drone mission. VBS3 has a database of about 40 different UAVs from armed forces around the world. It is possible to automatically launch a UAV. In order to control a UAV, it is necessary to deploy a ground control unit (GCU) and additionally establish a connection link between the UAV and its GCU. After that, it is possible to control the UAV manually by pushing the corresponding keys on the keyboard. The simulated optics are able to take photographs with a zoom while focusing on a ground object. It is also possible to give commands to a UAV, which are waypoints that the UAV has to traverse. Then the UAV will fly to these points in the designated ordering. Damages to the UAV are also simulated in a realistic manner. VBS3 distinguishes in those cases between various parts of the UAV, for example, damages to the instruments or damages in the engine. They result in disturbances of the flight behavior or malfunctions of the instruments, and are visually expressed by the formation of smoke around the UAV. A severe damage may result in a loss of the UAV, which leads to a signal loss at the GCU.

Artificial Intelligence for Solving Tactical Planning Problems

A drawback of the UAV simulation in VBS3 is the AI system that is only capable of directing the UAV automatically from waypoint to waypoint on a list that is specified manually in advance by the human user. There is no option to let the computer decide or assist on the ordering of these waypoints. However, the problem of first assigning waypoints to a fleet of UAVs and then deciding on the ordering is a very common one in reality, and the absence of a method that guides the user in both VBS3 and C:MA/NO initiated our research for developing an intelligent method that is capable of filling this gap.

We developed an algebraic formulation of the following UAV mission and trajectory planning problem. The model is well-suited for short-range UAVs whose operational radius is about 200 km and which can remain airborne for 8–12 hours. These UAVs are mostly used for surveillance and reconnaissance missions. Because of the range limitation, it is not necessary to take the curvature of the earth into account. It is enough to work on a flat projection of a fraction of the sphere that represents the region of interest.

Given is a set of UAVs, which can be inhomogeneous. Each UAV is represented in the model by its technical parameters: minimum and maximum velocity (true air speed) for horizontal flight maneuvers, limits on the maximum climb and descend rate, a maximum acceleration, a minimum and maximum flight altitude, and a fuel capacity. The fuel consumption rate depends on the velocity, the flight altitude, and the climb rate. Furthermore, a minimum safety distance between two airborne UAVs to avoid collisions is specified. The UAVs are radio-controlled (UHF/VHF band), so they must fly within a certain range to the ground control unit.

A mission is specified by the following parameters: the locations where the flight of UAVs start and end (which may be different spots) and the location of the ground control unit. A set of waypoints is given that need to be covered by one of the UAVs flying by. The relative importance of a waypoint is expressed by a score value. Each waypoint has a time window, in which a coverage has to be carried out, otherwise it would be invalid. Depending on the sensors on board a UAV and the desired image quality, a maximal distance for each waypoint and each potential UAV is determined. A set of restricted airspaces (no-flight zones) can be specified to model rectangular regions where the UAV is not allowed to enter. They represent mountains, cities, adversary radar or air defense systems, or blind zones with no radio connection to the ground control unit. As a further environmental parameter, a wind vector can be specified.

The model is formulated as a mixed-integer nonlinear programming problem. It takes into account the following constraints: The UAVs must start and end their flight trajectories at the desired locations, and they must stay within the range to the ground control unit(s). Once they are airborne, they must fly above the minimum and below the maximum altitude. The flight dynamics are governed by Newton's law of motion: The acceleration equals the first derivative of the velocity and also equals the second derivative of the location. The position of each airborne UAV is updated according to the formula, where the influence of wind is additionally taken into account. The underlying avionic model is that of the TC–TH–W vector triangle, where TH is the true heading vector, W is the wind vector, and TC is the true course vector, resulting from a vector addition TC = TH + W; for more details on this we refer to [9]. The horizontal velocity for airborne UAVs is limited by the given lower and upper bounds; the horizontal acceleration is bounded from above. To claim a waypoint by a certain UAV, it must fly by within a certain maximal distance; only one UAV can claim a waypoint. Restricted airspaces are modeled by additional binary variables that exclude a UAV from flying inside these spaces. For each three-dimensional rectangular restricted airspace, six binary variables are needed to check in which relative spatial position (left or right, below or above, in front of or behind the airspace) the UAV is situated. Collision avoidance is modeled in a similar way, here each UAV defines its own moving restricted airspace centered around it, and no other UAV is allowed to enter it. The fuel consumption is modeled by an ordinary differential equation taking into account the flight altitude and velocity. Here further binary variables are needed to find out within the model formulation in which altitude and velocity band the UAV is operating. The objective function has several components in it: The most important goal is to maximize the score for reaching the waypoints. On a subordinate level it is desired to finish the mission as soon as possible with the least amount of fuel flying in high altitude at low speed. Details of the algebraic model formulation can be found in [5]. Test data for this model was taken from C:MA/NO's library, which contains performance models for numerous UAVs.

The mathematical formulation of the UAV mission and trajectory planning problem falls into the category of mixed-integer nonlinear problems with ordinary differential equation constraints. Problems in this category are notoriously difficult to solve from a practical point of view (and NP complete from a theoretical point of view). Thus one cannot expect a kind of "simple" algorithm that is able to come up with optimal solutions after a short amount of computation (CPU) time. Instead, the solution of such problems is a highly active area of ongoing research. The continuous flow of time in the real world is discretized in the model by time steps. This transforms the infinite-dimensional problem to a finite-dimensional one. The differential equation constraints are discretized with a scheme from the literature. We tested several finite difference schemes, such as Euler, Euler–Cromer, Halfstep, Verlet, Leapfrog, Beeman–Schofield, Midpoint, and the second- and fourth-order Runge–Kutta methods. It turns out that the second-order Runge–Kutta is the best choice in terms of CPU time and accuracy of the solution. There are several nonlinear constraints, for example, the maximum distance between the UAV and the ground control unit is limited. Geometrically, this defines a circle around the GCU that the UAV must not leave. This nonlinear constraint is linearized by a piecewise linear reformulation. In the end, a finite-dimensional mixed-integer linear programming problem is obtained, which can be solved with numerical solvers, such as IBM ILOG CPLEX, GUROBI, or XPRESS. Such solvers apply an intelligent search strategy in the search for the global optimum using a divide-and-conquer approach. The search procedure is significantly accelerated by automatically exploiting knowledge about the problem structure, and multiple artificial intelligence strategies were developed to this end over the last decade, see [1] for a survey. Using these methods, the solvers are able to solve instances with two to three UAVs and up to 30 waypoints to proven global optimality within one hour on a modern laptop or desktop PC. A sample output is shown in Figure 1. For a detailed computational study we refer to [5].

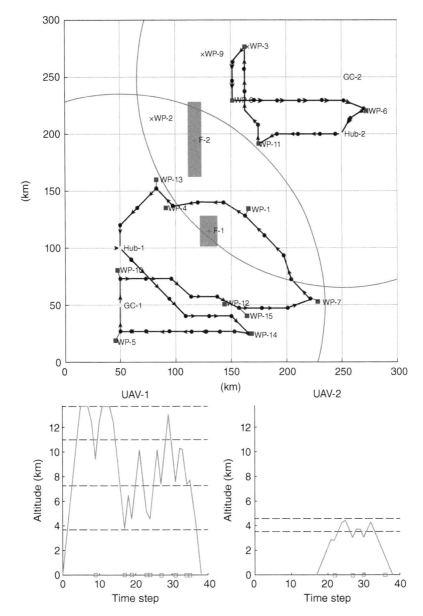

Figure 1 A solution to a test instance. (a) Map with two UAV trajectories (horizontal view), covering 13 of 15 waypoints. (b) Vertical view of same trajectories.

Wargaming Support

Having proven the global optimality of a solution at hand is only one part of the story, and in fact, it is not even its most important side. When demonstrating the capabilities of the above model and the solution method to experienced users (from a German battalion that regularly operates with UAVs), they were reluctant to use such a software as a black-box, whose internal "intelligence" may solve the problem but does not answer the question, why the given solution is in fact optimal, and how it was determined. To circumvent this opinion, we created a wargame in the form of a board game. This wargame can be played first without the support from a computer, so that the human operator can test his or her own ability and creativity in solving the problem. In a second step, a computer support is offered, and the operator can compare his or her decisions with that of the computer. Although the computer still operates as a black-box and gives no explanations for its moves, this approach seems to create some more appreciation for the help offered. In order to make the experience more interesting and viable, the game sets up a dynamic environment in which irregular forces driving in a vehicle must be detected by a UAV within a certain time frame.

The game implements a scenario familiar from the ISAF mission in Afghanistan. Again and again, irregular forces had made attacks there, and despite the clues and warnings from the intelligence services and the military intelligence, they rarely succeeded in preventing these attacks. In most cases, it was only possible to specify with certain probabilities where the attackers were approaching the attack target. As a result, it has always been sufficiently complex to detect the irregular forces before they reach the point of attack and, as a result, prevent them from carrying out the attack.

The game is designed as a board game, where a region of interest is discretized by hexagonal tiles. Each tile represents either a road, a city, or open/rural area. The game player takes the role of a drone operator of a single UAV, a LUNA drone. Due to safety regulations, it is only allowed to fly over road tiles and open/rural area tiles, whereas city tiles and airfield tiles are forbidden, see Figure 2a. The LUNA system is linked to a single ground control unit (GCU) with an operational range of 100 km. Since the area is flat, the UAV will never lose its contact to the GCU. Thus the GCU can be placed anywhere, and there is no necessity to move the GCU around later. In reconnaissance mode, the UAV can fly with 70 km/h. It is assumed that the irregular forces vehicle drives at 35 km/h, half the speed of the UAV. The goal of the player/UAV operator is to detect the irregular forces as early as possible, and thus prevent their successful attack. We assume that their destination is known (a bridge in hex tile 106 in Figure 2a), but that it is not possible to determine the path on which the irregular forces approach this goal. Since it is known that they use a vehicle, they have to take one of three possible road options. These options (A, B, C in Figure 2b) are shown on the umpires'

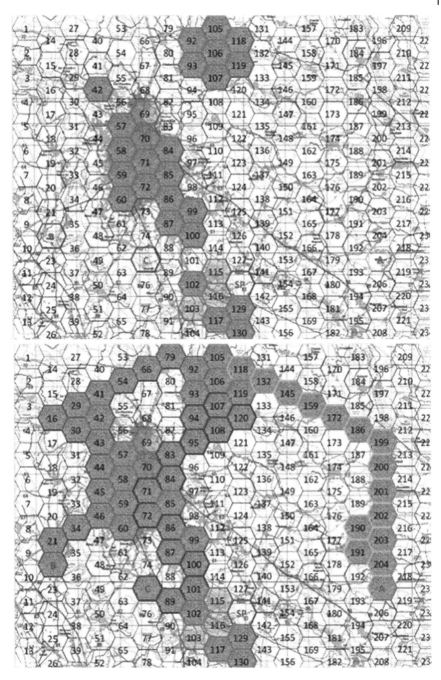

Figure 2 The playfield with its hex tiles. (a) Player view; dark tiles = no-fly zones; SP = start position of UAV; A, B, C = start of irregular forces. (b) Umpire view; potential roads A (blue), B (purple), C (green).

game map, who takes control of the irregular forces during the game play. The respective starting points A, B, C are also shown on the players' map in Figure 2a, but not the tiles that belong to the roads. It is part of the learning experience of the player to conclude, that flying along the road pays off most, whereas flying over non-road tiles should be avoided as much as possible.

When the game starts, the UAV player puts the drone on the SP tile. The umpire decides by rolling a die according to the given probabilities where the irregular forces actually start (and thus which path they take), and places them on his own map, which is hidden to the player. Within the wargame, time evolves in discrete chunks ("game rounds"). A round consists of the following actions: The player can move the UAV twice. Each of these two moves either ends in an adjacent tile to where it started, or the UAV can stay in the current tile (which means, the UAV is in loitering mode). After each move, the UAV observes the ground. Being a partly cloudy day, the sight is not always clear, but only in 50% of the cases. In the game this means that the umpire now tosses a die. If it shows 1, 2, 3, the player gets the message: clouds, no detection possible, and at 4, 5, 6 the player is informed that there are no clouds and a clear view to the ground is possible. Furthermore, he is informed about what to see: If the UAV is on the tile where the irregular forces are, then the player has fulfilled the mission successfully. Otherwise he is informed by the umpire that there are no irregular forces and the game continues. After the player has made two such moves, the umpire moves the irregular forces on his private map one tile closer to their destination (the bridge in tile 106). The game ends immediately when the irregular forces are detected by the player, or the irregular forces reach one of the tiles adjacent to tile 106, in which case the player has lost.

The game is played over at most 14 rounds, thus the player can make 28 decisions about how to move. When he is informed about clouds blocking the view, he needs to decide if his UAV should remain in the current position, having potentially more luck next time, or if he wants to move on and search somewhere else. In a certain way, this simple board game resembles the key ingredients of the real-life operational problem. In training sessions with real drone operators (on various levels of experience), this game was described as helpful, instructive, and also as realistic. When playing the game several times, one can see that the unexperienced newcomers behave more tactically from game to game.

In parallel to the game play, an operations research expert was feeding the game data as input to an optimization routine of an adapted version of the model described in Section Artificial Intelligence for Solving Lactial Planning Problems to the setting of the board game: The flight trajectory is aligned to the hex tiles, and there is a probabilistic model for the coverage of the waypoints, which are now all tiles of all three roads. For further details, we refer to Ref. [10]. The optimal solution of this model describes a trajectory for the UAV that leads to the highest probability for detecting the irregular forces; such trajectory has a detection probability of 88%. If the human player has no reconnaissance (and thus the game ended with a trajectory of 28 decisions), it is possible to evaluate this trajectory using the same mathematical model by simply fixing the discrete variables to the human decisions. In one such case using experienced UAV operators as players, it turned out that their trajectory has a reconnaissance probability of 53%. When revealing this difference to the player after having played several rounds, an understanding and appreciation for computer generated solutions was achieved. More details on the design of the experiment can be found in Ref. [11].

Conclusion

We analyzed two computer-based conflict simulation tools: The publicly available serious game C:MA/NO, and the army-tailored VBS3. Both allow the user to define two or more factions of a conflict, and endow them with military units, such as ships, aircrafts, and building structures (airfields, radars, defense systems). C:MA/NO allows the user to select one conflict party as his own and takes over the control of all others. The user can decide how much support he gets for his own troops. He can either micromanage, or put parts of the control also to the gaming AI routines. The tactical routines are considered as realistic by the group of students who assessed the game, and also the physics is represented correctly. VBS3 works similarly, but from the ego-perspective of an individual soldier. Here the AI takes control of other soldiers (or civilians). Here the students were observing many times an unrealistic behavior by the AI control, and when looking behind the effect of an impressive graphic, it became obvious that this system is not capable of simulating conflicts on the level of individual soldiers. A human observer is highly critical and suspicious when it comes to judging the behavior of other humans, in particular, when these other humans are simulated by the computer, an effect known as "uncanny valley" [12]. In C:MA/NO, this effect cannot be observed, since the abstraction level hides the view on the individual person.

However, in both simulation systems the user has still to take tactical and operational decisions on his own. As an example, we discussed a UAV mission planning problem, where a number of waypoints with time windows for their visits are given, and a fleet of UAVs has to be deployed to visit as many waypoints as possible during the specified time windows. Here the user of the software has to find a suitable ordering himself, and no help from the system is provided. One may of course argue that in C:MA/NO such tasks belong to the purpose of being a game and add to the fun of the game. For a realistic simulation there should be some computer-based intelligence that helps the operator defining the mission.

It was thus our goal to develop a method that solves the UAV mission planning problem. This method is based on a mathematical formulation of the problem, so that it becomes accessible for numerical solution methods, which are based on intelligent search techniques that are able to find global optimal solutions (at least, for instances having at most 20 waypoints and two UAVs). When presenting these solutions to drone operators, we experienced that they refrain from using them right away. To overcome their reluctance, we developed a board game, which is a simplified version of the UAV mission planning problem. For instance, fuel calculations were taken out, because it is a board game that is a simplified version of the UAV mission planning problem. Because it is a board game with clear and easy rules (which can be stated on less than one page), we believe that the user is better able to understand and appreciate the game mechanisms. Although it is possible to implement that board game one-to-one as a computer game, we believe that such implementation would again create a barrier between the player (who is more a learner at the time).

What we implemented instead was a method (an algorithm) that solves this particular version of the mission planning problem. It allows to compare the player's result with an optimal solution, so that the player can see what he could do better, or where he took the wrong decision that led to a lower probability for a successful mission accomplishment, although there is still no explanation given, why the computer generated solution is optimal. We think that this mixture strategy of a board game simulation and a optimization routine (based on intelligent search techniques to find optimal solutions) is able to close the gap between current optimization-free computer-simulations (C:MA/NO, VBS3) and real-world military planning problems on a strategical and tactical level. The approach of using serious gaming simulation allows a first quick allows a first quick assessment in order to solve an operator's task to some extent, instead of just being told by an expert that the modeling of the problem. "modeling of the problem might last at least 6–9 month." But it creates also enough complexity beyond the operators' gut feeling capability in order to discuss the necessity of a follow-up in-depth research, thus bridging the world of operators and experts through the use of serious gaming.

References

1 Bengio, Y., A. Lodi, and A. Prouvost. Machine Learning for Combinatorial Optimization: a Methodological Tour d'Horizon. *Technical report*, arXiv:1811.06128, 2018.

2 Curry, J., T. Price, and P. Sabin. Commercial-Off-the-Shelf-Technology in UK Military Training. *Simulation & Gaming*, 47(1):7–30, 2016.

3 Fügenschuh, A. Fallstudien zur Rechnergestützten Konfliktanalyse am Beispiel von Virtual Battlespace 3. *Technical report*, Angewandte Mathematik und Optimierung Schriftenreihe AMOS#55, Helmut Schmidt University/University of the Federal Armed Forces Hamburg, 2017.

4 Fügenschuh, A. and D. Müllenstedt. Rechnergestützte Konfliktanalyse am Beispiel von Command: Modern Air/Naval Operations. *Technical report*, Angewandte Mathematik und Optimierung Schriftenreihe AMOS#36, Helmut Schmidt University/University of the Federal Armed Forces Hamburg, 2015.

5 Fügenschuh, A., D. Müllenstedt, and J. Schmidt. Mission Planning for Unmanned Aerial Vehicles. *Technical report*, Cottbus Mathematical Preprints COMP#6, Brandenburg University of Technology Cottbus-Senftenberg, 2019.

6 Fügenschuh, A. and F. Scholz. Bewaffnete Konflikte: Geschichte, Dynamik, Simulation und Analyse. *Technical report*, Angewandte Mathematik und Optimierung Schriftenreihe AMOS#43, Helmut Schmidt University/University of the Federal Armed Forces Hamburg, 2016.

7 Fügenschuh, A. and I. Vierhaus. Rechnergestutzte Konfliktanalyse am Beispiel von Virtual Battlespace 3. *Technical report*, Angewandte Mathematik und Optimierung Schriftenreihe AMOS#47, Helmut Schmidt University/University of the Federal Armed Forces Hamburg, 2016.

8 Fügenschuh, A., I. Vierhaus, S. Fleischmann, T. Löffler, T. Diefenbach, U. Lechner, K. Knobel, and S. Marahrens. VBS3 as an Analytical Tool - Potentialities, Feasibilities and Limitations. *Technical report*, Angewandte Mathematik und Optimierung Schriftenreihe AMOS#49, Helmut Schmidt University/University of the Federal Armed Forces Hamburg, 2016.

9 Jameson, T. A Fuel Consumption Algorithm for Unmanned Aircraft Systems. *Technical report*, Army Research Laboratory ARL-TR-4803, 2009.

10 Johannsmann, L. Optimierte Flugroutenplanung des Wargames Enhanced LUNA Warrior. *Technical report*, Angewandte Mathematik und Optimierung Schriftenreihe AMOS#52, Helmut Schmidt University/University of the Federal Armed Forces Hamburg, 2017.

11 Matuszewski, S. Bewährtes zu Neuem verknüpfen – Wissenschaftliche Methoden für die Streitkrafte des 21. Jahrhunderts. *Technical report*, Angewandte Mathematik und Optimierung Schriftenreihe AMOS#64, Helmut Schmidt University/University of the Federal Armed Forces Hamburg, 2017.

12 Mori, M., K. F. MacDorman, and N. Kageki. The Uncanny Valley. *IEEE Robotics & Automation Magazine*, 19:98–100, 2012.

Part V

Emerging Technologies

12

Virtual Worlds and the Cycle of Research

Enhancing Information Flow Between Simulationists and Wargamers

Paul Vebber and Steven Aguiar

Naval Undersea Warfare Center, Newport, Rhode Island, USA

In both business and military contexts, simulation and wargaming often serve the same master: understanding a competitive decision space and how to achieve goals within it. Yet practitioners of each bring different sets of beliefs, assumptions and tools to bear in the effort to, "comprehend and cope," as Col John Boyd characterized the struggle to survive amid competitors[1].

Simulationists are keen empirical observers striving to comprehend and navigate competitive landscapes, used here more generally than typical in business analysis[2], as an analogy between trying to understand complex multidimensional interactions and the idea of visualizing movement over a topographic map. There are "hills" where advantageous circumstances lie, and likewise, disadvantageous "valleys." Ensuring we have the high ground – an advantageous position relative to our competitors – requires understanding both how to recognize positions of advantage, as well as what it takes to get you there. To find paths to the high ground, simulationists use bits – creating and using computer or mathematical models, typically iterated over time, to gain insight into the character of a competitive landscape.

Wargamers are storytellers at heart and architects of coping strategies. They use wits – acting as narrators and editors of an emergent story written by the players

---

1 John Boyd: Destruction and Creation, unpublished paper, 1976, available at: http://danford. net/boyd/destruct.htm
2 "A form of analysis that helps a business identify its primary online and offline rivals."http:// www.businessdictionary.com/definition/competitive-landscape.html

Simulation and Wargaming, First Edition. Edited by Charles Turnitsa, Curtis Blais, and Andreas Tolk.

about the nature of what the struggle over high and low ground might look like. They foster the creation of hypotheses regarding relationships and interactions with each other and their real-world analogs. Accordingly, they are expert at facilitating a dialectic conversation between players through the shared experience of gameplay.

The complementary relationship between the two sets of practitioners is more often seen as a dichotomy, with the simulationists garnering more respect and billions more dollars. This is particularly true in the US military. Despite the art of war being taught using games back to antiquity and through "kriegspiel" since its advent in the early nineteenth century[3], the siren song of the computer simulation lead to wargaming being nearly abandoned but for a few think tanks and boardrooms. The rise of computers during the early Cold War period led to a situation where the "problem" was static for the most part and shrouded in the cloud of potential nuclear Armageddon. The nature of the competitive landscape was thought to be well characterized. The West looked to technological quality as the foil to Soviet weight of numbers. The effect of myriad variables on the outcome of battles on the actual landscape of central Europe where the meat and potatoes of simulationists and their ever more complex models, algorithms and simulations. The wargamers continued in the traditional role of training headquarters' staff for battle, but the increasingly bureaucratic community that made decisions on weapons to buy was mesmerized by the simulationists whose comparative tools provided firm evidence regarding the advantages of "A vs. B" purchasing decisions. This justifiably trusted set of comparative tools increasingly passed itself off as predictive and ventured precariously far onto analytic thin ice[4].

Then the Cold War ended and the era of analyzing a decades-old, infinitely dissected problem ended with it. For a moment there was celebration that the era of such problems was over, but a competitive landscape never remains devoid of competitors forever. It took a couple of decades, but in the mid-2010s a new crop of competitors, both global and regional, became problematic. The competitive landscape had changed and in 2014, US Secretary of Defense Chuck Hagel penned a memo on a Defense Innovation Initiative that called for a reinvigoration of wargaming as a cornerstone in the effort to understand the new competitive landscape and re-establish control of its high ground[5]. Hagel's deputy, Robert Work, penned a memo to the senior leadership of the Department of Defense

---

3 See Mathew Caffrey "On Wargaming"; U.S. Naval War College Newport Paper 43 for history of wargaming: https://digital-commons.usnwc.edu/newport-papers/43/
4 One anecdote of many illustrating the problems using military simulations to predict conflict outcomes. https://www.theatlantic.com/technology/archive/2017/10/the-computer-that-predicted-the-us-would-win-the-vietnam-war/542046/
5 https://dod.defense.gov/Portals/1/Documents/pubs/OSD013411-14.pdf

describing in more detail what this reinvigoration of wargaming would look like.[6] Some simulationists saw this as a loss of confidence in their discipline, exacerbating the strain in the relationship with wargamers, while others re-enforced the idea as a return to the balance between wargaming, operations analysis, and live exercises of the past[7]. This "cycle of research"[8] provides a framework that can help inform how each type of practitioner approaches the problems related to understanding competition to inform decision-making. The structure it provides can help map out the types of information that must pass between them and can help with understanding the roles of simulationists and wargamers in achieving success, how they are often misunderstood by each other, and in particular, how emerging technology can help bring them closer together.

The Cycle of Research as a Communications Framework

If the simulationists are driven to understand the details of a competitive landscape's topography and how to navigate it, then wargamers are driven to contemplate what the idea of the map itself represents and how one derives meaning from it. Without simulationists to ground their abstract representations, wargamers' theoretical destinations can never be found. Without wargamers to inform what constitutes an advantageous destination, simulationists will wander the landscape gaining increasingly accurate representations of it, but without any idea of where it is they should be going or why. Together they form a formidable team that can iterate information back and forth between them about what good ground is, and how to most efficiently get from one such place to another. This relationship extends to the conduct of experiments and exercises to form what Dr. Peter Perla termed the Cycle of Research. Breaking exercise and experimentation out separately and defining inputs to and outputs from it, has been termed the "Cycle of Assessment[9]." In either case, as our analogy above helps to shed light on the complementary nature of what simulationists and wargamers do, it is the contention of the authors that the emergence of methods and technologies that create

---

6 https://news.usni.org/2015/03/18/document-memo-to-pentagon-leadership-on-wargaming
7 See CDR Phil Pournelle, "Preparing For War, Keeping the Peace", U. S. Naval Institute Proceedings, September 2014, Vol 140/9/1339: https://www.usni.org/magazines/proceedings/2014/september/preparing-war-keeping-peace
8 Dr. Peter Perla: *The Art of Wargaming*, March 1990 and May 2012
9 Paul Vebber: *Building Boyd Snowmobiles: Matrix Games as a Creative Catalyst for Developing Innovative Technology* in Curry, Engle and Perla (editors): *The Matrix Games Handbook: Professional Applications from Education to Analysis and Wargaming*, History of Wargaming Project, 2018

interactive virtual spaces can quite literally put simulationists and wargamers together, face-to-face, in the same competitive landscape, but now this landscape is further evolved into a *collaborative* landscape. Furthermore this landscape can now be adaptive to the "abstract representations and theoretical destinations" called for by simulations and wargamers, creating worlds that support problem genesis, landscape evolution, and culminating in the ability to inform decision-making.

Dr. Perla's cycle (Figure 1) depicts an interactional relationship between wargames, analysis, and exercises. Wargames deal with human decision-making and exercises with human activity. Analysis connects the two by providing information on the consequences of decision-making and the efficacy of activity. It implies a need for an effective collaborative environment where wargames can be played, exercises can be planned and rehearsed, and the information involved collected, analyzed, and used to inform the models and simulations that extrapolate from and form analogies to that competitive landscape (i.e. the representation of *Reality*) described above.

Moving information between the arcs of the cycle is seriously hampered by the lack of methods of exchange between practitioners of each of the three major activities. Just focusing on the interaction between wargamers and simulationists (and analysts more broadly) is exacerbated by the underlying conflicting philosophical mindsets of the two – one focused on the quantifiable, the other on the qualitative. This makes agreeing on a methodology for defining and bounding

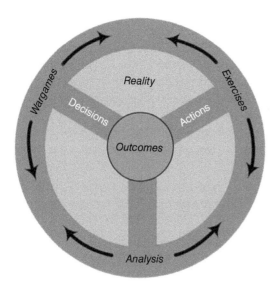

Figure 1 Cycle of Research.

the decision space, let alone forming a consensus on what a collaboration space to explore it might look like. What most wargamers and simulationists can agree upon is the need for a common meeting ground that can host the wide range of information while visually representing the current state of their interactive collaboration, together with a history of the collaboration that has occurred. Contentious issues include:

- Establishing a common level of abstraction (i.e. where to stop zooming in on the fractal pattern or opening black boxes to analyze their contents).
- The complementary nature of decomposing known characteristics and behaviors of objects and entities (at the agreed level of abstraction) and reassembling combinations that successfully complete tasks and achieve goals.
- The conflation of detail with realism by simulationists in their zeal to achieve rigor, validity, and fidelity, not wanting to let a good story get in the way of the facts.
- And conflation of insight with truth by wargamers in their zeal to achieve a "sum is greater than the parts" comprehension in "ah-ha moment" fashion, not wanting the facts to get in the way of a good story.

Why do wargamers chaff at the simulationists' preoccupation with validity[10], fidelity,[11] and rigor[12]? They feel it is too easy to miss the forest for the trees when trying to treat a "man in the loop simulation" as a "wargame." Increased validity does not necessarily follow from increased complicatedness. In an early form of his oft quoted aphorism, George E.P Box warns 'correct' one by excessive elaboration"[13] and continues "the scientist must be alert to what is importantly wrong. It is inappropriate to be concerned about mice when there are tigers abroad.[14]" What the wargamer sees as "excessive elaboration" (and to the simulationist is still insufficient to achieve the minimum "wrongness") often conceals the cause and effect relations from the player that allows him to sort mice from tigers and properly evaluate the risk of mistaking one for the other. The very validity, fidelity, and rigor that simulationists strive for are not usable by the

---

10 Validity being how well the outputs of the simulation match expectation. A low-fidelity simulation can have valid outputs.
11 Fidelity being how accurately the characteristics and interactions between entities and the environment within the simulation match data collected from real-world sources. A high-fidelity simulation can have invalid outputs.
12 Rigor being the degree of traceability and reproducibility there is in the simulation process. Wargames are by their nature not reproducible. There can, however, be traceability in the process by which analysis of them is done.
13 Box, G. E. P. (1976), "Science and statistics" (PDF), *Journal of the American Statistical Association*, **71**: 791–799.
14 ibid

player who is left to base decisions on heuristics derived from past experience, mostly unrelated to the context of the game decision at hand. They have what sometimes is the correct answer to the wrong question and lose confidence in the simulation results when it provides feedback that does not match the decision-makers' general expectation. Simulationists would argue that because of the validity of their simulation, real-world heuristics should be effective and any mismatch in expectations should lead to paying more attention to simulation-based decision-making, not less.

Why do simulationists dismiss wargamers' preoccupation with constructing narrative, engaging in dialectic argument, and paying little more than lip service to validity, fidelity, and rigor? They often feel it is little more than "making stuff up" at worst and at best presenting players with dangerously incomplete data that undermines the credibility of decisions that use it for risk assessment. It is like skiing down a mountain with a vague understanding there are these "tree entities" out there without worrying about where or how big they are. If wargamers can quote Box, simulationists can as well: "Remember that all models are wrong; the practical question is how wrong do they have to be to not be useful?"[15] With models as rudimentary as wargames commonly have, they are sufficiently wrong to have no usefulness and, in some circumstances, negative usefulness – they lead the player to misunderstanding the games' analogy to the real world contributing to basing future real-world decision on false premises acquired through oversimplified wargames.

So how can tabletop wargaming be effective even though to a simulationist it can be embarrassingly primitive? Can a hearty game of *Battleship*™ have any value (beyond hours of fun with your 6 year old) without a multi-million-dollar simulation to drive it? Of course it can and that is partially true because the game's abstraction layer compartmentalizes information and presents it to the player in an understandable manner without unnecessary informational (and visual) clutter. A wargame is not intended to provide the players synthetic experience skiing down a mountain, it is intended to inform the players about the characteristics of mountains that make for good ski resorts. Extraneous entities, characteristics, and interactions have been removed and the crux of the risk calculus underlying player decision-making allows cognitive energies to be focused on the game objective – find and sink the battleship. Our brain fills in the information gaps, creates a cognitive operational picture, and collectively the players exchange their beliefs about why they fill those gaps with the information they do and why they believe their decisions will advance them toward their desired end state and victory – hearing the long sought-after words from your opponent . . . "you sank my battleship."

---

15 Box, G. E. P.; Draper, N. R. (1987), *Empirical Model-Building and Response Surfaces*, John Wiley & Sons.

Bridging the Wargaming – Simulation Gap

Exacerbating the challenge is the traditional manner in which simulationists approach linking validity, fidelity, and rigor to the simulation environment. They see the degree of abstraction used in creating a simulation as something running either "top-down" or "bottom-up."

Either starting from broad generalities and deductively determining specifics, or starting from data and inductively aggregating them into generalizable rulesets, wargamers think more in terms of looking for insight from observed patterns – adductive reasoning. They form analogies from other domains thinking "sideways" across the spectrum of how things can be represented in greater or lesser degrees of abstraction.

Notice in Figure 2 how the three major layers of wargaming design representations, from the abstract to the concrete (i.e. live), are characterized as a spectrum. This mirrors the two extremes of wargaming merging in with analysis (simulation) at one end and live exercises or experiments at the other. The lack of readily accessible information about how different aspects of a game's representation should appropriately be abstracted results in many games being developed in a limited "bandwidth" (or stovepipe to use a bad word) requiring an "all or nothing" approach to represent a particular landscape with tactical-level games naturally moving right, while more strategic games go off to the left.

A paper-and-pencil "kriegspiel" with a spreadsheet-simulated table of attrition seems natural for a battle involving divisions fighting for a city. If one is going to look at how to takedown a particular building, then playing paintball in a mock-up of the actual building is about as high fidelity as you can get short of the real thing. A simulationist will focus on ensuring the ballistics of the paintball guns match that of actual bullets as closely as possible. High fidelity! Or will look out beyond

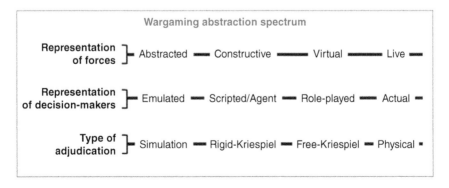

Figure 2 The wargaming abstraction spectrum.

the building to see how the surrounding neighborhood affects the ability to infiltrate and exfiltrate from the building before and after the takedown. The church steeple three blocks away must be held as an observation point to ensure the coast is clear. The times when school lets out and children crowd the streets need to be known. As you increase the number of such things, it becomes too complex and a new level of abstraction needs to be created further to the left. But how? The simulationist will box up the lower priority of the modeled city based on some criteria and retain high fidelity in some things but may take certain complicating factors totally out of the game (like the school kids and other civilians getting in the way).

The wargamer will complain that these limitations are arbitrary and capricious and drive players to stay in playing the game on the rails the simulation is prioritized in the things it retains at high fidelity or not. You should play the game first to see what the priorities are, not make them based on what is easy to depict at high fidelity, or involves pesky humans interacting and not equations. The wargamer will on what is important to a particular issue about the competitive landscape the players are interested in, not in creating a representation of the landscape itself. A wargame may even use a surrogate for the actual landscape in cases where the objective is assembling process components or is about identifying and dealing with certain conditions or constraints. You do not need a "realistic" landscape but one representative enough to be useful.

Too often, however, this is done by wargamers "winging it" with aphorisms such as "3-1 odds are required to win an assault" or "a unit either loses strength or has to retreat." The wargamer who wants to combine different levels of abstraction, or dynamically adjust it to the level of player decision-making, runs afoul of the "blurred lines" making things like shifting between free and rigid kriegspiel adjudication while combining live and constructive forces nearly impossible to manage. Like artists who limit themselves to impressionistic watercolors, wargamers can become increasingly narrow in their approach in their own way while chiding simulationists for doing the same. Each can learn immensely from the other if they would interact during the problem formulation stage and not when they meet to brief out their completed projects.

The US Navy has many examples of technologies created for a particular purpose, for example training or situational awareness, and then leveraged for wargaming. Indeed, it is really good at leveraging technology; but many applications successful for their original purpose fall short when extended too far beyond that purpose since the vertical solutions could not adequately meet the required horizontal adaptations.

About a decade ago the Navy's research and development community recognized the growing need for technology associating horizontal landscapes that cover a broader spectrum by including those higher fidelity simulations and stimulations that are available (courtesy of simulationists), and conceptually filling in the missing gaps to complete the narrative (desired by the wargamers). With so many to choose from, the Navy began a systematic exploration of synthetic environments with a focus on collaboration between the wargamer and simulationists' communities, and innovation via innate flexibility as important traits, features that would bring the users and the simulations together into a common virtual landscape.

Virtual World Beginnings

In 2008 one such technology making headlines was *SecondLife*™ – developed by Linden Lab, a small company in San Francisco, CA. In the time of massively multiplayer games like *World of Warcraft*, *SecondLife* was more than a game – it was a new type of virtual reality called Virtual Worlds[16]. More accurately, it was not even a game (it had no conflict or end-objective built in) but rather an infinitely flexible platform in which the users themselves create . . . well actually anything they want. It was marketed as a free 3D Virtual World where individual users (called residents) from across a wide internet-based community conceive and create spaces called *sims* – for shopping, entertainment, training and education, job fairs, research and concept testing, simulation and prototyping, communication, philanthropy and fundraising, political organization, brand promotion and yes . . . even gaming. But for this author, the "ah-ha" moment came when I discovered *Virtual Nantucket* in *Second Life* – a virtual community of sailing enthusiasts that created their own "game space" complete with public shops, private homes, live music on weekends, and a virtual yacht club that hosted daily sailboat racing. It was here that I purchased (with Linden Dollars – real currency of the virtual realm) my first virtual sailboat.

The is made possible because *SecondLife* offers simplified physics that is close enough to real-world behaviors that it can be used as an inexpensive and flexible simulation tool for small to medium-fidelity projects. In particular, *SecondLife* provides a built in 3D model creation tool. Users are able to rapidly and collaborately prototype complex systems with simple internal behavior models, or for increased fidelity, connect to external, sophisticated, simulation. For example, my

---

16 A Virtual World is a persistent *virtual environment* populated by many networked users, each represented by a personal *avatar*. Each user can simultaneously and independently explore the environment, participate in its activities, and communicate with others through text, voice, video, or other forms. The user can manipulate elements of the modeled world and thus experience a degree of *presence*. Journal of Virtual Worlds Research: http://jvwresearch.org/index.php/2011-07-30-02-51-41/overview

Figure 3 Virtual sailboat in SecondLife™.

virtual sailboat shown in Figure 3 is a Tetra35 modeled after a physical J35 class sailboat. It has a simulation script that responds to *SecondLife* wind and allows helm and sheet control through the user interface. My virtual sailboat allows me (seen in Figure 3 at the helm) to control mainsail, jib and spinnaker placement, sheet control and full rudder control . . . all in real time. It is this ability to visually create the landscape that is needed and then add the right level of behavior fidelity that launched 10 years of Virtual Worlds exploration and application.

Today the US Navy employs Virtual Worlds technology across secure military networks for a variety of innovative and collaborative applications including recent focus on composing adaptive environments for various levels of wargaming. This exploits the flexibility of Virtual Worlds to control fidelity from the simplest visual depiction to the most complex, often remotely driven, simulation spaces. An example of the former might be a 3D board game with virtual dice and 3D playing pieces with Virtual Worlds replicating the simplicity of the game play (without any additional features). In an example of the latter case, however, Virtual Worlds act as a visual and collaborative experiential layer between the players (which could be remote and distributed) and the complex synthetic environment that is actually producing the simulated game space. Many wargamers are attracted to the former because of a leaning toward simplicity (minimal cognitive clutter) while simulationists are drawn to the latter. Nevertheless, it is this flexibility to support collaboration between the two in a common space that gives Virtual Worlds the ability to be that bridge between the two communities.

Elgin Marbles – An Analytic Game

The first of several Virtual Worlds game spaces is *Elgin Marbles*[17]. Figure 4 shows the statue of Athena as an example for this collection. From a military perspective, the objective is to explore aspects of anti-submarine warfare (ASW) in regard to allocation of various types of sensors, each with unique characteristics (coverage, duration, fidelity, etc.) and cost (time to deploy, availability of assets, etc.). At its core, it is a game of strategy with focus on asset optimization. From a gameplay perspective, the approach is NOT to create a real-world operational space with each military platform simulated with full sensor suites (current and future). Frankly, that would be cost prohibitive if not technically impossible. Rather a different approach was considered – distill the wargame objective to an optimization of value vs. cost premise and then construct an abstracted game

Figure 4   Athena, Greek Goddess of Wisdom.

---

17 The Parthenon Marbles, also known as the Elgin Marbles, are a collection of Classical Greek marble sculptures made under the supervision of the architect and sculptor Phidias and his assistants. They were originally part of the temple of the Parthenon and other buildings on the Acropolis of Athens. https://en.wikipedia.org/wiki/Elgin_Marbles

Figure 5   Elgin Marbles strategy game board.

space with adjudication through simple simulation. The result was a virtual game of strategy called *Elgin Marbles* (see Figure 5).

The basic game board is modeled after the Greek Parthenon and is a 100 × 100 gridded square arena. Players control Zeus, King of the Gods, and aid in his task to find hidden deities (represented as transparent marbles) as they traverse the game board. The goal is omniscience (i.e. situational awareness) defined as knowing exact positions of all marbles. At the start of each game three red marbles with random location and velocity have been seeded in the arena; however, they are transparent and only become visible if intersected by one of the player's Elgin marbles (i.e. sensors). The player, as an individual or more often as part of a small team, has a selection of their own Elgin marbles in which to deploy and search for the red marbles. Each marble (named after a different Greek God or Goddess) has a unique set of characteristics (affecting value and cost). For example, Athena, Goddess of Wisdom, can be placed anywhere and exposes 10 grids in all directions for 10 seconds. The cost is 20 Pomegranate Seeds (PS). Nike, the God of Victory, has stamina and will expose three squares visible for the full game (three minutes). The cost is five PS. Demeter, Goddess of Agriculture, preplaces a 20 square line of five marbles each with a one square radius at a cost of 10 PS. Mercury . . . I think you get the point. Game play is timed and lasts for five minutes with a score generated based on how long it takes to expose all three marbles simultaneously (multiplying in a cost factor). If Zeus cannot expose all three marbles, then the objective is not achieved.

It should be obvious that with proper characterization, each marble represents an actual military sensor capability that would be available to a Theater ASW Commander. The balance of value versus cost is such that the trade-off decision-making with care should be able to translate from the abstract space to the military space (just like playing the Chinese game of *Go* can produce better Chinese generals).

To the simulationist, the game play is simplistic (even dismissively comical) since each marble has minimal physics defined by rudimentary behaviors.

Nevertheless, to the wargamer, the approach is useful in its simplicity. Since each marble can be characterized and simulated at a level that is quickly understood and is therefore predictable, the complexity of the game board space has been minimized to infuse trust in the simulation (to avoid second-guessing the results) while allowing cognitive energies to focus on strategy development. With guidance provided in real time from US Navy experts, designers were able to use Virtual Worlds to rapidly prototype this game in days with a working prototype in weeks. When viewed against the earlier wargame spectrum in Figure 2, Elgin Marbles has an <u>abstracted</u> representation of forces, <u>role-played</u> representation of decision-makers, and supports <u>rigid-Kriespiel</u> type of adjudication. The use of Virtual Worlds is helping to break the stovepipe by supporting greater horizontal fluidity.

Elgin Marbles is an example of what can be called a "man-in-the-loop simulation" game. Some call those "analytic games," but to a wargamer, the analytic nature of a game stems far more from what it is used for, than the manner in which it is constructed. The interaction of the marbles with the game board and other marbles is a simulation based on the laws of physics. The players make choices about what marbles to use and when to use them and the simulation tells them if the victory conditions have been achieved. The game rules are procedural – which marbles can the player chose and when can they introduce them onto the game board. The game has a predetermined degree of structural complexity depending on the fidelity of the physics model underlying the simulation engine. The inner workings of the game could be well-nigh incomprehensible with a multitude of characteristics determining the movement and interactions of the marbles. It has on the other hand, very low interactional complexity – the players collaboratively discuss what to do outside the simulation and once a plan of action is decided on, there is just the timing of the release of marbles based on information revealed during the execution of the simulation.

Analytical vs. Narrative Games

Games that use simulation engines to govern the physics of interactions and provide a means for players to make decisions affecting simulation behavior are often termed "man-in-the- loop" simulations. The way in which the simulation engine and player interface can capture information about how the player plays the game for analysis is what makes them "analytical." A data capture wrapper is created around the game play so as to understand what the player does, how it is done and why. It is about what the player does with the things in the game environment – the nouns and told in the first person by the players. *Elgin Marbles* game can be thought of as such an analytical game. Other types of games are more reflective, trying to stimulate abductive reasoning and generate hypotheses – the "ah-ha" moments. They are about why the player believes actions taken in the game will achieve the desired ends. They are about the verbs, not the nouns. These are "narrative" games and are more like collaborative writing, or a team debate. The discussion can go back forth about actions the players will take to gain this or that advantage and others describe pitfalls and countermeasures to be taken to thwart them. Whereas the Elgin marble game cannot be played by simply "talking about what you would do if you played the game" a narrative style game is often all about telling the story about why "playing the game like this will result in me winning." Role-playing games like *Dungeons and Dragons* are commercial examples of narrative style games. For professional purposes, there is often the need for some framework of simple rules to frame the interactions the game is about. A game similar in purpose to *Elgin Marbles* – uses capabilities to find hidden submarines – but one that is more narrative in implementation is *Subhunter*.[18]

In *Subhunter,* three enemy subs and three false contacts are hidden on a chessboard by an umpire who rolls two 10-sided dice and places the markers in the appropriate cell. If a 9 or 10 (0) is rolled, an additional false contact is added. Three players play as sub hunters. One controls two aircraft that each search a row or column. One controls two ships that sit on the vertex shared by four spaces. Players have five minutes to collaborate and make their moves. The aircraft covers a large area, looking for periscopes with its radar, but requires the sub to be recharging its batteries on the surface to have a chance, something it only does about 20% of the time. It carries sonobuoys that can help classify contacts detected by other units. The ship uses active sonar and so has a good detection capability but is poor at classifying correctly if it detects something. The sub has a short-range passive sonar (limited to the single square) but is quite good at determining a contact is in fact a sub.

A table of die rolls for each of those circumstances is the "simulation" and the players understand it explicitly. They then take turns moving their game pieces to find the three hidden enemy subs that remain relatively stationary, over the course of an average game only moving a square or two over the course of a game (six moves). They cannot move diagonally. The players must collectively agree on the locations of the three enemy subs at the same time before the end of the sixth turn to win. Once familiar with the game, the players may choose "advanced technology cards" that provide the player with the appropriate platform type modifiers to their die rolls, new ways to move, or other subtle rule changes reflecting the new capability.

---

18 *Subhunter* is a derivative of *Scudhunt,* used by the Center for Naval Analyses in a command and control experiment in conjunction with Thoughtlink, Inc. in 2003. (see http://www.dodccrp.org/events/8th_ICCRTS/pdf/077.pdf)yyCNA currently uses a game called *Subhunt,* that is also a SCUDHUNT derivative, but a different game.

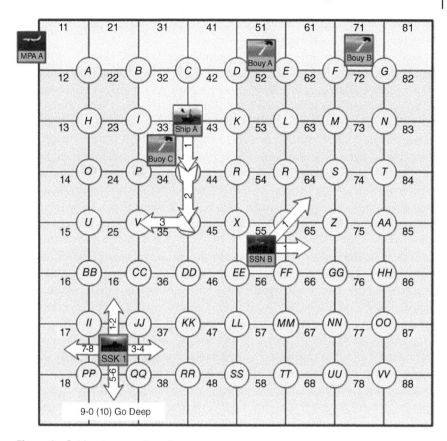

Figure 6 Subhunter gameboard.

If *Elgin Marbles* gameplay is thought by a simulationist to be "dismissively comical" then the gameplay of *Subhunter* would result in apoplexy (see Figure 6). A rudimentary familiarity with search theory would give the most advantageous moves to make each turn and there would appear to be no point whatsoever for humans to play as a computer program could easily "solve the game" and determine the optimum strategy. The actual purpose of the game, however, is not really about most efficiently finding the subs, but providing insight to the players regarding broad relationships between platform types and the information they provide. Players quickly find themselves with one platform classifying a target as a sub, while another classifies it a false contact. Aircraft are unreliable detectors, but invaluable in using sonobouys to correctly classify the detections. Subs are slow but sure classifiers but have to steer clear of their own ships to avoid being confused with the enemy. The uncertainty in the information they possess and what to do about it leads to a large degree of interactional complexity between the players in a game with minimal structural complexity (i.e. rules).

Using a time constraint has the players quite animated and insistent as to what to do by the third or fourth turn, each certain their course of action is the best one. The obvious lesson being that if such a simple game leads to such a confused state of affairs, heaven help those doing it "for real." They have the benefit of a lot more structural complexity in the technology of their sensors and information management systems, leading to a hierarchy of levels of fidelity in wargaming these issues. Extremely simple and elegant games can raise questions, analysis of which leads to more complicated, less abstract games that produce information with more fidelity that helps narrow down nuances of the problems and insight into the efficacy of proposed solutions. The cycle of research can spiral inward to higher levels of fidelity and complexity, or upward to lower fidelity and higher abstraction.

Adding an element of "coop-etition" to the game can provide insight into a different set of issues. If you give each player only the information his platform acquires and the winner of the game is the one who can claim that their platform type found the most subs, then – while it is an extremely simple game based on structural complexity – it becomes exceedingly high in interactional complexity. Players have to balance EVERYONE losing the game if they conceal too much information or fail to cooperate but being totally transparent will likely lead to someone playing a more cunning information control game winning. In the context of the Pentagon acquisition system, the player who wins individual games gets investment in new technology for their platform type, making it better and more capable of continued wins. If a platform type is becoming too dominant, the other two players will deny it the information it requires and "drag it back down." The simulation has almost zero fidelity, but the outcome is quite valid. Players repeatedly getting off to a good start who are denied their rightful victory will adjust their behavior making it impossible to reproduce any sequence of games. It becomes about the stories the players tell themselves about the synthetic history of their little simple decision-support contest in an artificial bubble. Now what if they had an actual virtual world? And simulationists joined them there!

This gets at one of the cruxes of the information sharing dilemma between simulationists and wargamers. Simulationists go to great pains to make available and capture as much information as they deem relevant, and given time and inclination err on the side of collecting too much information about the state-space frequently. Wargamers often carefully manage the information available about the game's state-space and how the abstraction level of the game is carefully chosen to limit the information available. The size of spaces on the game board allows anything happening in the space to be opaque to the players. A representative risk is assigned to performing a task in the space and you accept the risk and roll the die, or not. IF you roll the die and lose your piece, the details of how it happened are immaterial to a wargamer. IF the player felt the risk correctly captured the nature of the task, and accepted it knowing the consequences, that fact is captured and given to the analyst. If a certain interaction happened a lot in the game and the players accepted the risk when it was a "7 or less" to be successful, and these decisions directly lead to winning, then it behooves the student of the cycle of research to have analysts dig deeply into the factors that contribute assigning a risk value to that interaction. It was not critical in terms of the gameplay that the risk was portrayed with high fidelity, it was critical that the type of interaction was identified as being in the critical path to victory and thus needed far greater scrutiny.

If the simulationist likes the quote about models being wrong, but useful, the wargamer loves the quote from Einstein's office about "not everything that can be counted counts, and not everything that counts can be counted[19]." If a picture can speak 1000 words, then a number can hide 1000 sins; but we are drawn to them nonetheless thanks to their implied precision and factualness. Words are clumsy and imprecise when sparsely arranged but when well-crafted, stories unite people, as Tyrion Lannister in the final episode of HBO's Game of Thrones extols: "There's nothing in the world more powerful than a good story. Nothing can stop it. No enemy can defeat it." Which makes them threatening to those who want to have a factual basis to justify their decisions. Stories can have a firm foundation in facts, both qualitative and quantitative, and it is in the ability to present wargamers and simulationists with both the numbers and the stories that holds the potential of Virtual Worlds to bridge the gap and provide access to the different forms of information that is required to produce a complete assessment of options for consideration by decision-makers. We are comfortable with analytic approaches and associated tools, BUT NOT with "Narrative" approaches – "lacks rigor."

Virtual Worlds as a Virtual Reality

Let us pause our discussion on analytic versus narrative games and return to technology's ability to create game spaces. As mentioned earlier, Virtual Worlds hold promise in being able to create the virtual landscapes that both the wargamers and simulations require. However, when turning to virtual reality as a simulation technology, we must confront the reign of "first-person shooters," which dominate much of the entertainment gameplay, as compared to other environments in which the need for situational awareness prevails. First-person view implies that the experience presented to that player is carefully constructed to provide as much

---

19 While often misattributed to Albert Einstein, it is actually by William Bruce Cameron in "Informal Sociology: A casual introduction to sociological thinking".

and as little situational awareness as required. Think of a captain on the deck of a ship looking out over the horizon – atmosphere and distance affect how well he or she might see a passing ship over the horizon. We identify this as a Role Player Simulation (RPS)[20] viewpoint. Now think of a rousing game of *Battleship* in which players look down onto the scene as gods-eye view. The player is not an admiral commanding a fleet in the strictest sense of gameplay, but rather an observer within the virtual reality. We identify this as an Observer Situational Awareness (OSA)[21] viewpoint. Both views support the development of situational awareness; but the Captain on deck sees information filtered by realism and the *Battleship* player sees a space filtered to remove all informational clutter. Closer examination reveals that experience fundamentally falls into these two categories, dependent upon the purpose/objective of the viewer within that immersive space. To a wargamer this differentiation might not seem relevant, but to the simulationist it is perhaps the biggest influence on creating the landscape . . . because to adequately play as a role-player, there is an implication of maximizing presence . . . which primarily occurs by maximizing realism . . . which implies maximizing fidelity. Virtual worlds, again, stand out among technology in its ability to support both modes (RPS and OSA viewpoints).

Like most tabletop game boards, *Elgin Marbles* experience is through an observer. Zeus overlooks via the "Gods-eye" view and makes decisions based on all that is visible and known a priori. However, during the design phase in the next example, there was notable tension between the wargamers initially wanting a role-player simulation experience (with sufficient fidelity to support tactical decision-making) and an observer situational awareness experience, supportive of operational decision-making at a more abstracted level.

Operational Wargames

As a rule, submarine commanders are very well trained, especially on tactics (i.e. how to drive a submarine) and necessarily use high-fidelity, RPS-based, team

---

20 If the viewer is a part of the battle space (i.e. is a role-player within simulation), often depicted as an avatar, then the objective is categorized as *Role Player Simulation (RPS)*. The 3D virtual space is created to provide a simulation of real world and the reaction of the user to that virtual world is a key driver of effectiveness. Typically, the more realistic the virtual world the better, but that realism often comes with a cost in time and money to create.

21 If the viewer is not part of the scene itself and acts, instead, as an observer (with or without an avatar), then the application is categorized as an *Observer Situational Awareness* space (OSA). In these spaces, the objective of the observer is to maximize his/her situational awareness. This may be a full operational picture as one might find on a tactical chart; or it might be a complex data set in which 3D visualization provides insight into geospatial data relationships. This type often requires less fidelity (less costly) but more flexibility, simpler, and rapidly produced.

simulation systems to maximize realism and therefore training effectiveness. Nevertheless, these same tactical simulations fall short when supporting operational war games that require behaviors beyond that of an individual platform. Operational wargames in this context are not about the operational level of war (though they could be) but are about "operational" in the sense that they are exploring the competitions associated with military operations – naval operations in this case. The tactical simulation replicates the experience of a submarine's command and control function down to each button, knob, and display. This implies that within this simulation space the users can do nearly as much as a real submarine, but NO MORE. For example, a player cannot "look out the window" and see the bad guy unless that is a capability on a real sub. There must be an actual submarine sensor modeled and available within the context of the tactical scenario for it to be in play. An operational simulation assumes a submarine captain understands the correct tactics to "fight the ship" but lacks the situational awareness to support operational decision-making.

To explore these operational-level requirements several Virtual Worlds-based wargames were developed. Beyond the actual game itself, a secondary objective was to determine what features a simulation space would require above and beyond the existing submarine tactical capabilities. The following list of new abilities was identified:

- "Fast forward" simulation time to an event of interest (like a sonar detection).
- Pause gameplay for discussion for non-time critical decision making.
- Capture decision points (effects and logic) for geo-time synchronized after-action reviews.
- Support scenario replan by single-click waypoint insertion on to a 2D chart.
- Use "quick buttons" for ship maneuvers like *Come to Periscope Depth* (PD) and *Return to PD*.
- Generate simple target solutions based on range, target strength, and geometry to support weapon firing position.
- Depict merchant traffic through Automated Intelligence System (AIS) heat maps (instead of individually simulated contacts).
- Auto-suppress visualization of threats in game space (except during detection or while at PD).
- Play multiple scenario variations in a single session (e.g. Monte Carlo).
- Evaluate situational awareness and make gameplay decisions as a team (versus single player).
- Display other types of supporting information not necessarily available within current submarines.
- Represent large quantities and class diversity of possible threats.
- Represent future systems for both friendly and threat platforms.

● Represent the complexity of the battlespace beyond the immediate submarine-centric view.

While technically possible to add these features to the high-fidelity training system, a combination of many factors including the necessity to maintain support for core training requirements, system validation, performance, availability and of course . . . cost, led to selection of a Virtual Worlds solution. This approach would create several submarine operational wargaming spaces designed through rapid prototyping for the purpose of concept of operations (CONOPS) evaluation and training. In other words, the *simulationists* had a solution but it was not practical; therefore, the *wargamers* needed to innovate. Virtual world technology allowed the solution to be rapidly prototyped (innovation) with frequent input and review between Fleet wargamers and Virtual Worlds simulationists (collaboration). Furthermore, the Virtual Worlds' ability to support remote and distributed real-time development meant this rapid-prototyping primarily occurred even when the team was an ocean apart!

Like *Elgin Marbles*, the design approach is always to start with warfighter objectives (i.e. the purpose of the wargame), which in this case was identified as training submarine commanders to exercise operational decision-making skills when faced with the cost versus value of pursuing numerous threats of differing levels of capability. The next step was determining where in the wargaming continuum did this fall. Could a tabletop instantiation meet the objective? . . . maybe but the players tended to higher realism. Are high-fidelity submarine systems available? . . . no, they were being used for existing obligations and as is, they would drive gameplay to tactical decision-making at the expense of operational decision-making. Could we use Virtual Worlds to compose in several months an effective game space? . . . Yes, it was immediately accessible by all players, it could depict all necessary entities at a sufficient level of fidelity, and it could quickly prototype additional operational information layers to ensure the wargame exercised operational skills. A prototype of the resulting gameplay and interface is shown in Figure 7.

The game space is the Hawaiian Islands with terrain, imagery, and bathymetry at a level of accuracy defined by warfighter input. The game pieces are US and threat submarines with representative clutter via merchant shipping. The wargamers provided various scenarios, a priori, and then animated them within the virtual world. Several control and situational awareness (SA) heads-up displays (HUDs) were developed for both the instructor and student players. One critical feature was the ability to start, pause, continue, and stop the scenario in order to react and inject operational decision-making. During a pause, student discussion was encouraged with a result often being a change in submarine course, speed, and depth. These changes were recorded as a "Decision Point"

Figure 7   Virtual World operational wargaming toolkit.

available for post-event review. Various layers of information were available on demand and included: Automated Intelligence System (AIS) merchant historical tracks, high-fidelity navigational charts, and custom threat maps. SA HUDs included Ownship submarine readout and threat management boards. It is important to note that these did not pre-exist but were rather composed (sometimes in real time) per the wargamers' requirements. The result was a custom interface that reflected what the wargamers needed to meet their training objectives balanced by what the simulationist could reasonably provide.

Additionally, fidelity was provided during the prototyping and test phases on a "as needed" basis. Using Virtual Worlds models that emulate core platform behaviors such as movement, command and control, power, sensors, communications, and weapon launching, the players' submarine included basic behaviors resulting in a simulated submarine that met the expectations of the wargamers (it did enough but no more than necessary), while maintaining the flexibility to meet the wargaming requirements. However, tension points occurred several times when a wargamer requested features that deemed "tactical" (and hence already taught in the high-fidelity team trainer) and not essential for this "operational" decision-making game space.

Sometimes an increase in fidelity was deemed essential to the gameplay. For example, during a practice run, an instructor indicated that the decision to traverse the operational area at a high speed (to meet the objective of getting into an optimal position quickly) needed to have a cost impact of dramatically decreasing sensor range performance. A modification to the sonar model (i.e. increase in fidelity) was necessary to account for this specific behavior and was available in under an hour for the next practice run. In this example, tension was productive and resulted in a better capability as well as a clearer understanding of the wargamer needs.

In regard to the wargaming spectrum in Figure 2, this operational wargame example used <u>abstracted</u> and <u>constructive</u> representation of forces, <u>actual</u> representation of decision-makers, and supports <u>free-Kriespiel</u> type of adjudication. It was designed to fill the gap between traditional paper-based planning and desktop wargaming (i.e. chart with markers) and full live or virtual Modeling and Simulation (M&S) (e.g. Team Trainer). Since it allowed rapid creation for these particular blue-on-red scenarios, and supported variable speed game play, it traded the fidelity of a team trainer for the flexibility and availability of a constructive simulation. In addition, compared to the tabletop wargame that is often based on cognitive gymnastics with modest use of paper and white boards, Virtual Worlds provided an intuitive 3D game board that plays out the scenario and "adequately" depicts user moves and decisions. As further evidence of achieving the right tension, after the training event wargamers declared, "even with limitations it was still pretty awesome." This made the simulationists very happy ☺

Distributed LVC Wargames

Simulationists often think of the ideal game space as a single, beautifully crafted, synthetic environment that will ultimately, if one invests enough time and money, do whatever you need it to do. And in fairness (as a simulationist myself) I get up every day working toward that vision. Nevertheless, experience shows that synthetic environments themselves follow a fidelity continuum as shown in Figure 8. At its most basic level, the game board can be a static "single snapshot" of the game space. In military context, this is referred to as an "Operational View" (OV) and usually depicts the set of all possible systems (e.g. ships and aircraft) and connections (e.g. ship detects threat, ship communicates with satellite) that are play in the scenario. Though not particularly useful for actual game-play, it can be very useful during game design, in particular when first identifying system and subsystem requirements. Importantly, to the simulationist and wargamer alike, this view acts as a communication tool and helps reduce tension through the creation of a common vision.

The next level represents a series of OV events depicted as an "Activity Flow." What was static now has become a discrete series of steps, both temporal and geospatial, creating a storyboard view. Our game of *Battleship* could be described

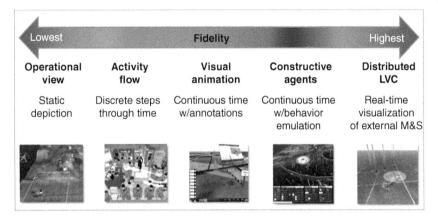

Figure 8 Game space fidelity across the live virtual constructive continuum.

as an activity flow since each move is discrete with game play fundamentally independent of time but very dependent on order of execution. Many tabletop war games could also be captured as a series of discrete moves in a geospatial context. Applying either a tactical or an operational context, an activity flow depicts State-1 (with systems$_1$ and communications$_1$), then State-2 (with systems$_2$ and communications$_2$), etc.

Note that while simulationists typically depict larger scale game spaces (like the Pacific Ocean), scale is entirely flexible within the Virtual World, which can create a game space that includes theater level (e.g. Pacific Ocean), platform level (20 miles around the submarine), and even human level (e.g. the submarine control center complete with human operators) all within the same environment. This is particularly useful when moves affect human communications (e.g. Captain orders torpedo launch), which results in an effect at platform and theater scale (e.g. torpedo launches and sinks ship). In this example, to tell the story sufficiently all scales are required.

If gameplay requires elements of time, then a third level of fidelity is introduced – "Visual Animation." In a game of *chess* analogy, this is the difference between simply representing the discrete moves to get to the end state (checkmate!); and the inclusion of players assessing the board to make a move with a timer going. Player time and a piece's motion are temporally non-discrete and part of the simulation. Likewise, our first example *Elgin Marbles* falls into a Visual Animation category since the overall score was based on time and the marbles had time-dependent motion. Traditionally animation is used extensively in gaming to tell a backstory or as a cut-scene to provide filler. It can weave together player moves providing the action–reaction necessary in gameplay whereas a decision causes a result. Strictly speaking, every animation needs a simulation to determine the motion. In this context, we differentiate animated movement as predetermined while a model interacting with its environment drives simulated movement (next level of fidelity). Animation is used extensively in scenario depiction with the high-fidelity calculations performed by analysists a priori, with the virtual environment simply reflecting the desired effect of those analysts; however, sole use of visual animation in wargaming, which by its very nature requires decisions, is less common.

In the Navy's Virtual Worlds wargaming toolbox, the next level of fidelity is most common – use of models and/or "Constructive Agents" to simulate expected behavior. Whereas an animation tells a story, use of models (a description of characteristics and behaviors) and simulations (dynamic interactions of models over time) creates a story of "consequences" based on a set of "actions." As an example, the previous submarine wargame example employed constructive agents to simulate sensor detection. Warfighter representatives identified the basic detection range of a particular sensor system in a particular environment against a particular threat – and came up with a number representing range. This particular agent requires about 1000 lines of code to create a binary detection– threat within range. The key was representing the intended behavior at the right level of fidelity to achieve the objective. The same sensor detection in a high-fidelity team trainer might require 100,000 lines of code since it requires a full modeling of acoustic sound propagation from a fully modeled source (bad guy sub) to a fully modeled receiver (good guy sub). As discussed earlier, for the submarine operational wargame, constructive agent fidelity was good enough.

At the right of the continuum scale, the highest fidelity level discussed introduces distributed "Live, Virtual Constructive (LVC)" Modeling and Simulation (M&S) capabilities as a method to more accurately create the ideal game space replicating all required aspects of the physical world within a single synthetic environment through the composition of distributed models and simulations. Done correctly, it can achieve both a simulationist's vision and a wargamer's need to hit the "I believe" button. LVC relies on a system-of-systems approach versus a monolithic simulation solution and is enabled and made practical through the coordination of usually high-fidelity, authoritative, distributed resources. Simply speaking in a US Navy context, LVC is characterized as:

- Live – which includes a Live submarine, ship, aircraft, vehicle, or personnel on range.
- Virtual – which includes both real people operating hardware-in-the-loop (HWIL) and real people operating or automated Virtualized Hardware (emulations).
- Constructive – which includes real-time, physics-based ("compact" or "synopsized") models, real-time, non-physics based ("behavioral" or "parametric") models and non-real-time (faster or slower), physics-based models.

Building upon the previous example where certain systems and subsystems (e.g. ships and sensors) were simulated through agent behaviors, distributed LVC allows those elements (and hence the entire wargame) to increase in fidelity by replacing those agents with external models and simulations. Depending upon the overall wargame objective, the game space can now be augmented until all requirements are met. Wargames that maximize fidelity tend toward the far right of the wargaming spectrum in Figure 2. By definition, LVC wargames rely on <u>live</u>, <u>virtual</u>, and <u>constructive</u> representation of forces[22]; however representation of decision-makers is usually <u>actual</u> to align the realism of the environment with the realism of the human component. Adjudication is usually built into the LVC event through <u>physical</u> means – usually data driven and analyzed post event to determine whether blue or red objectives were met.

The Future

Part of the recent success in getting M&S to the wargamers is through a crawl–walk–run approach of understanding the particular needs of the wargamer and providing small, incremental capabilities. In particular, the use of Virtual Worlds to visually "fill the gaps" and then later increase fidelity through the connectivity to external M&S is an approach that is showing success. From the wargaming side, one of the biggest issues is the representation to players of the risks they are accepting by choosing a particular course of action. Absolute knowledge of risk is never available in the real world, but many times the objectives of a wargame are to provide different amounts of information, of a variety of levels of accuracy, and see how the players use it. More information is not always better as the *Subhunter* game shows. It can cause more confusion than it resolves. So how much is enough and how do you decide when enough is enough? Current methods make it very difficult to control the information players have available and the different types of decision errors players can make based on it. The use of Virtual Worlds by simulationists and wargamers together has the potential to form a more controllable research environment to look at how information availability and its reliability affect military decision-making. This requires bringing multiple types of simulations, wargaming techniques, and data capture and analysis techniques together to get to the "meta" level the future informationized warfare will require. The line between "playing a wargame about a military operation to learn about the most effective course of action" and "planning an actual military operation"

---

22 Abstract representation is also possible and becoming more common when gaming future capabilities. If the LVC simulation cannot support the concept, it can be "white carded" so as not to interrupt gameplay.

may become blurred to the point of an "Ender's Game" situation where, to an observer, it is unclear whether a game is being played, or an actual military operation is being conducted. The Virtual World enables a seamless transition (in principle) along any of the three aspects of the wargame abstraction spectrum – decision-makers, forces, and adjudication. That poses a host of thorny issues, not the least of which is, in a cyberspace of fake news, bot-induced opinion memes, and adversaries attempting to manipulate and call into questions our sources of information, creating "bubbles" within which we can explore this new competitive landscape would seem a necessity. A firm alliance between the wargamers and simulationist (together with the operational military personnel whose decisions are being supported and enabled by it all) that work in that bubble will be vital to gaining the sort of advantages we will require to deter our enemies and persevere against them if that deterrence fails.

13

Visualization Support to Strategic Decision-Making

Richard J. Haberlin and Ernest H. Page

The MITRE Corporation, McLean, Virginia, USA

Introduction

Government agencies are facing unprecedented pressure on their budgets and increased demand for justification of their funding priorities. Across the Department of Defense (DoD) and broader government, organizations are challenged to understand the multifaceted impacts to cost, schedule, quality, and risk as limited resources are allocated to competing activities. A new approach and tools are needed to help leaders make informed decisions on resource allocation. Generally, decision support systems (DSS) provide decision-makers (DMs) with information relevant to a particular decision in an easily ingestible presentation. This chapter introduces a novel DSS solution – Interactive and Immersive Visualization (I2V) [1] – that transforms raw data into multilayered informative visualizations while also enabling real-time course of action (COA) analysis.

The traditional role of visualization in wargaming is primarily supportive in nature. It facilitates communication of results and comprehension of the underlying model behavior to strengthen confidence in representation of the domain. However, faster computing and lightweight software have opened new opportunities for a more interactive role in wargame execution through customization of an interactive layer that allows the user to transform wargaming analytics into collaborative decision-making tools. The overarching concept is an integrated association of parameters, data, and models with an interface that allows rapid input adjustment and real-time rough order of magnitude (ROM) results. While a ROM is does not provide the rigor of a full analysis, the trade-off in execution time provides wargamers with a scoping tool to narrow down the pool of

Simulation and Wargaming, First Edition. Edited by Charles Turnitsa, Curtis Blais, and Andreas Tolk.
© 2022 John Wiley & Sons, Inc. Published 2022 by John Wiley & Sons, Inc.

excursions requiring full assessment, as well as a flexible platform for the communication of results.

The Interactive and Immersive Visualization (I2V) DSS solution combines organizational data and analytic tools in an interactive, collaborative environment that provides visualization of decision impacts on capability in real time. I2V improves the quality and timeliness of decision-making by enabling leaders to visualize alternative courses of action and their impacts on operational metrics with a rough order of magnitude estimate. The I2V environment facilitates better decision-making processes and provides DMs a visual, interactive environment to facilitate making trade-offs, and understanding the multiple dimensions of available options. In this manner, DMs are able to identify resource allocations that better align with strategic goals and with greater transparency. I2V enables stakeholders to hold a collaborative decision-making session in which authoritative data is displayed and real-time excursions are executed using existing analytic tools. The key component of this methodology is the integration of the sponsor organization's own tools through the use of metamodeling and/or software wrapping. While I2V does not replace detailed analytics, it allows ROM estimates of solutions within the decision session.

Impact/Capabilities

I2V provides a more effective, engaging, and defendable decision-making approach that sponsor organizations can use to arrive at group solutions to challenging strategic decisions. Some specific applications of I2V to government strategic planning and acquisition decisions are described below.

Strategic Planning
- Optimize selection of actions to improve performance against targets/goals.
- Visualize effects of including/excluding specific actions on performance, timeline, and budget.
- Exercise "what-if" excursions to understand decision space.
- Forecast anticipated schedule and outcomes.
- Visualize overview of current performance status and drill down into underlying data.
- Compare historic performance data with current status and future projection.

Acquisitions
- Evaluate alternate courses of action for allocation of resources applied to acquisition programs to increase capability in the context of a scenario.
- Visualize budget changes from a budget baseline to multiple appropriation categories for alternate modernization strategies.

- Perform cost–benefit trades across acquisition portfolios in the context of a scenario and its required capital investment.
- Illustrate gap analysis and mitigation strategies, filtered by scenario, organization, or function.
- Recognize violations to budget, contracts, and industrial base with highlighting flags.

Spectrum of Visualizations

Visualizations are tools meant to aid the decision-maker in understanding data across a broad range of sources. The data may have been collected from sensors, experiments, surveys, studies, analytics, and so forth – each source varying in complexity. On the other side of the equation, the decision-makers will have a varying comfort level with data and analytics. Each visualization must be tailored not only to the decision to be made, but also to the interest and experience of the decision-maker.

Compounding the question of what is "right" visualization for a given decision, there exists a spectrum of visualization available to support decision-makers (Figure 1). On the low end are static displays of raw data, such as may be produced by a survey, spreadsheet, or simulation. On the high end, virtual reality allows the decision-maker to interact with the decision space to understand multidimensional relationships. The majority of visualizations lie somewhere between.

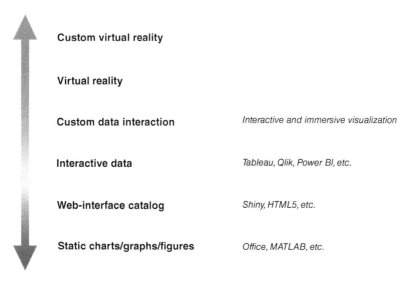

Figure 1 Spectrum of visualizations.

As decision-makers become more technologically adept, expectations rise regarding the presentation of data. A simple chart, that may have sufficed in the past, may no longer provide necessary stimulation to draw and maintain focus. Today, producing quality analyses is no longer sufficient; analytic results must be presented in an interactive, intuitive, and accessible manner.

Interactive Visualizations

There are many products, frameworks, and code bases available to the analyst to present results. Focusing on interactive visualization leads to two major categories: commercial and custom. Each is briefly introduced below.

Commercial Interactive Data Visualization

Commercial interactive data tools specialize in easily configured display of data, along with some business analytics functions. The emphasis here is to provide the user with many display options, and for those options to be easy to create. Three leaders in this domain include Tableau [2], Qlik Sense [3], and Microsoft Power BI [4]. Customization in this domain exists in the development of dashboards or storyboards, which are aggregations of several standard views into a single aggregate page focused on a common issue.

Visualization is an integral part of self-service BI tools, and they therefore share common feature sets. They enable the user to share and collaborate over data, investigate trends, and perform interactive reporting. There is also a strong user community and online third-party resources available technical support and to advise new users on project implementation. In many cases, use of commercial tools is sufficient for data display.

Custom Data and Analytics Visualization

Within custom interactive data and analytics tools, there also exists a spectrum of capability ranging from simple custom visualization to interactive and manipulatable dashboards, the focus of this chapter.

Simply viewing existing data in a customized visualization can be accomplished in a web-consumable format using software tools such as the Angular [5] or D3.js [6] JavaScript libraries. The general methodology is to bind data to a Document Object Model (DOM), and then transform that data into the visualization(s) desired within a web page. There are extensive open-source libraries of examples available online.

A more advanced solution involves integration of data and custom analytic tools with an interactive Graphical User Interface (GUI) to allow real-time collaborative decision-making. The GUI is built using the tools described above. This latter approach – using the I2V framework – is the focus of this chapter.

Leveraging the visualization described above, the additional benefit of this version is to allow the user to interact with not only the data, but also the underlying analytic tools that transform the data. Each of these analytic tools can be thought of as a black box to the web page. The model operates as follows:

1) A web-enabled dashboard of visualizations is created to inform a decision process (Figure 2). Adjustable variables are identified and "levers" assigned for interaction.
2) Data are bound to the visualization to produce the baseline position. The baseline data are often associated with the "current plan" or current budget.
3) Each of the "levers" identified above is connected via custom software coding to one or more analytic tools acting as "black boxes." Any adjustment to the lever causes recalculation of the analytic tool into a new data set.
4) The updated (calculated) data are displayed on the dashboard along with the baseline data to show the new situation resulting from the alternate course of action.

It is the custom coding between the web interface (also called software wrapping) that provides the additional power of this methodology.

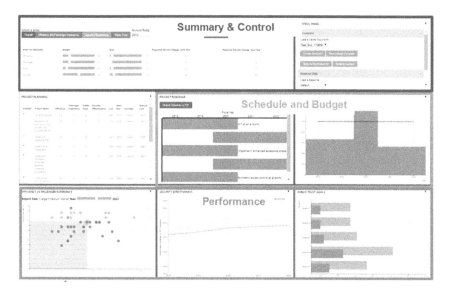

Figure 2 Example custom dashboard.

Methodology

Model Elicitation

Decision support tools are most effective when aligned to one or more specific decisions rather than simply displaying available data. While decision-makers differ greatly, there are commonalities among decisions that lend to a simple methodology for development of visualizations, summarized in Figure 3. Conceptually, this process can be thought of as building a specific decision support visualization by extracting a decision thread from the decision-maker one step at a time. Specifically:

- What decision must be made?
- What visualizations do you currently employ to make that decision?
- What data populates that visualization?
- What analytics interact with that data?

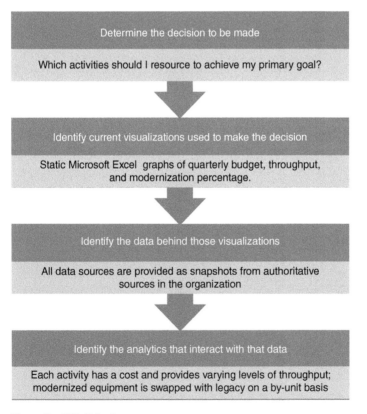

Figure 3 I2V elicitation process.

This process is summarized in Figure 3. The process step is shown in blue with an example in the associated grey box.

It is not uncommon for decision-makers to have difficulty specifying a unique decision to make without bleeding into other problems. Many decision-makers will begin describing a decision only to lead to a second or third without being able to articulate the first. Following the process articulated in Figure 3 helps to focus the decision-maker on a single problem at a time.

The I2V environment runs across several platforms, is distributed, scalable, and highly composable. Working closely with each sponsor organization provides clarity on how their enterprise data are utilized within the business processes, which provides the structure to be supported by the DSS. It also provides opportunity to understand the operational environment that will support the system.

In an ideal implementation, any analytic tools, data, and the visualization integration software will reside on the same server and be accessed via one or more client instantiations of the front-end web page. However, in some cases all calculations must reside within the client for security reasons with reach back to the data. Careful examination of the environment, the analytics to be integrated, the data sources, and the required visualizations inform a best of breed implementation customized to each sponsor that will provide the DM with greater perspective across the assigned portfolio.

Framework

Each decision space is built on a generalized framework that allows customization to data applicable to the DM's domain and preferences. By standardizing this framework, developers are able to scale the I2V environment to specific classes of decision problems across multiple domains. The target of each DSS is to inform COA comparison exercises.

All components of the I2V framework are written in JavaScript. This simplicity allows a single developer or small team to build a complete web application much faster. Enterprise data, stored in an easily accessible repository, requires minimal transformation from retrieval in the database to rendering in the browser, yields fast performance of visualization and interaction.

Considerations

While each decision support system is unique based on the decision to be made, there are commonalities across the domain, especially in the realm of government decision-making. There is a significant role of expectation management to be played with the decision-maker during the initial formulation of the data support system to ensure the "homework" has been done to enable the desired solution.

Data

It may sound obvious, but visualizations are only as good as the data behind them. This seems obvious, yet frequently data is not available to support the decision being made or is in a format that takes significant wrangling to be ready for consumption. A frequently used rule of thumb estimates that 80% of analytic effort is associated with data preparation [7]. The data must be accessible in a manner supportable by the business processes available to the organization. For example, if the decision-maker desires a real-time feed of data for the dashboard, there must be a technical solution to provide that data. On the other hand, if the dashboard will rely on data "snapshots" taken of authoritative data sources, who will perform those queries, and with what periodicity? Do the update rates of the data sources align in a manner that provides a useful decision? Without considering these items early, the output of the dashboard may not provide a valid solution.

Analytic Tools

It is also not uncommon for leaders to mistake the visualization provided with the analytic tools that are actually producing the trade-off. An important part of expectation management is ensuring leaders understand that the visualization/ dashboard is only the front end and that beneath it there must be a tool or tools performing calculations of benefit, risk, trade-off, budget, etc. This is especially true with the interactive dashboards introduced in section "Custom Data and Analytics Visualization." Many decision-makers see demonstrations or mock-ups of dashboards and immediately see benefit for their own organizations. This is good. However, if the decision support system is to be more than just a data display, there must be some analytic tool that performs the trade-off analysis, whether that trade-off is of cost, risk, performance, modernization, etc.

For the purposes of this effort, an analytic tool can be anything from a simple look-up table, to a multinomial equation, a response surface, a neural network, or fast-running simulation. To achieve maximum benefit, output from the tool must be near real time. If the existing tool is a long-running simulation, then meta-modeling may be necessary to provide a representative equation or surface that allows real-time exploration of the decision space.

Colors of Money

When working with the DoD, there are five major categorizations of money appropriated by Congress: Procurement (PROC); Military Construction (MILCON); Manpower; Research, Development, Test and Evaluation (RDTE); and Operations and Maintenance (O&M). Dollars within an appropriations category typically cannot be spent in another category, so it is important to understand the allocation of available dollars across the portfolio when it comes time to make budgetary trades. For example, if the decision-maker desires to speed up the delivery of an acquisition system, resources to accomplish this must come from PROC funds, thereby slowing down one or more other programs that would be used as "bill payers."

Of immediate importance to the decision-maker is a clear understanding of the assigned budget. The decision-maker must be able to view the budget across the entire domain, visualize both discretionary and nondiscretionary appropriations, identify major investments across the Future Year Defense Plan (FYDP), and anticipate programs at risk.

An I2V environment may provide a number of views that illustrate the contribution of the appropriation categories including an overall Budget Summary, Mission Support, Program Support and Program Element. From these views, the decision-maker is informed about the effect of moving appropriations between programs.

Courses of Action

Course of action evaluation is a common application of decision support, and the described I2V solution allows what-if analyses of multiple COAs in real time based on identified behavioral interdependencies of the underlying analytics. I2V is domain agnostic and can be applied in support of enterprise system solutions that support COA evaluations for all domains, starting with providing the option to analyze alternative COAs and describing what-if scenarios within other models. A SysML [8] implementation is described in section "Enterprise Integration," below.

Entities manipulated to describe the alternative COAs are drawn from enterprise data. This also implies that the results of the evaluation are also entities or transformations of the enterprise data. These entities have the same properties, but they will have different value content as the analytic tools provide their results. In this way, the same presentation schema may be used to display current and alternative predictions side by side. It also enables prediction of a future state by projecting the output trend.

Model Construction

An iterative, collaborative approach has proven successful to not only elicit the decision-maker's area of concern, but then to implement an operational solution in an agile fashion. Figure 4 summarizes the approach. The remainder of this section provides examples of strategic, budget, and risk dashboards developed for various government organizations.

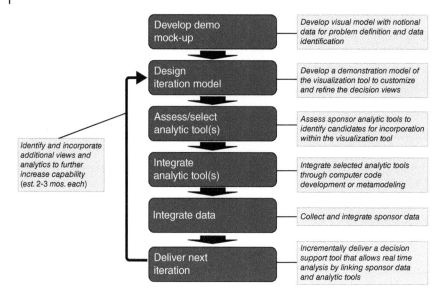

Figure 4 Development approach.

An initial mock-up is created in a JavaScript-enabled web page. This demonstration has the look and feel of the final product, including all necessary visualization graphics, levers, and filters. However, interaction between the levers and the output are purely mathematical and representative. At this point, the aim is for the directionality of change due to lever adjustment to be intuitive. For example, increasing resources for a project should increase output in the visualization. The mock-up process is iterative and collaborative with the decision-maker to ensure the proper decision is captured as well as the necessary graphics to inform that decision.

The remainder of the development follows an iterative process in which the analytic tools are integrated into the dashboard sequentially. Each analytic tool is a "black box" to the DSS, so a JavaScript software wrapper must be developed to capture "static inputs" that describe the problem space, "dynamic inputs" that are adjusted interactively through levers on the tool, and "outputs" that will become the data behind the visualizations. Additionally, if more than one analytic tool informs the DSS, integration between analytic tools may also be necessary to capture multi-order effects resulting from cascading outputs. A pairwise determination of tools must be performed to establish this additional integration.

Strategic

Strategic decisions are often characterized by shifting timelines of execution and/ or trading resources among several activities. In government, it is often the

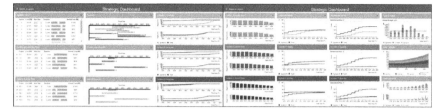

Figure 5   Strategic planning example.

timeline of an activity that is in flux and can be traded between activities. In this case, cost is an output variable, as is performance based on the selected activities. Figure 5 shows an example strategic dashboard for integrated planning across three domains.

In strategic planning, the primary variable under the control of the decision-maker is time. The decision-maker can observe, and sometimes control the sequencing and scheduling of events and activities. Therefore, the primary interactive levers for this DSS are the Gantt charts representing the start and execution time of each activity. The Gantt bars may be adjusted by sliding them horizontally on screen and/or lengthening/shortening the individual bar. Each adjustment to a single bar executes a recalculation of the underlying analytic tools that may provide a metric for one of three domains, or for the entire enterprise.

The decision-maker is also concerned about the influence that accelerating or delaying a program will have on mission effectiveness. A technology roadmap view provides an illustration of a capability to be delivered over time, categorized by appropriation. In the I2V budget framework, the program budget is related to the technology roadmap through an algorithm coded in JavaScript such that adjustments to the program budget result in an acceleration or deceleration of the technology delivery, resulting in a change in capability. Similarly, adjustments to the technology roadmap shift resources on the program budget. These adjustments are all collated in an overall budget visualization.

Budget

At a more detailed level, budget trade-offs between activities or programs are a common decision to be made. Within a decision-maker's portfolio, dollars are shifted from one activity to another. The output becomes delivery timeline, as well as performance over time. Figure 6 shows an example budget dashboard for resource planning of multiple activities.

Budget planning is primarily an exercise in shifting finite resources from one activity to another under the assumption of a fixed budget. Alternatively, an increase/decrease in funding may be anticipated and the budget DSS used to

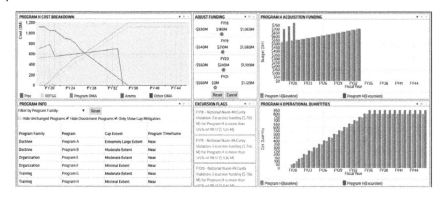

Figure 6   Budget visualization example.

identify the best activity to plus-up or reduce to maximize capability across the enterprise. While integration of algorithmic support to optimize this decision is possible (and has been integrated with I2V), it is not uncommon for the DM to drive toward a suboptimal solution. The primary levers for budget planning are individual spending by program within a period (e.g. years, quarters, months). Visualizations illustrate changes from the baseline program, as well as changes to overall capability caused by inclusion/exclusion of activities, or quantities of items updated due to the resource change. It is also appropriate to capture other discrete events such as penalties due to changes in contracts or delays between purchase in fielding.

Anticipating a budget adjustment, the decision-maker must identify which program or programs will have to absorb the reduction in resources. The decision-maker may have an idea of the relative importance of each mission. However, how much of the budget is tied up in each mission may be unclear and therefore the budget may not align to stated priorities. If not, then misalignments are targets for anticipated adjustments.

By displaying multiple programs support summaries simultaneously, the decision-maker can perform real-time trade-offs between programs while keeping the overall portfolio below the total obligation authority (TOA). Similarly, given a known amount for a budget adjustment or anticipated program, programs can be adjusted to free resources.

Risk Identification and Mitigation

This framework allows DMs to visualize technical, programmatic, and business risks, as well as illustrate the effect of mitigation strategies over time. It is important to note that this framework is a visualization mechanism of a risk analysis/gap analysis already completed. Visualization of a risk assessment

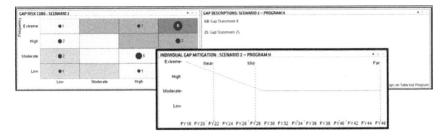

Figure 7   Risk visualization example.

allows decision-makers to drill down into the details of an assessment rather than be saddled with a summary chosen by the analytic team. An example is shown in Figure 7, below.

Here a standard DoD Risk Cube is used as a summary of the risk space, with frequency on the vertical axis and severity on the horizontal axis. High-risk areas are highlighted in red and low-risk areas are in green. The size and label of the circle indicate the number of risks at the confluence of each frequency-severity pair. Because the dashboard is interactive, selecting a circle via mouse click allows the user to drill down into the specific risk domain, as shown in the inset box.

Example: The MITRE Simulation, Experimentation and Analytics Lab (SEAL)

Inspired by the Center for Innovation at Lockheed Martin [9], the Decision Theater Network at Arizona State University [10] and the McCain Institute [11], MITRE's Simulation, Experimentation and Analytics Lab (SEAL) (SEAL) is a state-of-the-art facility designed to showcase modeling and simulation technologies in support of analysis, experimentation, training, mission rehearsal, integration, and testing in a uniquely compelling manner (see Figure 8). SEAL serves as the developmental testbed for I2V at large scale. Some of the key features of the lab are reviewed below.

Audio Visual Support

Many of the use cases for SEAL – particularly those involving human-in-the loop experimentation and facilitated analytics such as those supported by I2V – establish the need for a data wall. Some use cases (e.g. command center design) suggest the requirement that the data wall and associated A/V be *reconfigurable* in order to support design alternatives and analysis of workflows and ergonomics. Additionally, flexibly arranged and ubiquitous A/V assets should be utilized to increase the immersive experience of SEAL applications.

Figure 8 SEAL at the MITRE Corporation McLean, VA Campus.

To accommodate the flexible arrangement of potentially large numbers of displays within the lab, the MITRE 4 architecture team, led by SKA Studio [12], conceived a design that provides an open ceiling and A/V grid superstructure. The A/V grid, custom developed for SEAL by Trak-Kit [13], provides nine tracks (30 feet in length) running in one direction of main lab floor, abutting six tracks (10 feet in length) running in the perpendicular direction. Each of the tracks supports one or more panel-like structures, each of which is capable of supporting multiple displays (see Figure 9).

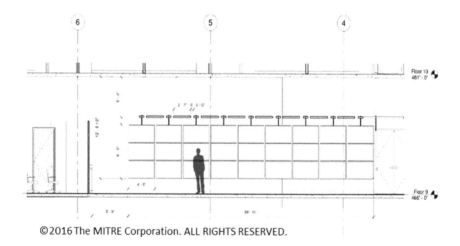

Figure 9 SEAL A/V grid concept.

The tracks are fully programmable – enabling lab users to orchestrate complex combinations of display panel movements, and quickly reconfigure display panels according to desired pre-sets. To control the heat load associated with very large numbers of displays, the A/V grid is designed to accommodate low-voltage displays. At full capacity, the A/V grid could support up to 99 displays. Currently the data wall consists of 45 55" high-definition planar displays. Through ICS Technologies switching solution, users are able to create up to 18 scalable and rotatable windows across a single or multiple display configurations. Any source is routable anywhere within the lab.

Multi-Level Security

SEAL is based on a zero-client architecture – all computing is extrinsic to the lab. Computing sources are routed into the lab by fiber connections to a 6.25 Gbps ThinkLogical KVM switch. Switch ports are assigned a classification status (e.g. unclassified, secret, collateral top secret), and SEAL systems automatically detect and display lab classification based on incoming signals. Executing an accredited CONOPS involving a few simple steps, the KVM can quickly switch between unclassified, classified, and mixed modes of operation.

Enterprise Integration

An I2V DSS also provides an excellent interface with an enterprise federation of data models and analytic tools. In this construct (Figure 10), parametric data is stored within an enterprise architecture tool in a language such as

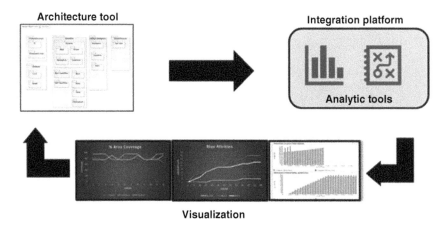

Figure 10   Enterprise model integration with I2V.

SysML. This provides a single location for updating overall performance parameters of the underlying analytic tools. Next, an integration platform may be used to facilitate the interaction and execution of a federation of analytic tools. When available, this simplifies input/output as well as integration between models. Finally, the I2V DSS provides an overall visualization of the enterprise, but also a means to perform course of action analysis through adjustment of levers.

This simple construct enables integration of multiple complex models that are undergoing continuous update to underlying parameters without needing to adjust software code. In a similar manner, all of the environmental parameters for a given decision may be stored in the architecture tool, making it the foundation on which all decisions are made, and assumptions are captured.

Community of Practice

As noted in section "Introduction," there is an ever-expanding spectrum of visualization capability available to display and interact with data. While this work has been primarily focused on interactive visualization for decision support, it is beneficial to share ideas and best practices across the greater visualization community. Figure 11 shows how four major categories of visualization skill sets may come together in a mutually supportive community.

In a similar manner, members of the visualization subgroups will attract members with differing skills. For example, the Interactive and Immersive Visualization group may be more interesting to software code developers, the Visual Business Intelligence group to the finance community, and the Visual Display of Complex

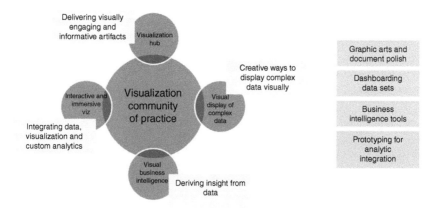

Figure 11   Visualization community of practice.

Data group to data analysts. These groups all have their strengths, so meeting periodically to share their experiences will serve to raise the entire community. A robust community of practice raises the overall quality of members' products through shared learning and experience.

Summary

Decision-makers face a daunting number of critical decisions and are frequently deluged with information, not all of which is relevant to the current decision to be made. The I2V construct allows creation of "decision spaces" that are customized to a particular DM and decision type. I2V provides a more effective, engaging, and defendable decision-making approach that sponsor organizations can use to arrive at group solutions to challenging strategic decisions. Trading fidelity for execution time provides wargamers with a real-time scoping tool to narrow down the list of excursions requiring full assessment, as well as a flexible platform to deliver results from a detailed simulation. By leveraging faster computing and lightweight software, today's wargames are more accessible through customization of an interactive layer that allows the user to transform wargaming analytics into collaborative decision-making tools.

References

1 Glazner, C., Chin, S., and Haberlin, R. (2016). *Immersive Visualization*. McLean: The MITRE Corporation.
2 Tableau. 2019. "Tableau Software". www.tableau.com.
3 "Qlik Sense Desktop," Qlik, 12 June 2019. [Online]. Available: www.qlik.com. [Accessed 12 June 2019].
4 "Business Intelligence Like Never Before," Microsoft 12 June 2019. [Online]. Available: https://powerbi.microsoft.com. [Accessed 12 June 2019]
5 "Angular," Google 12 June 2019. [Online]. Available: https://angular.io. [Accessed 12 June 2019]
6 M. Bostock "Data-Driven Documents," 12 June 2019. [Online]. Available: https://d3js.org. [Accessed 12 June 2019]
7 CrowdFlower "2016 Data Science Report," CrowdFlower 2016
8 "OMG Systems Modeling Language," Object Management Group, 12 June 2019. [Online]. Available: http://www.omgsysml.org/. [Accessed 12 June 2019].
9 Lockheed Martin "Center for Innovation (The Lighthouse)," [Online]. Available: https://www.lockheedmartin.com/en-us/capabilities/research-labs/center-for-innovation.html. [Accessed 11 June 2019]

10 Arizona State University "Decision Theater," [Online]. Available: https://dt.asu.edu/. [Accessed 11 June 2019]

11 McCain Institute "About the Decision Theater," [Online]. Available: https://www.mccaininstitute.org/decision-theater-mission/. [Accessed 11 June 2019]

12 Steven Kahle Architects Inc. "SKA Studio," [Online]. Available: https://www.skastudio.com/. [Accessed 11 June 2019]

13 Trak-Kit "The Design Studio," [Online]. Available: http://www.trak-kit.com/. [Accessed 11 June 2019]

14

Using an Ontology to Design a Wargame/Simulation System

Dean S. Hartley, III

Hartley Consulting, Oak Ridge, Tennessee, USA

Motivation and Overview

Both wargames and simulations require a model (in the most general sense of the word) of the domain of interest. For a wargame/simulation system, the model must be consistent throughout the system. This commonality is difficult to achieve when the model resides only in the minds of the creators. An ontology provides a significant starting point for a well-defined model by identifying and defining all of the potentially needed components of the model and their relationships. For some domains, the ontology may also identify the processes that connect causes to results; however, for domains involving human conflict, many of these processes are ill-defined. These ill-defined processes must be supplied by human wargamers or by calls to theoretical or probabilistic algorithms in the simulation. Thus, in human conflict domains, the ontology supplies a framework for the model definition and identifies the places where decisions are needed on wargaming or simulation algorithms.

In this chapter, I will describe the Modern Conflict Ontology and show how it provides the components that apply to a (generally) theater-level conflict. Then I will show how the ontology specifies the places where calls to either the human decisions or algorithmic decisions are needed. I will conclude with observations on modeling human conflict at different levels of granularity.

Simulation and Wargaming, First Edition. Edited by Charles Turnitsa, Curtis Blais, and Andreas Tolk.

Introduction

We will begin with an ontology for modern conflict. This ontology is derived from the ontology for unconventional conflict described a book with that name (Hartley D. S., An Ontology for Unconventional Conflict, 2018). Unconventional conflict is the union of operations other than war (OOTW), which consists of operations other than garrison duty and warfare, and irregular warfare, which consists of conflicts between and among human opponents larger than the individual scale but smaller than conventional state-on-state scale. This means that unconventional conflict includes situations in which one "opponent" is a natural or man-made disaster and those involving guerrilla war and terrorism (Hartley D. S., Unconventional Conflict: A Modeling Perspective, 2017). Modern conflict simply adds conventional conflict up to, but not including, nuclear war. The use of "dirty" bombs, chemical and biological agents, and cyber warfare are included. The extension of the unconventional conflict ontology to encompass modern conflict was relatively easy because the "conventional combat" envisioned from World War II through the cold war is no longer the norm. No longer can we assume that there will only be two sides to a conflict or that defeating the armed forces of one side solely through use of armed forces will be sufficient to win the conflict. The entire set of diplomatic, informational, military, and economic (DIME) instruments of power (or some similar taxonomy) may be needed. Similarly, the entire set of state variables will be needed to determine the results: political, military, economic, social, information, and infrastructure (PMESII) (or similar taxonomy). Impacts on the social fabric can no longer be relegated to the category of collateral damage – and ignored.

An ontology is merely a tool for codifying what we know, could know, or should know about a domain of knowledge. A good introduction to ontologies can be found in Lacy's book (Lacy, 2005). A more detailed description is found in a book by Fensel (2004). A very detailed description may be found in the book about the Basic Formal Ontology (BFO) (Arp, Smith, & Spear, 2015). Basically, an ontology consists of names of the classes of things included in the domain and the various relations among these things, with definitions. When instantiations are added, the result is often called a knowledge base instead of an ontology. For example, imagine an ontology that contained the pertinent knowledge about World War II in Europe. (Note, "pertinent" is an important adjective. It indicates that the definition of the ontology will generally depend on the use to which it will be put.) In this case, suppose there was a class of thing called *supremeAlliedCommander*. The ontology might define this class as part of the *army*, with some sort of relationship that also specified that this class was the leader. An instantiation of this class would be a particular person, with the *name* attribute "Dwight D. Eisenhower." The full knowledge base would include the names of the units making up this *army* and so forth. It should also be noted that often there are multiple ways of designing an ontology, some more useful for one purpose and others more useful for another purpose. For example, rather than making *supremeAlliedCommander* a class of thing with leader as an afterthought, *individualKeyLeader* might be the class, with *supremeAlliedCommander* being a role that might attach to the class.

A Modern Conflict Ontology

This ontology codifies modern conflict roughly at the theater level. The word "roughly" is used because the level of granularity is not completely consistent. Theoretically, an ontology could include all levels of granularity, from grand armies down to the decision processes in a human mind. This is not practical, so a decision has to be made concerning the target level of granularity. The target here is based on a desire to include the entire scope of a theater-level conflict. Then, with the given scope, an estimate can be made as to what the size of an ontology would be at successively finer levels of granularity. For example, suppose that the level of granularity is set at division-level units. This level of granularity would produce a relatively small ontology of things and relations among them. Setting the granularity level to brigade-level units would result in a larger ontology and so on. The "roughness" comes in when you consider that diplomacy, economics, etc., are part of the domain. At any given granularity, what is the appropriate unit for diplomats, corporations, nongovernmental organizations (NGOs), etc.?

An Introduction to the MCO

The full description of the Modern Conflict Ontology (MCO) requires an entire book. However, for the purpose of this chapter, only portions of the ontology will be mentioned and they will be sketched, rather than fully defined. We start with a simple context diagram (Figure 1).

The *Operational Environment* (undefined) includes an *Actor* class, an *Object* class, and an *Action* class. The lowest-level classes of these classes are collectively called element classes, as distinguished from state variables. An *Actor* performs an *Action*. The *Action* affects an *Object* and/or an *Actor*. Everything in the *Operational Environment* is described by a state variable (generally a vector of variables). This means that the states of *Actors*, *Objects*, and *Actions* can be perceived by an *Actor*, which may then continue the cycle.

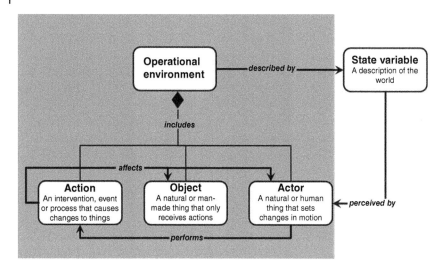

Figure 1 A simple context diagram of the MCO.

The elements of the MCO can be in only one of these classes. This was not true in earlier versions. For example, a hurricane might be modeled as an active thing that moves and causes damage – an *Actor*. On the other hand, it might be modeled as part of the environment with trafficability or vision properties – an *Object*. Similarly, the wind of the hurricane might be modeled as an *Action*. Earlier versions of the ontology provided a single class for a hurricane, allowing its use to determine the categorization. In the MCO, a hurricane can still be modeled as an *Actor*, *Object,* or Action; however, there are separate *Actor*, *Object*, and *Action* classes to clearly distinguish the role being modeled.

Actors

Actors are the simplest class. Most *Actors* are humans or groups of people. The superclass is named *genericActor* and has five subclasses:

- *individualActor* is a *genericActor* who is a particular person and is subdivided into
  - key leaders, further subdivided by occupation.
  - other individuals, further subdivided by occupation.

- *significantGroupActor* is a *genericActor* that is an organization composed of people and small enough that it does not require a density distribution and is subdivided into
  - social organizations (such as a religious faction).
  - economic organizations.
  - armed forces.

- unarmed political organizations.
- armed political organizations (such as a police force).

- *demographicGroupActor* is a *genericActor* that is composed of enough people that it is best described by a density distribution over a geographical area and is subdivided into
  - static populations (such as a religious population).
  - mobile populations (such as a class of migrants).

- *nonhumanActor* is a *genericActor* that is not a human being and is subdivided into
  - civilian vehicles.
  - military systems (such as a military ship).
  - environmental actors (such as a hurricane).
  - animals.

- *compositeActor* is a *genericActor* that is composed of other actors and is subdivided into
  - simple composite actors (such as a side in a conflict).
  - hierarchical actors (such as an armed force brigade).

These classes collectively have 15 subclasses and 114 *Actor* classes.

Objects

Objects are also fairly simple. They are the inanimate concrete or intangible things in the environment that can only be the recipients of actions. The superclass is named *object* and has seven subclasses:

- *infrastructureObject* is an *object* that is a portion of infrastructure and is subdivided into
  - water infrastructure
  - transportation infrastructure
  - government (including military) infrastructure
  - energy infrastructure
  - business infrastructure
  - social infrastructure.

- *neededThingObject* is an *object* that is something needed by people and is subdivided into
  - business objects (such as an insurance system).
  - immediate needs (such as a food supply).
  - service needs (such as trash disposal).

- *naturalObject* is an *object* that is a portion of the natural environment and is subdivided into

- conditions (such as a tempest).
- geographical objects (such as a natural or human made feature).

• *conflictObject* is an *object* that is a portion of the conflict environment and is subdivided into
- hot conflict objects (such as the piracy environment (as opposed to a piracy *Action*).
- warm conflict objects (such as stress migration).
- cool conflict objects (such as political power-sharing).

• *governingObject* is an *object* that describes the governing situation and is subdivided into
- government objects (such as the type of government).
- economic objects (such as taxation structures and policies).
- criminal objects (such as the state of drug use).
- intervention objects (such as the relations with the intervenors).

• *conceptualObject* is an *object* that is largely intangible and is subdivided into
- rights objects (such as freedom of movement).
- cognitive objects (such as opinions).
- documents (such as a military plan).

• *compositeObject* is an *object* that is composed of other objects and is subdivided into
- simple composite objects (such as the general economy).
- hierarchical objects (such as a geographical area).
- complex composite objects (such as a geopolitical unit).

These classes collectively have 24 subclasses and 193 *Object* classes.

Actions

Actions are moderately complex. They are part of two ontological organizations, one semantic and one structural. The *Actions* are semantically connected to the DIME categories (not shown here); however, as a structural matter, the superclass is named *genericAction* and has nine subclasses:

• *strikeAction* is a *genericAction* that by its nature causes damage and injury and is subdivided into
- strikes on a target type (such as strikes on dam infrastructure).
- strikes using a particular method (such as chemical strikes).
- environmental strikes (such as earth shifts).

• *C2Action* is a *genericAction* that involves command and control (C2) and is subdivided into

– control actions (such as move oneself).
– C2 actions (such as organize a force).

- *informationAction* is a *genericAction* that relates to persuasion or observation and is subdivided into
  – persuasion actions (such as decrease the legitimacy of an actor).
  – monitoring actions (such as monitor human rights practices).
  – Intelligence actions (such as collect data).

- *conflictAction* is a *genericAction* that relates to general conflict and is subdivided into
  – sustainment actions (such as distribute equipment of material).
  – security actions (such as clear mines and improvised explosive devices [IEDs]).
  – general conflict actions (such as conduct a military exercise).

- *conflictOrgOrPersAction* is a *genericAction* that changes a conflict organization or personnel and is subdivided into
  – conflict organizational actions (such as decrease paramilitary organizations).
  – conflict personnel actions (such as decrease paramilitary force personnel).

- *humanAffairsAction* is a *genericAction* that affects human affairs and is subdivided into
  – social aid actions (such as resettle people).
  – civil training actions (such as train educators).
  – civil personnel actions (such as increase law enforcement personnel).
  – change civil situation actions (such as change religious factions).

- *economicAction* is a *genericAction* that is economic in nature and is subdivided into
  – business economic actions (such as conduct labor strikes).
  – consume or produce actions (such as consume food).
  – civil building actions (such as rebuild manufacturing infrastructure).
  – business organization actions (such as decrease service businesses).

- *policingOrCriminalAction* is a *genericAction* that relates to police work or security and is subdivided into
  – criminal actions (such as conduct drug trade).
  – policing actions (such as interdict drugs).

- *civilGovernmentAction* is a *genericAction* that is governmental in nature, excluding economic and military actions and is subdivided into
  – policy or legal actions (such as change energy policy).
  – government economic actions (such as create new currency).

- government organizational actions (such as increase bureaucracy organizations).

These classes collectively have 26 subclasses and 475 *Action* classes.

Metrics or State Variables

State variables or *Metrics* are complex. They are part of two ontological organizations, one semantic and one structural. The *metrics* are semantically connected to the PMESII categories (not shown here); however, as a structural matter, the superclass is named *metricType* and has five subclasses:

- *metricVectorKey* is a *metricType* that contains the key information identifying the entire metric vector and is subdivided into
  - current situation identifying information
  - identity of the entity
  - time of the metric vector.

- *physical* is a *metricType* that describes physical attributes and is subdivided into
  - location data.
  - quantity (such as quantity of intelligence service personnel trained).
  - members (such as judicial branch members).
  - disaster indicators (such as the entity a disaster or condition, man-made or natural).
  - movable.
  - weaponry.
  - damage.

- *flow* is a *metricType* that describes flows and is subdivided into
  - capacity or flowrate
  - munition consumption
  - supply consumption
  - power consumption
  - money consumption.

- *relationship* is a *metricType* that describes relationships
  - affiliation.
  - hierarchy (such as authority level, name of superior).
  - owner.
  - initiator (such as initiator of an action).
  - object (such as object of an action).
  - activity (such as migrants activity level).
  - availability.

- *HSCB* is a *metricType* that describes human, social, culture behavior (HSCB) attributes and is subdivided into
  - decision-making
  - influence
  - fairness and corruption
  - effectiveness
  - efficiency
  - operating health
  - level rating
  - progress
  - professionalism
  - transparency.

These classes collectively have 32 subclasses and 5,953 *Metric* classes. The large number of classes results from the connection of each *Actor*, *Object*, and *Action* class to its own set of *Metric* classes. Each element class connects to the same three *metricVectorKey* classes. The other *Metric* classes are unique for each element class. However, each element class has only a subset of the lowest-level *Metric* classes shown above because only some of them pertain to the particular element class. This is illustrated in Figure 2.

There are many other parts to the ontology; however, these suffice for the moment. Other parts will be presented as they are needed. To ensure understanding before we proceed further, we will look at some examples.

MCO Examples

One of the subclasses of *individualActor* is *keyLeader*, defined as an *individualActor* who is an important, influential personage. And one of the *Actor* classes that are subclasses of *keyLeader* is *keyMilitaryIndividual*, defined as a *keyLeader* actor in an armed force. When we create an instance of *keyMilitaryIndividual*, we are thinking of a particular person, such as Dwight Eisenhower.

Figure 2 is not meant to be read; rather, it is the structure that is important here. The first two columns contain the five *genericActor* subclasses (listed above) and their subclasses. The first two rows contain the five *metricType* subclasses (listed above) and their subclasses. The crosstab cells are blacked out if each *Actor* class of a particular *genericActor* subclass has a *Metric* class of the corresponding *metricType* subclass. The white cells indicate that no *Actor* class has a *Metric* class of the corresponding type. The regularity of the resulting pattern shows that the subclass division relates to the kind of state variable (metric) information that is relevant for the Actor classes.

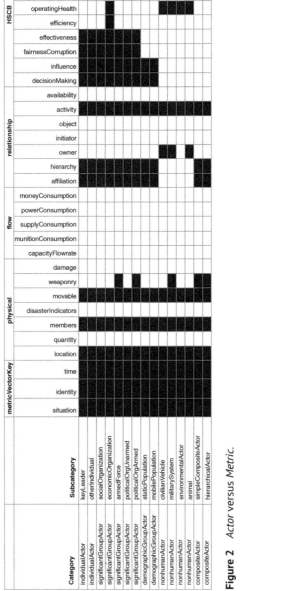

Figure 2 *Actor versus Metric.*

For example, the *keyMilitaryIndividual* class belongs in the first crosstab row. The first three crosstab columns comprise the *metricVectorKey* and are identified as follows:

- *currentSituation*: What is the identifying information on the current situation?
- *identity*: What is the name or other unique identification of the entity?
- *time*: What is the time data?

The instantiation of *currentSituation* would be "World War II," *identity* would be "Dwight D. Eisenhower," and *time* would be some date/time during the war. One other crosstab column is of interest here:

- *keyMilitaryIndividualHierarchy*: What are the Actor's authority level, name of superior, and type of distribution of authority (define hierarchy)? (If you are interested, this is the name of the *Metric* class for the *keyMilitaryIndividual* (a subdivision of *keyLeader* in the first crosstab row) that falls in the *hierarchy* column of the *relationship* class (the 17th crosstab column). You will see that almost all of the subcategories have blacked out cells in this column. The subcategories with white cells are the *civilianVehicle*, *militarySystem*, *environmentalActor*, and *animal* subcategories. *Hierarchy* is not appropriate for these. The next column, *owner*, is the appropriate Metric for all but *environmentalActor*, which does not have any corresponding concept.)

The instantiation of this class would differ depending on the time of this vector of *Metrics* entry. During the time when Eisenhower was Supreme Allied Commander, this fact would be included in the instantiation. This points out a fact about the state variable vector is actually a large list of vectors, with the *metricVectorKey* classes identifying the situation being described (for example base case, scenario 1, historical World War II, etc.), the identity of the entity, and the timestamp of the information. The rest of the vector entries consist of appropriate details. This example illustrates some of the details for the *Actors*; however, there are analogous crosstab connections for *Objects* and *Actions*.

As an example of *Object* classes, consider the class *drugCrimeOverall*. This class is a subclass of *simpleCompositeObject*, which is a subclass of *compositeObject* (listed above). The class *drugCrimeOverall* is a *simpleCompositeObject* that has *drugUse*, *drugCultivation*, *drugManufacture*, and *drugTransshipment* as components. Each of these components is a subclass of *criminalObject*, which is a subclass of *governingObject* (listed above). This construct permits the instantiation of more complex objects. In addition to *simpleCompositeObject* classes, there are *hierarchicalObject* classes that support the definition of hierarchies of objects (such as infrastructure at different political unit levels) and there are *complexCompositeObject* classes that support the definition of such things as geopolitical units and facilities. (There are also hierarchical actor classes that support the definition of hierarchical organizations.)

Provenance of the MCO

Before the MCO could be built, someone had to create knowledge about the domain. I have divided this into five parts: knowledge of warfare, knowledge of OOTWs, modeling issues, precursor ontologies, and early versions of the MCO.

Knowledge of Warfare

The discussions of warfare are nearly countless; however, I have chosen a set that describes theories of warfare. These begin with Sun Tzu's *The Art of War* (*circa* 400-320 B.C.) (Sun-Tzu, 1963), Machiavelli's *The Prince* (1513) (Machiavelli, 1966), and Clausewitz's *On War* (1832) (Clausewitz, 1993). Sun Tzu and Machiavelli couched their writings as precepts and aphorisms, with Machiavelli concentrating more on the political end of the spectrum and Sun Tzu more toward warfare. Clausewitz was at pains to analyze warfare, drawing conclusions from historical data. While he is sometimes mischaracterized as concentrating solely on combat, he actually put combat in the context of the wider political situation.

In more modern times we have Liddell Hart's *Strategy* (1954, revised 1967) (Liddell Hart, 1967); Wylie's *Military Strategy: A General Theory of Power Control* (1967) (Wylie, 2014); DuBois, Hughes, and Low's *A Concise Theory of Combat* (1967) (DuBois, Hughes, & Low, 1997); and the US Marine Corps' *Warfighting* (1967, revised 1997) (U.S. Marine Corps, 1997). Each of these authors drew on analyses in attempting to create theories of war. They also (variously) included the more political aspects, such as guerrilla warfare.

For quantitative theories, we have Lanchester's work (Lanchester, 1956), Helmbold's work (Helmbold, 1997), Hartley's *Predicting Combat Effects* (Hartley D. S., Predicting Combat Effects, 2001), Kuikka's work on stochastic attrition (Kuikka, 2015), and Paciencia et al., on modeling air attrition (Paciencia, Richmond, Schumacher, & Troy, 2018). These authors concentrated exclusively on combat effects.

Figure 3 illustrates the spread of the works of these authors, from politics to combat and from precepts to quantitative analysis.

Knowledge of OOTWs

There are also numerous sources that relate to OOTWs and the "softer" side of modern conflict, from which I have selected a few. For OOTW experience, there are Larry Wentz's *Lessons From Bosnia: The IFOR Experience* (Wentz, 1997) and Avruch, Narel, and Combelles-Siegel's *Information Campaigns for Peace Operations* (Avruch, Narel, & Combelles-Siegel, 1999). These authors presented details of past campaigns and drew lessons from them.

For analytical approaches, there are three collections of papers: Woodcock and Davis' *Analytic Approaches to the Study of Future Conflict* (Woodcock & Davis,

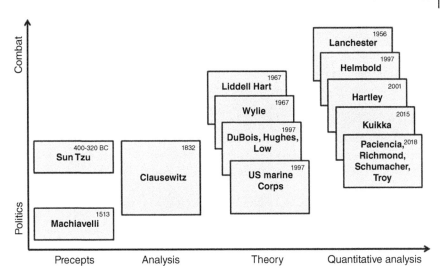

Figure 3 Knowledge of warfare.

Analytic Approaches to the Study of Future Conflict, 1996); Woodcock, Baranick, and Sciarretta's *The Human Social Culture Behavior Modeling Workshop* (Woodcock, Baranick, & Sciarretta, The Human Social Culture Behavior Modeling Workshop, 2010); and Nicholson and Schmorrow's *Advances in Design for Cross-Cultural Activities, Part II* (Nicholson & Schmorrow, 2013). Dell and Sparling examined logistics concerns for OOTW (Dell & Sparling, 2008). Each of these works presents analyses of different aspects of OOTWs.

Modeling Issues

For modeling issues, there are Hartley's *Operations Other Than War: Requirements for Analysis Tools* (Hartley D. S., Operations Other Than War: Requirements for Analysis Tools Research Report, K/DSRD-2098, 1996) and Kott and Citrenbaum's collection of papers, *Estimating Impact* (Kott & Citrenbaum, 2010). Additionally, there is the Gallagher *et al.*, rethinking of the hierarchy of models (Gallagher, Caswell, Hanlon, & Hill, 2014). An earlier paper by the author discusses using ontologies in modeling conflict (Hartley D. S., Ontology Structures for Modeling Irregular Warfare, 2012). These authors investigate the types of models that might be useful in investigating or supporting conflict.

Some particular issues in modeling require new concepts in social theory. Modeling perception, misperception, and deception is a difficult, yet important modeling issue (Mateski, Mazzuchi, & Sarkani, 2010). Hartley introduced methodologies for incorporating psychosocial issues in models (Hartley D. S., Modeling Psycho-Social Attributes in Conflict, 1995).

Some issues in modeling can be addressed by modeling techniques. Two sources discuss modeling for forecasting (Haken, Burbank, & Baker, 2010) and (Shearer & Marvin, 2010). Bernardoni, Deckro, and Robbins used social network analysis for stability operations (Bernardoni, Deckro, & Robbins, 2013).

Terrorism and insurgencies present (relatively) new conflict modeling issues. Coles and Zhuang used terrorist behavior and objectives to create archetypes to define descriptions of particular terrorist groups (Coles & Zhuang, 2016). Several authors worked on evaluating counterinsurgencies and their classification (King, Hering, & Newman, 2014), (Kucik & Paté-Cornell, 2012), (Paul, Clarke, & Grill, 2012), (Schroden, Thomasson, Foster, Lukens, & Bell, 2013), and (Ünal, 2016).

Precursor Ontologies

The original version of the MCO was derived from a set of lists, taxonomies, networks, and ontologies that described things and metrics that were needed to describe irregular warfare (IW). The lists were comprised of

- Haskins' list (Haskins, 2010).
- The Training and Doctrine Command (TRADOC) Analysis Center (TRAC) IW Decomposition list (TRAC, Jan 2009).

The taxonomies were comprised of

- The Department of State and U.S. Agency for International Development (USAID) Foreign Assistance Standardized Program (FASP) Structure and Definitions (Department of State, 2016).
- Hillson's taxonomy (Hillson et al., 2009).
- The TRAC Metrics v3 taxonomy (TRAC, 2010).
- The Office of the Coordinator for Reconstruction and Stabilization (OCRS) Essential Tasks taxonomy (Department of State, 2005).
- SRI International's Probative Rapid Interactive Modeling Environment (PRIME) taxonomy (Lowrance & Murdock, 2009).

The networks were comprised of

- Hayes and Sands network from *Doing Windows* (Hayes & Sands, 1997).
- Hartley's Interim Semi-static Stability Model (ISSM) (Hartley D. S., Interim Semi-static Stability Model, 2006).

The ontologies were comprised of

- Hartley's DIME/PMESII VV&A tool (Hartley D. S., DIME/PMESII VV&A Tool, 2008).
- Hartley's Corruption Model (Hartley D. S., Corruption in Afghanistan: Conceptual Model, 2010).

- The Measuring Progress in Conflict Environments (MPICE) ontology (Dziedzic, Sotirin, & Agoglia, 2008).

Early Versions of the MCO

The Modern Conflict Ontology has grown in scope and evolved over time.

- The first version of the ontology, version 1.0, was created as the IW Metrics Ontology for TRAC. As the name implies, it focused on irregular warfare (Hartley & Lacy, Irregular Warfare [IW] Metrics Ontology Final Report, TRAC-H-TR-13-020, 2011).
- Version 2.0 also focused on IW, but also introduced a generalized concept for agendas (Hartley & Lacy, IW Ontology Final Report, 2013) (Hartley & Lacy, Creating the Foundations for Modeling Irregular Warfare, 2013).
- Version 3.0 was renamed the Unconventional Conflict Ontology. In it several changes to the structure were introduced and new elements were created to increase consistency. Its use in modeling was described in *Unconventional Conflict: A Modeling Perspective* (Hartley D. S., Unconventional Conflict: A Modeling Perspective, 2017) and the ontology's structure was described in detail in *An Ontology for Unconventional Conflict* (Hartley D. S., An Ontology for Unconventional Conflict, 2018).
- The current version, version 4.0, has been renamed the Modern Conflict Ontology because it adds the elements of conventional combat to the ontology. Besides adding the ncccssary elements, the ontology has been restructured to be more compliant with general ontology principles.

Various versions of the ontology have been used in projects, resulting in improvements and increased credibility. These are discussed in (Hartley D. S., Unconventional Conflict: A Modeling Perspective, 2017) and (Hartley D. S., An Ontology for Unconventional Conflict, 2018).

Creating a Simulation/Wargame from the Ontology

Of course we do not know whether the MCO is or is not complete, consistent and correct; however, we strongly suspect that it is not perfect. In each new version that increased the scope, a few improvements have been made to the part that was within the old scope. This version should not be expected to be perfect when previous versions were not. However, let us suppose that the ontology is complete, consistent, and correct for the domain of modern conflict at the theater level of detail. What could we do with such a thing?

Tolk et al. defined "*a reference model as an explicit model of a real or imaginary referent, its attributes, capabilities, and relations, as well as governing assumptions and constraints under all relevant perceptions and interpretations.* The reference model captures what is known and assumed about a problem situation of interest (Tolk, Diallo, Padilla, & Herencia-Zapana, 2013)." They go on to say that a reference model provides the foundation for creating a conceptual that is needed to specify a simulation or wargame.

Model Building Steps

The perfect ontology would be just such a reference model. The imperfect, actual MCO, however, should still be pretty good at being a substitute. Figure 4 illustrates the "waterfall" style for creating a model: the requirements are used to create a design; the design is used to create the coded model (implementation); the implementation is subjected to verification, validation, and accreditation (VV&A); and the model is maintained throughout its life.

First, we make some changes at the requirements and design stages, as shown in Figure 5. Because we are dealing with modeling and simulation (M&S) software, rather than a systems engineering project, the requirements are made more flexible – "requirements" – and the problem to be solved by the software is added to the requirements stage as a driver of requirements. Then the ontology and conceptual model steps are added before reaching a design.

The "requirements" and problem statements are used to determine which parts of the ontology are relevant. (Note that scenario construction [discussed later] starts with the problem statement.) For example, in defining requirements, if no volcanic eruptions, hurricanes, or use of chemical, biological, or radiological weapons will be addressed, then those parts of the ontology can be omitted. The reduced ontology is then used to create the conceptual model (more on this later) and this is used to create the design specification.

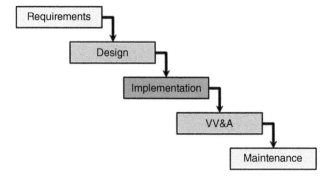

Figure 4   Old-style "Waterfall" modeling diagram.

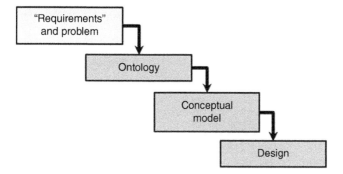

Figure 5 Modified requirements and design stages.

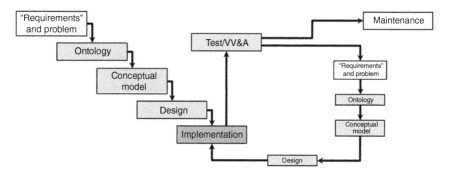

Figure 6 Ontology-based M&S software design.

Second, we make some changes to the "waterfall" structure. Because of the fluidity inherent in designing M&S software to solve a problem, a version of the "spiral" design technique has proven to be better than the "waterfall" structure. This is why Figure 6 contains a loop, with smaller versions of the "requirements" through design steps. As the developers learn from each implementation, the "requirements" and the problem may need modifications, including planned changes to add functionality. The selection from the ontology may need to be changed and the conceptual model may need changes. These changes result in a new design specification and a new implementation.

The other difference was reported by Hartley and Starr. It is a fact that the testing that is performed on software as it is being developed is a component of VV&A, making VV&A an integral part of the development process, not an "add-on" (Hartley & Starr, Verification and Validation, 2010).

There are several other activities required in building a simulation or wargame. The discussion above relates to establishing the software design process. However, there is some model infrastructure that is also required. This infrastructure includes deciding on the modeling language to be used, building or importing a simulation engine to drive the time progression of the model, defining the visualization support that will be needed, defining the analysis engine and data storage system, and defining the VV&A process and the code version control.

The building block concept, discussed in a later section, corresponds most closely with what most consider the heart of building a model, creating code. The metric model concept, discussed in the two sections after the building block section is necessary for adjudicating results in the model. Following those two sections, the VV&A section discusses VV&A and maintenance. Data requirements are discussed in a section on scenarios. These discussions are brief and, thus, incomplete. More detailed discussions are found in the modeling book (Hartley D. S., Unconventional Conflict: A Modeling Perspective, 2017).

Moving from the Ontology to the Conceptual Model

The first phase in moving from the ontology to the conceptual model consists of (tentatively) removing the elements that will not be used (based on the "requirements" and problem). We will use the *Actors* portion of the ontology as an example.

In step 1 (Figure 7), we consider the various categories to see if there is one in which we will be using none of the *actors*. Suppose we think we will not be using any *nonhumanActors*. We examine this category further and decide will be using some, but not all of them.

In step 2 (Figure 8), we consider the subcategories. For example, within *nonhumanActors* we determine that we will not be considering hurricanes, earthquakes, etc., as *actors*. (We may want to use them as environmental *objects*, however.) Similarly, we decide that the granularity of the model will be too large to consider

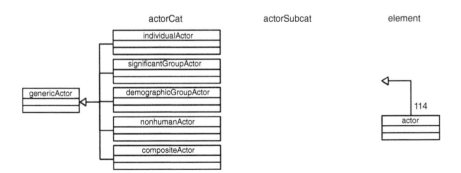

Figure 7 Reducing actors Step 1.

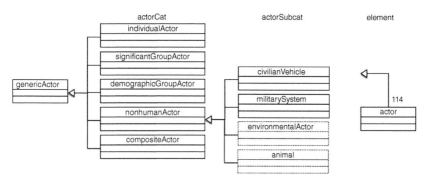

Figure 8 Reducing Actors Step 2.

animals as *actors*, such as drug-sniffing dogs, explosive-detecting dogs, or cadaver search dogs. Thus, we remove these subcategories and all of their *actor* classes.

In step 3, we examine the *actor* classes of the remaining subcategories to see if any will not be needed. Those will also be removed. We will also remove any *metrics* that are unique to the removed *actor* classes.

These steps are repeated for *objects* and *actions*.

In the discussion of the ontology earlier, I mentioned that there are other parts. We introduce two of the cross-connecting relationships here.

- The simplest relationship contains connections among elements (whether *actors*, *objects,* or *actions*) that are semantically similar. A bridge is related to a road because they both involve transportation. This relation is obvious to a human, but not to a computer. These are implemented in the ontology as semantic concepts.
- The second relationship has similarities but is more complex. A bridge is related to an action that damages bridges and to one that repairs bridges. This is a stocks-and-flows (SAF) type relationship. The bridge capacity is the stock and the damage and repairs are flows that change the capacity. Figure 9 illustrates a more complex stocks-and-flows relationship. The Host Nation *judiciarySAF* an *organizationOrientedClass* with a large number of classes participating in the relationship. The *judicialBranch* and *govtTypeUnit* are *actor* classes that are "organizations" in the figure. There are *action* classes to increase and decrease judicial organizations, to increase, decrease, and train the people in the organizations. There are other impacting *action* classes, such as changing the legal system, increase and decrease the legitimacy, and produce a constitution. The *keyJudicialLeader* class represents any key persons and the *govtPerson* class represents related *actors*. Related environment *objects* include the *legalSystemTradition* class and the international legitimacy of the organization. A *typicalAction* is the *conductJudicialAction* class.

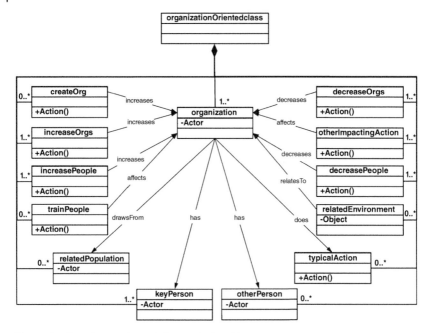

Figure 9   Complex stocks-and-flows relationship.

These two cross-connecting relationships identify elements that **may** be needed in the model as supporting elements, even if they had not been considered as necessary initially. Phase 2 involves reconsidering the tentative removals. Each of the selected elements should be reviewed using these relationships to see if additional elements should be added back into the conceptual model.

Building Block Concept

The building block concept rests on principles of object-oriented design. A simple example helps to explain the concept. Consider the *keyMilitaryIndividual* class discussed earlier. There have been many books written about object-oriented programming, design, and analysis and many programming languages have been built to support these processes. Grady Booch was one of the most influential early experts (Booch, 1991). Naturally, with so many books and authors and languages, the terminology is not uniform; however, the basic concepts are fairly well established. We will put object-oriented terms in quotes to distinguish them from similar ontology terms, which are in italics.

An "object" (as opposed to our *Object* in the MCO) is an instantiation of a "class" (similar to a class in the ontology, but basically only of either *Actor* or *Object* type).

To fully define a "class" named *keyMilitaryIndividual*, we would need to define "attributes" for the class, which are the same as our *Metrics* for *Actor* and *Object* classes. We would also need to define the "operations" of the class. These "operations" are taken from the *Action* classes and consist of those Action classes that it is appropriate for this "class" to perform. In addition, there would be technical "operations" that are needed for the model to operate on a computer, such as operations to import data into the "attributes" at the beginning of the simulation and operations to output the "attribute" values at proper times and to appropriate other "classes" throughout the execution of the model.

Inheritance is an important part of the object-oriented concept. The ontology provides the "is-a" structure of class and subclass, which defines which classes (or "classes") inherit from which class (or "class"). From a programming point of view, this means that the "attributes" and "operations" need only be defined at the highest level in which they appear and will then be inherited by the lower levels. This also means that these "classes" can be defined independently of the eventual implementation. Just as only some of the classes of the ontology will be used, only some of the "classes" will be used – the corresponding ones. This is the basic building block concept.

As a side note, another modeling decision might be to define "class" to include *Action* classes and make the "operations" of the other classes to be calls to the resultant *Action* "classes." This would preserve the current *Metric* vectors of the *Action* classes. The *Action* "classes" would then perform the actions implied by their definitions.

The MCO is actually much more complex than shown in Figure 1. As discussed earlier, there are several semantic relations that are contained in the ontology. Figure 10 illustrates the SAF relationship that includes the bridge/damage/repair relationship mentioned earlier. The classes shown here include meta-classes, which stand in for actual classes. This figure illustrates an environment-oriented stocks-and-flows (SAF) class (there are two other types of stocks-and-flows classes, which are much more complicated, one of which is shown in Figure 9).

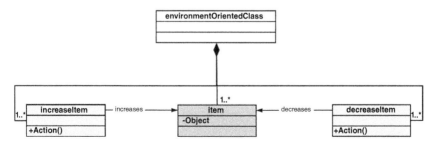

Figure 10   Environment-oriented stocks-and-flows class.

The structure derives from the fact that some items (all *Object* classes) have semantically associated *Action* classes, one of which acts to increase the item and the other acts to decrease the item. There are 49 of these *environmentOrientedClass* classes, each structured as shown, collectively relating 147 of the *Action* and *Object* classes. With respect to the building block concept, the point is that if an Object class is to be included and appears in the item position, then the corresponding *Action* classes should be strongly considered as needing to be included. These semantic relations are useful in ensuring that all of the classes that are needed for a complete and consistent model are considered.

For a more complete explanation of these structures and others, such as structures to model networks of *Actors* and ownership relations, see the ontology book (Hartley D. S., An Ontology for Unconventional Conflict, 2018).

Agendas and Implicit Metric Models

An important part of the MCO that has not been discussed is the concept of agendas. In the cold war version of conflict, there were two sides, each with its own agenda. Basically, these were identical, "beat the other side militarily." Therefore, there was not much need to dwell on the concept of agendas. In modern conflict, where two sides seems to be the exception, each side or party has an agenda. Even when parties are allied or in a strong coalition, their agendas have differences. For example, Germany contributed forces to the NATO Operation Resolute Support mission in Afghanistan, "to train, advise and assist the Afghan security forces in the ongoing conflict with groups such as the Taliban, the Haqqani network and the regional division of the Islamic State group (Lyneham, 2018)." This differs greatly from the US mission in that same operation that has included counter-terrorism operations (Wikipedia, 2018). In Modern Conflict, an agenda is can be created for each significant party to the conflict.

The content of an agenda depends on the situation; however, there is a common structure for all agendas.

- There is an owner of the agenda, some *Actor*, whether individual or group.
- There is a *goal* or a set of *goals* in the agenda.
- There is a *task* or a set of *tasks*, each connected to one *goal*, believed by the owner to be necessary (and perhaps sufficient) to accomplish the *goal*.
- Each *task/goal* pair is decomposed into a set of *subtask/subgoal* pairs.
  - The owner believes that the decomposition is complete and minimal.
  - The owner believes the *subtask* is necessary to accomplish its *subgoal*.
  - The owner believes that accomplishing all of the *subgoals* will accomplish the *goal*.
  - There is a final generic *subtask/subgoal* pair that consists of general support to the other *subtasks*, with its *subgoal* being the overall *goal*.

• Each *subtask/subgoal* is decomposed into unpaired sets of *Actions* and *Metrics*.

  – The owner believes that the collection of *Actions* is complete and minimal.
  – The owner believes that the collection of *Actions* that decompose a *subtask* is necessary to accomplish the associated *subgoal* and that the results of the collective *Actions* will be reflected in the collective *Metrics*.
  – The owner believes that the collection of *Metrics* is sufficient to determine the status of the *subgoal*.

In the MCO, this structure is implemented as a Goal-Task-Owner (GTO) Set. It is composed of meta-classes in much the same way as those of Figure 10. However, the meta-classes in that figure are resolved with real classes that are independent of the situation. (For example, the bridge/damage/repair SAF resolution is valid for all models. However, it may not be used in any given model.) The GTO Set meta-classes are only resolved when they are instantiated for a particular situation. That is, it is only when the situation has been defined that the parties requiring agendas can be known and owners are assigned. Similarly, the *task/goal* pairs, *subtask/subgoal* pairs, and *Actions* and *Metrics* can only be determined after the parties are defined.

It is after instantiation of the agendas that you can see the intrinsic metric models that exist. Each owner (at a given time, for agendas can change over time) believes that he can change the state variables in the desired directions through a set of *Actions*. This intrinsic metric model is a cause and effect chain. Each intrinsic metric model consists of the Actions, Metrics, subtask/subgoal pairs, task/goal pairs, and their connections in his GTO Set (agenda). *Each owner has a different intrinsic metric model, which is probably incomplete, inconsistent, and wrong – to some extent.*

Theoretical Metric Models

The owners' intrinsic metric models drive the owners' actions (in the modeling sense); however, they do not directly drive the results of those actions. Each *Action* (in the ontology), when coupled with a set of environmental conditions, produces results. (There is a part of the ontology that defines this; however, it will not be described here.) The critical issue for a simulation or wargame is that this result must be adjudicated. It may be adjudicated by calling a routine that is based on some scientific theory; it may be adjudicated by calling a routine that is based on an empirical set of probabilities; or it may be adjudicated by human decision-making in a wargame. This adjudication is external to the ontology. The part of the ontology mentioned above only explicitly defines the need for adjudication.

A second need for justification lies in the nature of the *Metric* classes. There are many *Metric* classes that are measures of merit at a very low level: measures of performance. For example, the amount of damage done and the amount of supplies delivered are measures of performance. There are other *Metric* classes at high levels (measures of political effectiveness) such as satisfaction of the people with the status quo. There are also *Metric* classes at intermediate levels. These *Metric* classes are not connected in the ontology. When creating a simulation or wargame, a theoretical metric model is required to connect the proper *Metric* classes with inferences. Ideally, these inferences would be based on scientific theory. Unfortunately, scientific theory is largely silent in this area and the modeler will be forced to rely on subject matter experts.

Together, the implicit cause and effect logic of adjudication and the explicit inference chain of *Metrics* constitute a theoretical metric model. This theoretical metric model is external to the ontology and must be created by the modeler; however, the ontology does provide the locations for its insertion. Ideally, the theoretical metric model would be completely and firmly based on scientific theory. Given the state of theory regarding the social domain, this ideal will not be met. The verification, validation, and accreditation process (VV&A) will supply information about the degree to which the ideal is met.

VV&A

Verification is the "process of determining that a model implementation accurately represents the developer's conceptual description and specifications." *Validation* is the "process of determining the degree to which a model is an accurate representation of the real world from the perspective of the intended uses of the model." *Accreditation* is an "official determination that a model is acceptable for a specific purpose." These definitions are taken from a US Department of Defense publication (Department of Defense, 2007). The words "intended uses" and "purpose" are there for a very good reason. "A model that is good for one purpose need not be good for another. A scale model airplane has the purpose of representing the visual and physical relationships of the model: most scale model airplanes cannot fly. A flying model airplane has the purpose of flying and usually has only a rough semblance of the appearance of the real airplane (Hartley & Starr, Verification and Validation, 2010)."

The more complicated a model is, the harder it is to verify it. Once a certain level of complication is reached, verification **must** be regarded as an ongoing process, rather than something that can be completed. Anyone who has experienced a bug in a commercial software product is, generally, a victim of this fact, rather than a victim of bad software creation processes.

For validation, "[t]he problem is that in many cases, M&S is used where a real system is not available, or not perfectly observable or where the problem is viewed with different perspectives, which leads to different solutions (human social behavior modelling for instance) (Tolk, Diallo, Padilla, & Herencia-Zapana, 2013)." Not only is a validation an ongoing process, as is verification, but also it must emphasize continued learning about the model. The phrases "degree," "accurate representation," and "intended uses" are critical in expressing what needs to be learned. We need to learn how much confidence we can place in the model over various ranges of variable values and conditions of use. "When we speak of validation, we must emphasize coverage of the appropriate domains of events and variables and expect only general directional and magnitude correspondence for output values (Hartley & Starr, Verification and Validation, 2010)."

Creating a model using the MCO has certain inherent advantages. The ontology decomposes the operational environment into a neat set of things that can be checked to see, first, if they are covered in the model and then how well they are covered, both with respect to verification and a validation. This ontological organization also supports an accreditation decision based on a holistic picture of the model.

At any given point in time, both verification and validation will be incomplete. This incompleteness requires periodic testing to supplement knowledge, which implies the need for good record keeping. The VV&A process recommended by Hartley and Starr contains four types of testing/VV&A: development, periodic, triggered, and M&S use.

- Development testing/VV&A consists of all of the testing, both for verification and validation, that takes place during the model development cycle, followed by an accreditation decision.
- Periodic testing/VV&A consists of a defined cycle of testing (perhaps annually) during the life of the model, in which new verification and validation tests are performed to increase the knowledge about the model, followed by an accreditation decision.
- Triggered testing/VV&A consists of a set of testing that occurs when triggered by some event, such as changes to the model or its data (perhaps a new scenario). This testing includes some old tests to make sure that previous competence has not been lost and that any new features meet the appropriate standards. It also is followed by an accreditation decision.
- M&S use testing/VV&A consists of evaluating (and recording) the results of using the software system. If the use results in no known problems and produces satisfactory results, this fact adds to the knowledge of the M&S and should be entered into the body of knowledge of the system because it improves the level of confidence. Any observed problems should be evaluated and entered into the record as needing attention. This attention may require immediate corrective action, leading to triggered testing or it may be added to the list to be

addressed in the next version of the system where it will be checked in development testing.

The discussion of the four types of testing/VV&A above omits management of residual risk, which actually precedes the accreditation decision. This is actually alluded to in the last type. Following each testing/VV&A type, verification and validation are still incomplete. The knowledge has just been improved. There will still be a residual risk that there is something wrong. This risk must be managed, either with verbal and written caveats, by restricting the domain of use, or by some other method.

Wargames introduce an additional consideration into the discussion, namely the actions of humans during the operation of the system. The computer code portion of the system still needs to undergo the testing/VV&A described above. The problem lies in the evaluation of results. Without the human component, the system operation is repeatable (within the limits of possible stochastic variables in the system). If the human interaction is restricted to simple choices at various points in the system operation, then it can be treated as a random choice with effects on the sequels. If the human interaction includes changing data, the problem becomes more complex. Within the context of the DIME/PMESII theoretical questions of any modern conflict model, it is not clear that the VV&A problem is noticeably harder for a wargame than for a simulation without internal human interactions.

The major difference between the VV&A processes for older combat models and modern conflict models lies in the metrics required for assessing validity. Modern conflict models rely heavily on the use of social theories in determining the results of actions, whereas older combat models relied more heavily on physical theories, with stronger validity. This requires both an emphasis on PMESII metrics and the understanding that modern conflict models will require much more care in their use and assessments of their results. In particular, the assessment of the theories used in a model becomes central to VV&A for models of modern conflict. Although risk assessment and mitigation – and the understanding of residual risks – are common to both cases, the likelihood of larger risks is inherent in modern conflict models.

The ontological underpinnings of this modeling approach ensure that the problem of testing for coverage is relatively simple. The things that should be covered are explicitly modeled (or explicitly not modeled). The object-oriented implementation makes it relatively simple to verify the implementation of the conceptual model and perform validation on the more mundane aspects of the model. The explicit exposure of the use of theories makes the assessment of the validity of the more difficult aspects of the model easier.

Constructing the Scenario

As mentioned earlier, the initial scenario and variants will be created as part of the definition of the problem that is a driver for the system requirements. Generally, if the model is successful – using some unspecified definition, it will be used with new scenarios, which will have to be created.

A scenario starts with an English (or other natural language) description of a situation and contains what is worrisome about the situation (i.e. that the volcano will explode and cause various kinds of damage to infrastructure, social structures, and psyches or that an insurgency will threaten a government, etc.). It will also include potential reactions to the situation and desired outcomes from the reactions. As good modeling practice, the scenario should also include a "road to war" narrative, with metric data. This "road to war" narrative is a prequel to the beginning of the actual simulation or wargame that motivates the situation and actions. The end of the "road to war" is the starting point of the model and contains all of the data needed for the simulation or wargame to run from that point.

When this narrative is converted to inputs to the system, it contains all of the instantiations of the "classes" (derived from the ontology classes) and the "attribute" values corresponding to the instantiations of the *Metric* classes.

Assuming the MCO is perfect, then every input type is predefined in the model and only needs to be instantiated – except for a relatively small number of control variables, such as time-step size, input and output file names, etc. (The ontology contains classes for model (scenario) name etc., so some control variables are also pre-existing.)

Model Infrastructure

Figure 11 recapitulates the model design process. The "requirements" and the scenario narrative are used to select the appropriate parts of the Modern Conflict Ontology (MCO). This is used to define the conceptual model. The conceptual model and the selected parts of the MCO are used to create the object-oriented (OO) design. This design is then coded into the simulation or wargame.

The simulation or wargame is not complete when the model is complete. The model will be coded in some computer language or set of languages. The model will need a simulation engine to drive its progress. Analysis and data storage systems will be required to support the use of the model. Almost certainly, a visualization system will be required. Finally, a strong version control system is required so that VV&A results and statements can be tied to the proper object.

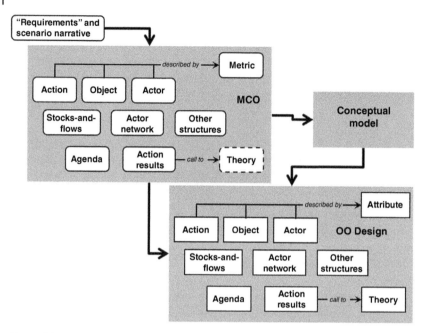

Figure 11 Model design diagram.

This model infrastructure is required; however, it is independent of the ontology. Any wargame or simulation requires this infrastructure. Therefore, it will not be covered in this chapter.

Conclusion

Using a (good) ontology to design a wargame or simulation is arguably the best way to do so. The ontology is independent of the wargame or simulation and is thus free to concentrate on the elements that could or should be covered by the model. Both wargames and simulations require a model (in the most general sense of the word) of the domain of interest. For a wargame/simulation system, the model must be consistent throughout the system. This commonality is difficult to achieve when the model resides only in the minds of the creators. An ontology provides a significant starting point for a well-defined model by identifying and defining all of the potentially needed components of the model and their relationships. For a wargame or simulation that has the granularity approximating a theater-level model, the MCO provides the best option. Not only does it provide the proper list of elements and attributes to populate object-oriented classes, but it also provides semantic relationships that offer tests for missing elements.

Further, the MCO supports the creation of agendas for the principal parties of the conflict in question. The creation of these agendas is important in defining the scenario and defining the calls for adjudication (important VV&A points). In human conflict domains, the cause and effect processes must be supplied by human wargamers or by calls to theoretical or probabilistic algorithms in the simulation. Thus, in human conflict domains, the ontology supplies a framework for the model definition that identifies the places where decisions are needed concerning whether to use the wargaming or the simulation part of the system. The implication of connections among the metrics also provides an explicit call for theoretical connections (important VV&A points).

The use of the MCO in designing the wargame or simulation enforces a division between the "easy parts" of the model (the choices of things to model, their relationships, and their implementations) and the "hard parts" of the model (the cause and effect relations and implications). This division supports good VV&A.

The MCO should be useful for modeling conflict at a coarser level than theater level by selecting only those elements that reflect the desired level of granularity. If the model will require a finer level of granularity, the MCO will not be appropriate. The process described in this chapter will need an ontology that is oriented to a finer level of conflict. However, it should be just as efficacious.

References

Arp, R., Smith, B., and Spear, A.D. (2015). *Building Ontologies with Basic Formal Ontology*. Cambridge, MA: The MIT Press.

Avruch, K., Narel, J.L., and Combelles-Siegel, P. (1999). *Information Campaigns for Peace Operations*. Washington, DC: CCRP.

Bernardoni, B.J., Deckro, R.F., and Robbins, M.J. (2013). Using social network analysis to inform stabilization efforts. *Military Operations Research* 18 (4): 37–60.

Booch, G. (1991). *Object Oriented Design with Applications*. Redwood City: The Benjamin/Cummings Publishing Company, Inc.

Clausewitz, C.v. (1993). *On War*. (M. Howard, & P. Paret, Trans.). New York: Alfred A. Knopf, Inc.

Coles, J. and Zhuang, J. (2016). Introducing terrorist archetypes: using terrorist objectives and behavior to predict new, complex and changing threats. *Military Operations Research* 21 (4): 47–62.

Dell, R. F., & Sparling, S. J. (2008). Optimal distribution of resources for non-combatant evacuation. *76th Military Operations Research Society Symposium*. New London, CT: MORS.

Department of Defense. (2007). "Department of Defense Directive Number 5000.59.". http://www.dtic.mil/whs/directives/corres/pdf/500059p.pdf

Department of State (2016). Updated foreign assistance standardized program structure and definitions. In:. http://www.state.gov/f/releases/other/255986.htm.

Department of State, O (2005). Post-conflict reconstruction. In:. http://peacebuildingcentre.com/pbc_documents/US_State_Department_Post_Conflict_Essential_Tasks_2005.pdf.

DuBois, E.L., Hughes, W.P., and Low, L.J. (1997). *A Concise Theory of Combat, NPS-IJWA-97-001*. Monterey: Institute for Joint Warfare Analysis, Naval Postgraduate School.

Dziedzic, M., Sotirin, B., & Agoglia, J. (2008). *Measuring Progress in Conflict Environments (MPICE) - A Metrics Framework for Assessing Conflict Transformation and Stabilization*. DTIC.

Fensel, D. (2004). *Ontologies: A Silver Bullet for Knowledge Management and Electronic Commerce*, Seconde. Berlin: Springer-Verlag.

Gallagher, M.A., Caswell, D.J., Hanlon, B., and Hill, J.M. (2014). Rethinking the Hiearchy of Analytic Models and Simulations for Conflicts. *Military Operations Research* 19 (4): 15–24.

Haken, N., Burbank, J., and Baker, P.H. (2010). Casting Globally: Using Content Analysis for Conflict Assessment and Forecasting. *Military Operations Research* 15 (2): 5–19.

Hartley, D.S. (1995). Modeling Psycho-Social Attributes in Conflict. In: *In AMORS, Third Asia-Pacific Military Operations Research Symposium: Operations Research for Regional Security Issues in Peace and Low Intensity Conflict*. Bangkok, Thailand: AMORS.

Hartley, D.S. (1996). *Operations Other Than War: Requirements for Analysis Tools Research Report, K/DSRD-2098*. Oak Ridge, TN: Lockheed Martin Energy Systems, Inc.

Hartley, D.S. (2001). *Predicting Combat Effects*. Linthicum, MD: INFORMS.

Hartley, D. S. (2006). *Interim Semi-static Stability Model*. Retrieved May 4, 2016, from Hartley Consulting: http://drdeanhartley.com/HartleyConsulting/TOOLBOX/issm.htm

Hartley, D. S. (2008). *DIME/PMESII VV&A Tool*. Retrieved May 4, 2016, from Hartley Consulting: http://drdeanhartley.com/HartleyConsulting/VVATool/VVA.htm

Hartley, D.S. (2010). *Corruption in Afghanistan: Conceptual Model. NDU Corruption Workshop*. Washington, DC: National Defense University.

Hartley, D.S. (2012). Ontology Structures for Modeling Irregular Warfare. *Military Operations Research* 17 (2): 5–18.

Hartley, D.S. (2017). *Unconventional Conflict: A Modeling Perspective*. New York: Springer.

Hartley, D.S. (2018). *An Ontology for Unconventional Conflict*. New York: Springer.

Hartley, D.S. and Lacy, L.W. (2011). *Irregular Warfare (IW) Metrics Ontology Final Report, TRAC-H-TR-13-020*. Ft Leavenworth, KS: US Army TRAC.

Hartley, D.S. and Lacy, L.W. (2013). Creating the Foundations for Modeling Irregular Warfare. In: *Advances in Design for Cross-Cultural Activities, Part II* (eds. D.M. Nicholson and D.D. Schmorrow), 13–23. Boca Raton, FL: CRC Press.

Hartley, D.S. and Lacy, L.W. (2013). *IW Ontology Final Report*. Ft Leavenworth, KS: US Army TRAC.

Hartley, D.S. and Starr, S. (2010). Verification and Validation. In: *Estimating Impact* (eds. A. Kott and G. Citrenbaum), 311–336. New York: Springer.

Haskins, C. (2010, Sept-Oct). A Practical Approach to Cultural Insight. *Military Review*.

Hayes, B.C. and Sands, J.I. (1997). *Doing Windows: Non-Traditional Military Responses to Complex Emergencies*. Washington, DC: CCRP.

Helmbold, R.L. (1997). *The Advantage Parameter: A Compilation of Phalanx Articles Dealing With the Motivation and Empirical Data Supporting Use of the Advantage Parameter as a General Measure of Combat Power*. Washington, DC: US Army Concepts Analysis Agency.

Hillson, R. et al. (2009). *Requirements for a Government Owned DIME/PMESII Model Suite*. Washington, DC: Office of the Secretary of Defense Modeling & Simulation Steering Committee.

King, M.L., Hering, A.S., and Newman, A.M. (2014). Evaluating counterinsurgency classification schemes. *Military Operations Research* 19 (3): 5–25.

Kott, A. and Citrenbaum, G. (eds.) (2010). *Estimating Impact*. New York: Springer.

Kucik, P. and Paté-Cornell, E. (2012). Counterinsurgency: a utility-based analysis of different strategies. *Military Operations Research* 17 (4): 5–23.

Kuikka, V. (2015). A combat equation derived from stochastic modeling of attrition data. *Military Operations Research* 20 (3): 49–69.

Lacy, L.W. (2005). *OWL: Representing Information Using the Web Ontology Language*. Victoria, BC: Trafford.

Lanchester, F.W. (1956). Mathematics in warfare. In: *The World of Mathematics*, vol. 4 (ed. J.R. Newman), 2138–2157. New York: Simon and Schuster.

Liddell Hart, B.H. (1967). *Strategy, Second Revised Edition*. New York: Meridian.

Lowrance, J.D. and Murdock, J.L. (2009). *Political, Military, Ecomomic, Social, Infrastructure, Information (PMESII) Effects Forecasting for Course of Action (COA) Evaluation*. Rome, NY: Air Force Research Laboratory.

Lyneham, C. (2018). *Bundeswehr in Afghanistan: What you need to know*. Retrieved August 4, 2018, from DW: https://www.dw.com/en/bundeswehr-in-afghanistan-what-you-need-to-know/a-43122424

Machiavelli, N. (1966). *The Prince*. (D. Donno, Trans.) New York: Bantam Books.

Mateski, M.E., Mazzuchi, T.A., and Sarkani, S. (2010). The Hypergame Perception Model: A Diagramatic Approach to Modeling Perception, Misperception, and Deception. *Military Operations Research* 15 (2): 21–37.

Nicholson, D.M. and Schmorrow, D.D. (eds.) (2013). *Advances in Design for Cross-Cultural Activities, Part II*. Boca Raton: CRC Press.

Paciencia, T.J., Richmond, D.J., Schumacher, J.J., and Troy, W.L. (2018). A Framework and System for Theater Air Attrition Modeling. *Military Operations Research* 23 (2): 41–59.

Paul, C., Clarke, C. P., & Grill, B. (2012). Qualitative Comparative Analysis of 30 Insurgencies, 1978-2008. *Military Operations Research, 17*(2), 19-40.

Schroden, J., Thomasson, R., Foster, R. et al. (2013). A New Paradigm for Assessment in Counterinsurgency. *Military Operations Research* 18 (3): 5–20.

Shearer, R. and Marvin, B. (2010). Recognizing Patterns of Nation-State Instability that Lead to Conflict. *Military Operations Research* 15 (3): 17–30.

Sun-Tzu. (1963). *The Art of War*. (S. B. Griffith, Trans.) New York: Oxford University Press.

Tolk, A., Diallo, S.Y., Padilla, J.J., and Herencia-Zapana, H. (2013). Reference modelling in support of M&S - foundations and applications. *Journal of Simulation* 7 (2): 69–82.

TRAC (2010). *Metrics v3.xls*. Ft Leavenworth, KS: TRAC.

TRAC (2009). *IW Decomposition Analytic Strategy: Overview Briefing for IW WG*. Ft Leavenworth, KS: TRAC.

Marine Corps, U.S. (1997). *Warfighting, MCDP 1*. Washington, DC: U.S. Marine Corps.

Ũnal, M.C. (2016). Counterinsurgency and Military Strategy: An Analysis of the Turkish Army's COIN Strategies/Doctrines. *Military Operations Research* 21 (1): 55–88.

Wentz, L. (ed.) (1997). *Lessons From Bosnia: The IFOR Experience*. Vienna, VA: CCRP.

Wikipedia. (2018, August 4). War in Afghanistan (2001-present). https://en.wikipedia.org/wiki/War_in_Afghanistan_(2001%E2%80%93present)

Woodcock, A. and Davis, D. (eds.) (1996). *Analytic Approaches to the Study of Future Conflict*. Clementsport: The Canadian Peacekeeping Press.

Woodcock, A., Baranick, M., and Sciarretta, A. (eds.) (2010). *The Human Social Culture Behavior Modeling Workshop*. Washington, DC: National Defense University.

Wylie, J.C. (2014). *Military Strategy: A General Theory of Power Control*. Annapolis: Naval Institute Press.

15

Agent-Driven End Game Analysis for Air Defense

M. Fatih Hocaoğlu

Istanbul Medeniyet University, Istanbul, Turkey

Motivation and Overview

End game analysis is basically a type of analysis used to understand missile defense capability, plan defense design, evaluate interoperability, and assess defense performance. Four types of analysis come under end game analysis: missile footprint analysis, operating area, defended area, and scenario view.

In this study, an agent-based solution is developed to automize end game analysis scenarios and to analyze them in run-time. The agent creates parallel and nonparallel footprint analysis scenarios automatically and executes them. Automated analysis is applied to all these types of scenarios and is also supported by a reasoning mechanism. Depending on the results of this reasoning, the scenario being executed is diverted to alternative cases such as changing search intervals for sensors, changing altitude and so on. The basic motivation is to use the intelligence capability of the agents to design the missile footprint analysis and enrich the scenario analysis by using automated analysis to achieve a standardized footprint analysis.

Introduction

The land-based missile defense system effectiveness analysis is composed of a set of so-called End Game Analyses. End game analysis aims to measure the defense success level of a surface to the air defense system success level from the conditions that are idealized to the real scenario conditions for both opponent missile

Simulation and Wargaming, First Edition. Edited by Charles Turnitsa, Curtis Blais, and Andreas Tolk.
© 2022 John Wiley & Sons, Inc. Published 2022 by John Wiley & Sons, Inc.

types and the types and configurations of defense systems. The idealized conditions consist of flat world assumptions, with no mountains or weather activity, and with perfect communication between units. The complex conditions are defined as real-world situations, such as a world with communication latency and loss, weather conditions (such as rain, snow, turbulent weather), real earth representation. The aims of the analysis are to determine the best engagement rules, defense missile site deployment strategy, missile flight profile selection, and C4ISR architectures and topologies for sensory systems and communication systems. With regard to their purposes and complexities, the scenarios are taken into consideration and placed under the categories of *Basic Missile Footprint Analysis*, *Defended Area Analysis*, *Operational Area Analysis*, and *Scenario View*. Using the basic missile footprint analysis that is implemented under the assumptions of a flat world and there being no mountain on it, the defense success criteria are measured against that of the opponent missiles in parallel and nonparallel fire patterns to determine system configurations and deployment. All of the analyses mentioned here have graded complexities, from measuring the defense success of a defense system that is in charge of an area to be defended, to making decisions regarding the deployment of defense systems on an area in order to defend a set of high-value assets.

In this study, an agent model named "*Missile Footprint Analysis Model*" is developed, based on the Agent driven Simulation Framework (AdSiF) for the analysis of land-based defense systems. The agent is not only in charge of the missile defense system's run-time analysis, but it also designs and manages simulation scenarios. Scenario management consists of creating grids for both the defense and opponent sides, deploying their sensors on defense sites (sensors and launchers are sometimes deployed on different grids), igniting opponent missiles on their destination grids, and collecting the statistics of defense success. The statistics include the grid defense ratio, calculation of the distances for target detection points, identification points, recognition points, and interception points. Additionally, the effects of the defense success of communication delays, detection cues between sensors, and C4ISR Architectural design decisions are also taken into consideration. The agent uses all the computed statistics as inputs for a rule-based decision-making process. Using its reasoning mechanism with predicates in its knowledge base, it makes the decisions that will guide the execution of the scenarios. The decisions may change the process of the execution of the scenario or the scenario's resources and deployment being used to create a new scenario that will result in better statistics.

The study is organized as follows. In section "Related Studies," a brief literature survey is given. The agent-directed simulation concept, endgame agent, command and control structures, and the simulation infrastructure AdSiF that the implementations are done on are summarized in section "Agent-Directed

Simulation and AdSiF." In section "Aims and Performance Measurement," the performance measurements widely used in an air defense simulation scenario are summarized. The model library necessary for such a simulation environment is explained in section "Types of End Game Analysis." The types of analysis used in end game analysis are summarized in section "Types of End Game Analysis." End game agent online analysis and scenario replication design capability are summarized in section "Online Analysis and Scenario Replication Design." A complete scenario design and analysis example are given in section "An Air defense Scenario: Scenario View." A conclusion part summarizes the idea in section "Discussions."

Related Studies

An air defense study has a cycle called *observe–orient–decide–act (OODA)*. The observation phase includes the collection of sensor detections from the sensors, while the orient phase is related to target identification using data fusion and rule-based decision-making. The phases "decide" and "act" are related to a target being selected, a weapon being assigned to it, and an engagement being initiated. The whole process is repeated until all targets or loosing defense units are coped with.

The purpose of an air defense system is to evaluate the air scenario and identify hostile aircrafts that represent a real threat to the assets under protection. For the finding of solutions for threat evaluation problems, rule-based systems [1], Dempster–Shafer based systems [2], and Bayesian network systems [3] probabilistic systems [4], or fuzzy logic-based reasoning technologies are widely used [5].

Weapon target assignment (WTA) optimization is taken both as a multi-objective decision-making (MODM) and a multiple-attribute (criteria) decision-making problem (MADM). Analytic hierarchy process (AHP), a commonly used MADM technique, is one of the most commonly used algorithms in the solving of this problem.

Goal programming is one of the most commonly used multi-objective optimization techniques and is also used to solve weapon target assignment optimization [6]. Some exact solutions for WTA problems have been proposed, such as integer programming [7] and network flow-based lower-bounding methods that are obtained using a branch-and-bound algorithm [8]. It is hence proved that heuristic search algorithms [9] such as swarm optimization [10], genetic algorithms [11, 12], ant colonies [13, 14], and chicken swarm algorithms give fast and accurate solutions for these problems.

In target evaluation, aerial target threat evaluation criteria such as flight distance, flight velocity, flight height, short cute route, and flight angle are represented by fuzzy membership functions in order for the threats to be arranged according to their threat score by using the Grey rational analysis method [15]. In the study, the target evaluation problem is taken to be a multi-criteria decision-making problem.

To cope with the shortcomings of both AHP and goal programming, there are studies that combine both MADM and MODM techniques.

In modern aviation, the layered defense approach is a central tenet. According to this doctrine, each layer compensates the other's weakness. In such a defense strategy, each layer is added to satisfy some specific security and safety requirements [16].

Jackson and Latourette put forward a set of aviation security measurements such as obscure target, detect target, harden target, interdict attack or dynamically mitigate effects, and respond to and recover from attack [16].

Some swarm optimization algorithms such as ant colonies [17] and hybrid algorithms consisting of improved artificial fish-swarm algorithm (AFSA) and improved harmony search (HS) [18] are applied to the WTA problem. Using the ant colonies algorithm, a solution is developed considering the WTA problem to be a bi-objective WTA (BOWTA) optimization model that maximizes the enemy's expected damage and minimizes the cost of missiles [13].

Intelligent agents are used in air defense analysis simulations for mission planning and execution, as well as for threat assessment and weapon selection. A model for autonomous intelligent agents in an air defense system, which uses the concept of meta-level plan reasoning to make decisions about threat assessments, countermeasures, and weapon allocations, is developed by Das [19]. The agent developed in this study achieves a logical and statistical evaluation. The agent's goals and their mutual conflicts are represented through rule-based logical evaluation and the agent's behavior is characterized by random statistical distributions.

Defining air defense behavioral structures and uncertain environments in fuzzy reasoning is widely studied domains in the existing literature available on the subject [20]. In this study, there is uncertainty present in information acquisition, command and control, operational effectiveness, and the natural environment of the battlefield.

Air defense missions are also seen as a scheduling problem. An air defense formation, surface/underwater warship, or land-based formations are defined using three basic concepts: resources, channels, and missions. Resources refers to the participating entities, which include targets, missiles (weapons), and sensors. A channel is a collection of air defense resources required for an air defense mission, including guidance resources and weapons resources. A mission is the time interval in which one attack is performed during the continuous air defense operations, including the channel, target, and the implementation of the attack time. The scheduling optimization is related to the question of which missile is to be fired toward which target and when, so that the scheduled actions succeed as a whole in a given mission goal [21]. With regard to this concept, assigning multitask to multi agents is also a task assignment optimization problem [22].

Communication between air defense units in order to be able to increase situational awareness is a fundamental component of C4ISR concept. Nowadays, information is enriched by space information and mainly includes imaging reconnaissance information, electronic reconnaissance information, early warning reconnaissance information, mapping information, ocean surveillance information, weather information, navigation orientation information, and all kinds of communication information [23].

For an air defense analysis, in the existing literature, there are two main tendencies, which are game theory solutions and decision tree-like, decision analysis-based solutions. In the decision tree approach, the action to be taken is chosen using a decision tree consisting of the choice alternatives of the attacker and attackee with their respective selection probabilities, effects, and after action consequences [24].

In the existing literature, the deployment of sensors and defense systems including fire schedule optimization are studied as a part of protocol determination. The protocols determined require the threat constraints of (i) perceived threat inventory, (ii) threat missile, and (iii) the ground assets being attacked [25].

Although end game analysis takes into consideration land-based air defense systems, some reliability analysis is also done for portable air defense systems [26]. Reliability, which is a measure of important weapons system performance indicators, is also an important factor that affects system performance.

Agent-Directed Simulation and AdSiF

In this study, all entities accommodated in a simulation scenario, such as missiles, weapon systems (launchers), communication devices, and sensors, are designed as agents. There are two intelligent agents: The Command and Control agent and End game analysis agent. While the C2 agent and End game agent consist of a set of goals that make them goal-based agents, with a knowledge base that has a set of predicates and facts, a reasoning mechanism that allows them to guide and choose behaviors with regard to reasoning results, the others are just reactive agents that respond to the events occurring in and the state changes in the environment in which they are situated. Although, while it seems as if the end game agent follows a well-defined procedure and reacts to events, it also changes the scenario depending on user-defined rules, which moves it beyond being a reactive agent. In contrast with what the reactive agents do, the C2 agents choose actions to satisfy their goal(s) by evaluating the situation. In this sense, it can be said that the C2 agent is a deliberative agent. In general, deliberative agents do not ignore or avoid reacting to the events happening around them, but choose behaviors that will take them closer to their goals.

A BDI agent has beliefs, desires, and intentions. Beliefs refers to the information that is assumed to be true of the world and which is unquestioned. For example, detections received from a sensor are considered as being true at the defined level of accuracy, and they are stored as time-labeled facts in the knowledge base. Desires represent the tasks that they wish to achieve, and a set of tasks that they have chosen to carry out in accordance with the situation is defined as their Intentions. In the case of air defense, for a commander to defend their assets with complete success by eliminating all the enemy missiles that are targeted at the units being defended is considered as a desire. During defense, the arranging of resources to defend relatively high-priority targets and expend less focus on the targets that have lower importance or to deal with a lack of resources can be defined as an intention.

An agent in an end game analysis scenario perceives the geographical environment and the other entities in the scenario as their environment. An agent environment is categorized as being:

- Accessible: The environment is an accessible environment because it is one in which the agent can obtain complete, accurate, up-to-date information about the environment's state. However, the rest of the agents such as command and control systems can also reach information via their sensors.
- Nondeterministic environment: The environment is detected by the C2 agent sensor sensitivity, and action selections are determined by a particular probability. For example, the target selection is performed according to the score calculated based on the sensor data obtained.
- Episodic: In an episodic environment, simulation executions consist of a repeating scene. The differences between the scenes are defined within the framework of the analysis carried out.
- Dynamic: Both the end game agent and the command and control agent work in and interact with the environment. Since the assets have a dynamic nature, they exhibit changes in a continuous form on the time axis.
- Hybrid (discrete and continuous): In a hybrid environment, a scenario includes simulation of assets being executed on a discrete and continuous event execution. For example, missiles are modeled to be continuous event simulation, but command control agents, sensors, and communication devices are modeled as discrete models.

End game analysis scenarios are dynamic, hybrid, episodic, accessible, and nondeterministic environments.

The interaction between agents is designed as event-driven architecture, and there is no dependency between software components. An agent interacts with any other agent for two purposes: *"information request"* and *"requesting the assignation of a task"* [27]. In the information request, an agent asks for the state information of any other agent, and while asking for a task to be assigned to them, an agent asks any other agent to make a specific task. The events trigger a behavior or a set of behaviors depending on the state the agent is in. The event handling, time management, and behavior management of the agents are provided by the abstract simulation interpreter. The simulation interpreter executes behavior as developed language components in the state automaton form. The interpreter does not contain any information on the area of application on which the entities operate.

An intelligent agent has a reasoning mechanism, consisting of a knowledge base that stores facts (beliefs) in a time-tagged way to be updated or populated during a simulation about a simulation environment, and a set of predicates that are constituted on the facts in horn clause form. The statistics are used as input parameters for the predicates, and output parameters are used as the rules.

AdSiF: Agent Driven Simulation Framework

Agent driven Simulation Framework (AdSiF) is a declarative simulation language, an environment for agent development, and in more general terms, a computation environment. It basically provides a state-oriented interpreter and a simulation layer to manage simulation execution algorithms for both discrete-event and continuous-event systems. In comparison with the currently used agent programming systems (e.g., RAP [28, 29]), from the simulation and modeling environment point of view (e.g., DEVS [30] and to CosMos [31]), AdSiF offers a single paradigm built on agent based, object oriented, aspect oriented, and logical programming (termed here as "state oriented"). The power of this paradigm stems from its ontological background. Its ontological commitment extends from describing what exists to include the modeling of mental abilities through the use of a reasoning mechanism, thereby driving behaviors.

Ontologically, AdSiF constructs a world composed of operations (behaviors, which are sequences of states), a set of events trigging each behavior, and a time-reference with a reasoning mechanism for managing the behaviors. The categories of the related behaviors are handled by the components termed as behavior containers, which represent different aspects of the system being simulated. All these components are organized within a layered architecture, in which the behaviors are defined on the basis of state-oriented paradigms and executed by a kernel. In this way, the services of event synchronization, time management, and data management are defined as being semantically independent of the kernel.

Since AdSiF is also based on agent-oriented programming, each model has its own dual-world representation. A dual-world representation is defined as an inner representation of an environment to be simulated and is constructed of time-stamped sensory data (or beliefs) obtained from that environment, even when this data contains errors. Time-stamped facts represent the models' own earlier values of state variables along with those of other models within that environment. In this way, dual-world representation is a good candidate for the representation of conflicting behaviors and knowledge as they occur in the real world.

End Game Agent

The end game agent is in charge of missile footprint analysis planning and the scheduling of events to fire the opponent side's missiles to their targets, and the scenario has at least one commander agent managing defense side, with the necessary assets such as sensors, weapon systems, and any other equipment required. The responsibilities consist of scenario design, execution management, collecting statistics, managing replications, and altering resources under given conditions. The conditions are determined using the graphical user interfaces (GUIs) of the agent and consist of missile types to be analyzed, selections for sensory systems and communication systems, and a series of preferences such as allowing sensor cueing and remote data launches. In run-time analysis, it undertakes the tracking of scenario dynamic parameters, computing statistics, and making scenario management decisions.

The end game agent implements footprint analysis, operating area analysis, and defended area analysis in a serial order. In scenario view analysis, with it being a more complex analysis that is different from the others, the end game agent collects execution data and calculates statistics, and applies the rules given by the scenario designer to divert the scenario. But the scenario design as a whole is achieved by the analysts.

Command and Control Agent

C2 agents are accommodated in an end game scenario as a defense side-command and control unit. The commander creates a tactical picture using sensory data and the information obtained from data fusion for target positions and target identities. The knowledge that is used to create the tactical picture is stored in the agent's knowledge base in time-labeled fact forms. In this sense, the commander agent is aware of the current and past tactical situations. The knowledge belongs to an earlier time and the current time as well, both of which are used in decision-making. For example, whether an attacker gets closer to its target is decided using its position trajectory, which is a time-labeled position information series. The examples of predicate and fact structure are shown in Table 15.1. The predicate

Table 15.1   The predicates.

Predicate	Definition
detection(Snsr, Stype, Trgt, PosX,PosY,PosZ0, Vel, Dr, Dst, W, L, Ac, DstR, T)	Formula (15.1)

Snsr: The sensor ID that sends the detection	DstR: Distance ratio
Stype: Sensor type (1: 2: radar)	Dst: Target distance
Trgt: Detected target Id	W: Target width (as pixel)
PosX, PosY, PosZ: Target position X, Y, Z components	L: Target length (as pixel)
	Ac: Sensor accuracy level
Vel: Target velocity	T: Detection time
Dr: Target direction	

Approaching target	It is decided that when an attacker approaches an attackee (a defense site or a defended asset), the distance between the attacker and the attackee gets closer by time. If it is decided that it is a threat that is approaching, the defense element starts the engagement

approaching(TargetId, DefenseSiteId):-distance(TargetId, DefenseSiteId, Distance0, Time0),
distance(TargetId, DefenseSiteId, Distance1, Time1),
distance(TargetId, DefenseSiteId, Distance2, Time2), Time0>Time1, Time1>Time2,
Distance0<Distance1, Distance1<Distance2.
approaching(_, −1).

Getting far away	It is the opposite of the predicate "Approaching target," and if there is an engagement for the target, it is terminated

gettingFarAway(Target,true):-detectionDistance(Target, Dist0, Time0),
detectionDistance(Target, Dist1, Time1),   detectionDistance(Target, Dist2, Time2),
Dist0>Dist1, Dist1>Dist2, Time0>Time1, Time1>Time2.
gettingFarAway(_,false).

Dead target	If the target has damage exceeding its limit, the predicate returns "true" truth value and the engagement already started is terminated or an engagement is not started

checkTarget(TargetId,0):- detection(Sensor,1,Target,Health,_,_,_, _,_, _, _, _, _,_,_),
Health==0.
checkTarget(TargetId,1).
targetDead(TargetId):- checkTarget(TargetId,1).

Identity fusion	The predicate returns a type of information regarding a given target type

decision(Target, plane):-detection(Sensor,1,Target,_,_,_,PosZ, Vel,_, _, _, _, Ac1,DistanceRatio0,_),
detection(Sensor, 2, Target,_, _,_,_, _, _, _, Width, Length, Ac2, DistanceRatio1,_),
Vel>=200, 800>=Vel, PosZ>=50, Width>=3, 10>=Width, Length>=4, 15>=Length,!.

(Continued)

Table 15.1 (Continued)

Predicate	Definition
Position fusion	The target position is calculated based on the weighted average of the position information received from the multi-sensors

list_sum([], 0).
list_sum([Head | Tail], TotalSum) :-list_sum(Tail, Sum1), TotalSum is Head + Sum1.
fusedPosition(A,Avg):-findall(X*T, (detection(S,A,X,T),member(S,[1, 4])),L), list_sum(L,Total),
findall(T, detection(S,A,X,T),L2), list_sum(L2,Total2), Avg is Total/Total2.

In control (responsibility) region	If the given target is in a responsibility region, it returns truth value and the region id of the target

inThisACO(TargetId,PosX,PosY,PosZ,What,1):-
inWhichACO(TargetId,PosX,PosY,PosZ,RegionId),regionType(RegionId,What),!.
inThisACO(_,_,_,_,_,0).
inWhichACO(_,_,PosY,_,ARegionId):-regionPoints(ARegionId,_,PosY0,_),
regionPoints(ARegionId,_,PosY1,_), (PosY0<PosY, PosY1>=PosY).
inWhichACO(_,PosX,PosY,_,ARegionId):-regionPoints(ARegionId,PosX0,PosY0,_),
regionPoints(ARegionId,PosX1,PosY1,_),
(PosY1<PosY, PosY0>=PosY), (PosX0 +(PosY-PosY0)/(PosY1-PosY0)*(PosX1-PosX0)<PosX).
inWhichACO(_,_,_,_,-1).

In control time interval	The predicate checks whether or not the target that is detected in the control region is in the control intervals

inTimeInterval(RegionId,Time):-timeInterval(RegionId, StartTime,EndTime),
Time>=StartTime, EndTime>=Time.
inTimeIntervalCont(RegionId,Time):-timeInterval(RegionId, StartTime,EndTime),
Time>=StartTime, EndTime>=Time.

"*approaching*" is used to decide whether a target is getting closer or not, and the predicate uses the earlier values and the latest value of the fact *distance*. For a given target id, the predicate returns a defense system showing that the target is approaching. The predicate gives the following results, depending on the call procedure:

• Given: Target Id (TargetId), output: to the defense site (DefenseSiteId) that the target (TargetId) is getting closer. If the DefenseSiteId is equal to −1, it means that there is no defense site that the opponent missile getting closer to.
• Given: Defense Site Id, Output: that the target id (TargetId) is getting closer to the defense site given,

- Given: the target id and defense site id (both parameters are given). The predicate gives a true or false result. If the result is true, it means that the target given is getting closer to the defense site given, and if the result is false, it is not getting closer.

The command control agent's knowledge base contains a set of predicates, such as the fire doctrine rules, target evaluation rules, target identification rules, decisions regarding whether a target is approaching or not, and a set of facts that predicates uses in time-labeled forms such as target detections. Reasoning is used to evaluate predicates and for obtaining a truth value representing whether the predicate is satisfied or not, and getting output parameters from the predicate. The truth value and the parameters are used to manage agent behaviors. Behavior management can be defined as activating, canceling, finishing, suspending, or resuming a behavior defined as a specialized finite state automaton in AdSiF representation. Making a decision on whether a detected entity is an enemy or not also involves reasoning, and if it is decided that it is an enemy, an engagement behavior is activated. The parameters necessary are obtained from the predicate, allowing the agent to execute the decision.

The agent's situation awareness is determined by the facts that it stores in its knowledge base, and it's decision capability is determined by the predicates. The fact named *detection* represents the sensor detections sent from the sensors to the related C2 units, and these detections are kept in the C2 knowledge base in time-labeled forms. Each fact declaration has a time frame, and any fact older than this time frame is accepted as being expired and is removed from the knowledge base. In end game scenarios, the predicates being used are given in Table 15.1.

The engagement behavior diagram is seen in Figure 15.1 for a target that has been detected and decided in a responsibility region. The guard *InControlRegion* controls whether or not the detection received is within the responsibility area and in control time (*inTimeInterval*). The guard also consists of an additional control to determine whether the detected target is getting away. The two rues are connected with "and" operator. If the guard is satisfied, the behavior *Fsa_TargetEvaluation* is executed.

The event detection also activates the behavior *Fsa_UpdateTarget*. The behavior populates the knowledge base with time-tagged detections (seen in Table 15.1). The facts that are older than given time limit are erased from the knowledgebase. If a target is getting away, it is decided it is not a target anymore and the engagement is ended (*Fsa_CancelEngagement).*

Defending the assets that the command control agent is in charge of is the main mission of the agent. Defense is not only against the missiles fired by the end game agent but also against the opponent side of the scenario view. The responsibilities of a C2 agent are summarized in the following subsections.

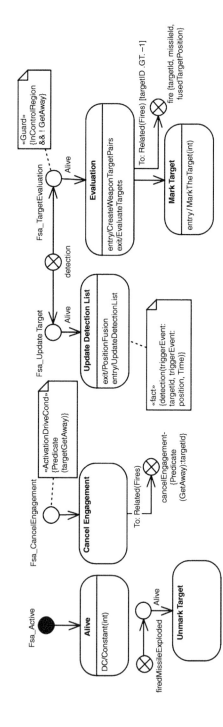

Figure 15.1 Engagement behavior.

C2 Architecture and Information Sharing

There are three types of architecture between command and control units. These are centralized C2 architecture, coordinated C2 architecture, and autonomous C2 architecture.

In centralized command and control architecture, a hierarchy exists between commanders. A commander manages a set of commanders of a layer below them, and each commander may manage a set of commanders at this layer. The commanders of the lower layer are in charge of conveying their detections and the target scores calculated by them to the commander they are connected with at the upper layer, and they do not take any decision on engagement.

In coordinated command and control architecture, commander models coordinate between themselves, share the target scores they compute with each other, and execute the commander's engagement with the highest score for their common goal.

In autonomous command and control architecture, each commander entity makes their own decisions, and thus it is possible to engage in the same target with a set of other commanders.

The C2 agents convey their detections to the commanders they coordinate with or to the commander they are controlled by within a centrally controlled command and control architecture. In the centralized C2 architecture, the purpose of conveying detections to the commander of a higher order is to enhance the awareness of the commander and to expect it to make a decision and send an order dictating what to do. But in coordinated C2 architecture, the commanders pass their target scores to the commanders they are coordinated with, in order to choose the best target for them to initiate an engagement with.

The coordination and the centralized command and control architecture are designed by the relation defined between the C2 units. The relations named "*Orders*" and "*Coordinated*" are defined for the centralized C2 architecture and coordination C2 architecture, respectively. When the relation "*Orders*" is established between two C2 units, the C2 unit located on the left side manages the C2 unit located on the right side of the relation. The C2 unit that is managed activates its "*send detections*" behavior list and deactivate the list "*Default.*" After breaking the relation, the C2 unit reactivates the list "*default*" and deactivates the "*send detections*" behavior list. Similarly, the relation "*Coordinated*" activates the behavior list "*InCoordination*" on both sides of the relation any time it is established. The behavior list consists of sharing detections, sharing target scores, and choosing the target with the highest score.

Target Evaluation

In target evaluation, a target selection score is calculated for each target weapon pair for all the defense sites, using target, weapon, and defense system parameters.

The parameters that are used frequently are summarized below and are also not limited by them.

- Arrival duration: It is defined as the duration of time taken for the attacker missile to reach the attackee. A shorter duration gives a higher score. A short arrival time indicates that the available engagement duration is also short, and this results in a high prioritization being given to the related target in the selection.
- The attacker's altitude: It expresses the vertical height of the target to the ground and the rising time of the missiles is longer for a high target.
- Attacker type: The target type is an integer selected from 0 to 10 and a high value indicates that the target criticality is high.
- The weapon type the attacker has (for aircrafts): It determines the criticality of the platform carrying the target platform with a high ammunition effect. The level of criticality of an aircraft platform is determined by the type of weapon of the highest criticality it carries.
- The attacker's damage level: The attackers that have lower damage levels have the higher precedence here.
- Probability of the kill: The probability of the kill is determined on the basis of a rule base. As in the target evaluation calculation, the P_k value is determined for the target and weapon pairs by using the rule sets created by the parameter values of the target and weapons system. The parameters to be taken from the weapons system and the target are determined by the user selection.

The target evaluation function is selected by the scenario designer, and the parameters to be used in target evaluation are also determined by the designer. The function selection is done from a set of plugin functions and the parameters are selected from the target parameters and the defense system parameters.

Fire Decision

Depending on the calculated weapon target pair scores, the C2 agent makes the engagement decision, following which it initiates the engagements. The fire order is given to the related launcher that has the highest score for the target being engaged. The order is sent to the commander that has the launcher, if the target is in the range. The range must be within the interval $[nr_e, xr_e]$.

where

xr_e: Maximum effective interception range;
nr_e: Minimum effective interception range.

The number n can be derived as:

$$n = \text{int}\left(\alpha T_D / t_L\right) \tag{15.1}$$

The fire decision range (r_{L1}) is calculated as below [32]:

$$r_{L1} = r_d, \quad \text{if} \quad r_d \leq xr_e \quad \text{else} \tag{15.2}$$

$$r_{L1} = xr_e \left(1 + V_{op} / V_{int}\right) + \left(\mu + t_L\right) V_{op}, \quad \text{if} \quad r_d > xr_e \tag{15.3}$$

where

μ: Initial reaction time
t_L: Inter-firing time
V_{int}: Interceptor missile (defense missile) velocity
V_{op}: Opponent missile velocity
r_d: Target detection range

If a target is detected out of the range of ($r_d > xr_e$), then the fire order is given after a wait duration, which is calculated as shown below [32]:

$$T_D = \left(r_{L1} - r_{min}\right) / V_{op}, \quad r_{min} = \max\left(D_r, n_{re}\right) \tag{15.4}$$

where
D_r is considered as the denial range, that is the range which the target opponent missile would not be allowed to cross.

Fire Doctrine
The fire doctrine is a rule set to determine what type of ammunitions should be used for what type of targets and in which fire pattern. The fire doctrine (F_i^k) is defined as < target i, weapon k > tuples. The fire doctrine is also defined as a salvo fire pattern, as $N \times T_0 \times T_1 \times M$.

Where N is the number of missiles to be fired and T_0 represents the duration between fires, M and T_1 represent repeated numbers and the duration between each repeat. In this sense, a single shot is represented as $1 \times 0 \times 0 \times 1$. Similarly, the no engagement rule is represented as $0 \times 0 \times 0 \times 0$, and this means that no missile k is fired to target I, which makes the probability of the kill equal to zero. As an example, $3 \times 1 \times 4 \times 2$ indicates that three missiles are fired one second apart from each other, and this cycle is repeated two times with intervals of four seconds each. In total, 6 missiles are fired, taking 10 seconds. The total duration (Ft_i^k) is calculated as $N*T_0*M + (M - 1)*T_1$.

In this representation, N is given and the number of missile necessary to kill a target is calculated depending on its probability of kill value as below:

$$\beta = \left(\log\left(1 - \left(1 - p_a\right) / p_a\right)\right) \Big/ \log\left(1 - p_k\right) \tag{15.5}$$

and the number of cycle

$$M = (\text{int}) \max \left((\text{int}) \beta \, / \, N, 1 \right) \tag{15.6}$$

M is casted to the integer.

Where p_a is the acceptable leakage probability, and p_k is the interceptor single shot kill probability.

A sample rule set is given in Code 15.1. The rule representation consists of target type, ammunition to e fired, salvo pattern, and the preference coefficient. A rule given a target-weapon (ammunition) pair overrides the calculation given above, and if there is no entry regarding the cycle number in a rule, it is determined by the calculation.

As seen from the rule, a salvo fire with the pattern 2-1-10-3 is applied for the target F16. The cycle number is given as 3, as a default value. Its current value is calculated through the use of Formula (15.6). There is no rule for the F104 type of fighter and thus the rule written for "Plane" is used instead.

Decision-Level Data Fusion

Data fusion is the process of integrating multiple data sources to produce more consistent, accurate, and useful information than that provided by any individual data source [33]. In this study, decision-level data fusion is considered rather than sensory signal level, and it is used for two purposes. The first purpose is to estimate the detected target position using the position information collected from different sensor detections. Each of the sensory data may give different but close

---

Code 15.1   Fire Doctrine Rule Base

```
! fireDoctirne(Target, Ammunition, Salvo, PreferanceFactor)
fireDoctrine(f16, harpoon, [0,0,0,0,0],0) ! no engagement to
F16 by Harpoon
fireDoctrine(f16, ciws, [10,3,1,2,3],1) ! Salvo for F16
by Ciws with parameters given.
fireDoctrine(f16, patriot, [1,0,0,0,1],1)
fireDoctrine(Plane,Patriot,[1,0,0,1],1) ! Single shot
for F16 by Patriot
fireDoctrine(Patriot,Chaff,[1,0,0,1],2)
fireDoctrine(Helicopter,Stinger,[1,0,0,1],)
fireDoctrine(Plane,Stinger,[2,200,200,2],1)
fireDoctrine(Ship,Nike,[2,200,200,2],1)
fireDoctrine(Helicopter,Bullet,[2,200,200,2],1)
fireDoctrine(Helicopter,AAALauncher,[5,500,1000,3],1)
```

position information, mostly with differing time labels. The idea is to combine the sensory data that are received from the sensors and weighed by the importance and reliability of the sensor they are received from. The combining function is selected on the basis of being a weighted average function. The combined or in other words, fused position information is used as the target position information. The second purpose is to identify the target. A rule-based identification reasoning mechanism based on first-order reasoning is developed. The detection parameters, such as the target velocity, target altitude, and target physical properties (detected using day-cams or thermal cams) are used as input parameters for the identification of the predicate. The target detection parameters are stored in a knowledge base in the form of facts and are populated by the detections received from the sensors. The fused target position information is used in the engagement as the target position for the target to be engaged in and as guidance information for a missile that has already been fired at a target. The C2 agent, which is in charge of target evaluation and weapon selection, is the owner of the knowledge base. The detection information and fusion algorithms are defined as facts and predicates, and as the components of logical programming as well.

Depending on the sensor type, a detection consists of the information given below:

- Radar detection: The direction, velocity in 3D, position.
- Day-cam/thermal cam detection: The pixel value that supports the estimation of the target physical size.
- Laser distance measurer: The measure of the target distance.

The target position is combined with the sensor reliability value for the different position information of the target from the same type of sensors (radars) in fusing, by taking a weighted average. The weighing average of the detection information is stored in a timed manner and is provided to create a trace with the previous detections of the target. The detections are kept in a time tagged manner in the knowledge base and the detection exceeding a given time limit are removed from the knowledge base. A detection is represented in the standard prolog fact format and is shown in Formula (15.1). How old the data used is determined by the covariance of the detection array. It is noted that the detection information taken on average is within the proximity of the defined time, which means that relatively older detention may not have been fused.

The position fusion rule code is shown in Code 15.2.

Identity fusion makes a decision about the identity of the target by using defined rules, and the information contained within the detections are collected from different sensors. The sensor detection information is used as entries for the identity rules (Table 15.1).

Code 15.2   Position Fusion Predicate

```
list_sum([], 0).
list_sum([Head | Tail], TotalSum) :-list_sum(Tail,
Sum1), TotalSum is Head + Sum1.
fusedPosition(A,Avg):-findall(X*T, (detection(S,A,X,T),
member(S,[1, 4])),L), list_sum(L,Total),
findall(T, detection(S,A,X,T),L2), list_
sum(L2,Total2), Avg is Total/Total2.
```

The target position, direction, and identity of the three results obtained from the position and identity fusions are used in fire control. The target position, speed, and direction information are used in the interception calculation and for firing time, and the target identification information is used for weapon selection.

Aims and Performance Measurement

The agent-driven end game analysis aims to create standard end game analysis scenarios, execute them, calculate performance measures in runtime, and manage scenario executions by making decisions based on the analysis results.

While creating an end game scenario, the agent provides a wizard to support the deployment of the defense assets, selections for data sharing, C2 structures, and fire doctrines. Scenario management consists of creating grids for the defense and opponent sides on the map, deploying defense sites with their sensors (sensors and launchers are sometimes deployed on different grids), igniting opponent missiles to their destination grids, and collecting defense success statistics. The statistics collected are used as inputs for the online analysis that is to be conducted. The analysis consists of central tendency measurements such as mean, standard deviations for any performance criteria, and statistically difference tests such as chisquire, F test. For any value calculated, it is possible to write a rule and manage behaviors or schedule events depending on the rule truth value. For example, if the average value of the detected target altitudes exceeds a given value, or an ARM missile (Anti-radiation missile) is detected, the stop event can be sent to the sensors.

Real-time analysis is taken into consideration in this study to drive scenarios in run time. Mainly, real data is collected in real time and four types of scenarios are executed. These are Tactical Ballistic Missile (TBM) active defense, TBM passive defense, attack operations, and standard defensive counter-air [34].

Measures of Performance (MOPs) are being computed for each of the four operational missions, such as:

- Percentage of TBMs killed.
- Timeliness and accuracy of TBM impact point predictions (IPPs).
- Percentage of TELs (transporter erector launchers) killed.
- Percentage of penetrating enemy aircraft killed.
- Timeliness and completeness of sensor detection and reporting.
- Timeliness and completeness of resource allocations.

The performance measurements differ for both the defense side and opponent sides. While the opponent's side aims to cause as much damage as possible to its targeted assets and defense units, the defense side aims to minimize its damage or maximize the probability of survival, and also cause as much as damage as it can to the opponent sides. There are some practical calculations for the air defense performance measurements in the existing literature [35]. In Table 15.2, the performance measurements for both the defense and offense sides and the relation between the two sides are given. The calculations for the same are also given, based on the parameters presented in the table. The performance criteria are categorized with regard to active air defense, passive air defense, and the cost.

A typical example of a measure of effectiveness covering both active and passive defenses is the survival probability of a target point.

$$Q_k = \Pi \left\{ 1 - K_k \left(1 - E_{jk} \right) \left[0.5^{\left(r_{jk}/e_j \right)^2 \left(X_j - \sum_i Z_i U_{ij} P_{ij} \right)} \right] \right\} \quad n_{ij} \leq 1 \tag{15.7}$$

A typical measure of the effectiveness, disregarding the cost considerations, is that of the threat destructive potential surviving the combined action of all the active defenses:

$$T = \sum_j X_j Y_j^a - \sum_i \sum_j Z_i Y_j^a U_{ij} P_{ij} \quad n_{ij} \leq 1 \tag{15.8}$$

This sample expression can be transformed into a measure of cost-effectiveness through the incorporation of the defense system's total and unit costs.

$$T = \sum_j X_j Y_j^a - \sum_i \sum_j \frac{A_i}{C_i} Y_j^a U_{ij} P_{ij} \quad n_{ij} \leq 1 \tag{15.9}$$

Some of the commonly used achievement goals of air defense systems are of the following types:

- Maximize threat attrition: to maximize the number destroyed or neutralize that of the attackers, the nuclear yield they are carrying, or, in general, of the pertinent threat potential.

Table 15.2 Air defense effectiveness measure inputs.

	Relative to		
	Attackee	Attacker	Interaction
Attackee's kill effectiveness		P_{ij}	Active defense
Attackee's commitment ratio		n_{ij}	
Attackee's weapons inventory	Z_i		
Attackee's weapons utilization factor		U_{ij}	
Attacker's strike strength	X_j		
Attacker's yield per strike unit	Y_j		
Attacker's yield exponent	a		
Attackee's active defense unit cost	C_i		Cost
Attackee's active defense total cost	A_i		
Attacker's offense weapon unit cost	C_j		
Attacker's offense weapon total cost	A_j		
Attackee's target value.	T_k		Passive defense
Attackee's passive defense total cost	B_k		
Attacker's weapon CEP	e_j		
Attacker's weapon lethal radius		r_{ij}	
Attacker's probability of target location knowledge		K_k	
Attackee's probability of successful target removal		E_{jk}	

Subscripts: i, Active defense, j, Offense, ki, Targets

$$\text{Max}\left[\sum_i\sum_j Z_i Y_j^a U_{ij} P_{ij}\right] \quad n_{ij} \le 1 \tag{15.10}$$

- Maximize probability of non-penetration, i.e. maximize the probability that not one of the attackers will succeed in penetrating the defense.

$$\text{Max}\left\{\prod_j\left[1-\prod_i\left(1-P_{ij}\right)\right]^{X_j}\right\} \quad n \le 1 \tag{15.11}$$

- Maximize probability of ground target survival, i.e. maximize the probability that a protected target will not be destroyed by any of the attackers, e.g. Eq. (15.7).
- Minimize active defense cost, i.e. minimize the cost of active defenses per kill of onie attacker or, more generally, per unit of threat potential destroyed or neutralized.

$$\text{Min}\left\{\frac{\sum_i A_i}{\sum_i\sum_j Z_i U_{ij} Y_j^a P_{ij}}\right\} \tag{15.12}$$

- Maximize offense cost, i.e. maximize the attacker's cost per unit value of attackee's targets destroyed.

$$\text{Max}\left\{\frac{\sum_j A_j}{\sum_k T_k\left(1-q_k\right)}\right\} \quad q_k \text{ is target survival probability.} \tag{15.13}$$

- Minimize defense/offense cost ratio, i.e. minimize defense total cost per attacker's cost to destroy one value unit of attackee's targets.

$$\text{Min}\left\{\frac{\sum_i Ai + \sum_k B_k}{\sum_j A_j / \sum_k T_k\left(1-q_k\right)}\right\} \tag{15.14}$$

The mathematically exact expression for the fraction of attackers surviving engagement with a defense system is

$$\left(1-n_{ij}P_{ij}\right) \quad \text{for} \quad n_{ij} \le 1 \tag{15.15}$$

$$\left(1-P_{ij}\right)^{n_{ij}} \quad \text{for} \quad n_{ij} \ge 1$$

Types of End Game Analysis

End game analysis essentially aims to determine the defense success of a land-based air defense system from the simplest cases to the more complicated and real-world situations. The simplest analysis assumes that all the conditions present are ideal, such as the flat land, with no mountains present, as well as ideal communication (zero latency and no data loss), a launch on remote data, and data sharing between sensors. In this respect, there are four types of analysis to be taken into consideration here, in the order of the simplest to the most complicated one: footprint, operational area, defended area, and scenario view.

Footprint analysis is aimed at understanding the defense capability of defense units against a specific type of missile. Operational area analysis is used to determine where a movable defense unit must be placed in order to engage in a specific trajectory. Analysts can look at several potential trajectories and determine whether or not there is one movable operating or maneuver area that can handle a group of launch locations. Defended area analysis provides the analysts and others with a complete defense architectural design. This analysis lets the user place defensive units on the map, examine at the risk of the defended asset list against a specified threat, connect the defensive units on a network, set up cueing between platforms, and understand which assets can be protected by the defense. Scenario view analysis is related with a complete defense scenario. The first three analyses prepare the input data and deployment information to be used for the scenario view analysis. The deployment of defense sites, ammunition selection, selection of target evaluation criteria, launches based on remote data decisions, etc. are done with regard to the previously conducted analyses.

The air defense simulation model library consists of sensory systems, wired and wireless communication devices, a series of air platforms such as fixed wing fighters, rotary wing platforms, and Surface to Air Missile (SAM) sites. A SAM site has one or more launchers, missiles, radars, and a command and control unit. Optical sensors and microwave radars are used as defense site sensors. Each sensor connected with the defense site sends their detection in two different ways. The first one is to send the detections directly to the C2 unit, and this is known as the perfect communication assumption. The second one is to use communication devices. In this case, communication latencies and losses are taken into consideration.

In any type of analysis, different scenario designs are obtained by differentiating between the following configurations for a scenario. According to the criteria defined by the operation of scenarios, relative rankings are made between the scenarios.

The scenario parameters are defined as follows:

- Whether or not there is perfect communication between the sensors and command and control systems (whether communication devices are used or not):

Perfect communication means that there is direct communication established between sensors and command and control units. There are no communication loss between communication devices. The communication succeeds with no loss, no time latency, and with zero error. A relation between the sensors and command and control units is established to make the sensors send their detections to the respective command and control units. In this study, the relation is named "Sends detections." The same relation is established between the sensor and the communication device if the detection is to be sent via a communication device. The communication between sensors and C2 units is achieved through wireless communication devices, wired communication devices, link communication, and any other types of communication devices that are allowed to be used in conceptual models.

- Whether or not cueing between sensors is allowed: If cueing is allowed between sensors, a sensor will share its detections with other sensors within a given distance. Otherwise each sensor possesses only its own detections and transmits this information to the associated assets (for example, to a C2 unit with which it has a relation). If the sensor deployments are positioned so as to provide access to the C2 centers on a network, all of the detections are transmitted to the centers of the C2, which means creating a larger pool of information for the decision makers.
- Whether or not a launch on remote data is allowed: This determines whether a weapon system will be able to decide whether to engage with a target by using only its own detections or by using the detection information received from other sensors.
- Command and control architecture: One of the three types of C2 architecture is defined. These are autonomous C2 architecture, coordinated C2 architecture, and centralized C2 architecture. A scenario may have more than one architecture defined between the commanders in the scenario. For example, a specific region may be defended by a group of coordinated C2s and any other asset may be defended by a group of centralized C2s.
- Fire doctrine: The fire doctrine to be applied is chosen for each C2. It provides the answers to the questions of which weapon is to be used against which type of target, and in which pattern.

The collecting of performance statistics from a scenario execution depends on the purposes for the execution of the scenario. Since end game scenarios are prepared for a set of specific purposes, a performance measurement set and a set of statistics related to the performance measurements are collected. Additionally, it is possible to define some other performance measurements as well. The performance measurements given below are considered to be fundamental:

1) Detection distances: For each detected target, its first detection distance is traced.
2) Detection time: For each detected target, the first detection time is traced.

3) Target tracking duration: This indicates how long a threat has been detected and traced by defense sensors throughout its flight.
4) Engagement distances: This represents how closely engagement is achieved with the attackers. Two types of distance are considered. The first type is the distance between the attacker and the attacked (the asset the defense unit defends) and the second type is that between the defense site and the attacker.
5) Communication delay (if it is not assumed there is perfect communication): The communication time taken in the communication network, such as the transmission of detections, transmission of engagement orders, is calculated.
6) Number of grid cells defended: The number of defended grids in a gridded defense area.
7) Flight duration of opponent air platforms.
8) Number of opponent missile suppressed.

For the criteria from 1 to 4, the average value is calculated for the attackers and for criterion number 5, the average density is calculated by weighing the intensity of the message. On the other hand, the total number for criteria 6 and 7 is taken as the basis of an air defense analysis. For the first eight criteria, basic statistical calculations such as minimum value, maximum value, and standard deviation are carried out. A simple scenario consists of an air defense system with a sensor and as many defense missiles as the opponent missiles.

Footprint Analysis

In footprint analysis, a single system is usually placed at the origin and ballistic missiles are flown against it. The footprint result is then taken to be the area on the ground that the defense system can protect [36]. The defense footprint shows the area on the ground that can be defended against ballistic missiles. The probability of negation or the level of protection, and the depth-of-fire (or number of engagements) to indicate where the defensive system may have shoot-look-shoot capability are computed during the execution. Detection and interception points are taken into consideration in this analysis. Analysts can perform the calculation with multiple launchers, radars, or systems, but generally, a single collocated system serves as the basis of the calculation. Footprints are useful in understanding how a system works and for doing trade within a system. Varying the interception altitude and changing the interceptor logic or radar capability are all examples of system trades. Footprints are also helpful in understanding how a system capability varies from different threats.

In footprint analysis, the scenario is designed on the basis of parallel and non-parallel missile defense strategies. Analysts determine the analysis grid that represents the analysis resolution by entering grid sizes and numbers. The end game agent draws defense and opponent side grids, puts a mission on each opponent grid cell for parallel footprint analysis, and deploys defense units (SAM sites) at the center of area to be defended. The defense system that consists of sensors, launchers, missiles, and a C2 system (at least a commander) to give fire orders at the beginning of the analysis is placed on the center of the grid. The agent determines the distance between the grids of the opponent and defense side, according to the opponent's effective missile range (Figure 15.2). After the scenario entity deployment, consecutive scenario designs are created. Some specific design parameters for footprint analysis are as follows:

- The land is assumed to be flat: It is assumed that the land is flat. In this case, the targets that remain undetected because of the roundness of the world will now be detected. The main goal behind the creation of such an assumption is to carry detection success to the maximum level and eliminate the problems related to the line of sight from the analysis.
- The area with mountains: It is assumed that there are no mountains in the flat world. The main purpose is to create an analysis level that lies between the ideal flat world and the real world.
- The rounded world: The rounded real world with mountains is used. All the environment variables are used as they are.

In the parallel footprint analysis, opponent missiles are deployed in accordance with the grids to be defended. The distance between the two grids is equal to the range of the opponent missiles. A defense system is located at a user-selected

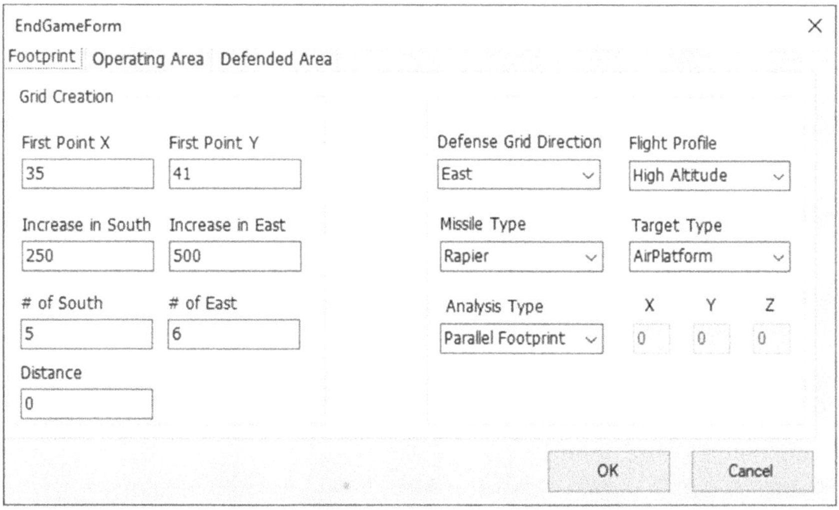

Figure 15.2 Footprint analysis design interface.

position on the defense grid. The defense system has defense missiles, sensor(s), and a command and control unit. In a nonparallel footprint analysis, the opponent's weapon system is placed by user selection. The number of defense systems, the number of sensors, and the command and control architecture are determined by the scenario design phase through user selection (from the model library). The other scenario parameters selected during the scenario design phase will be explained later and are enumerated in Table 15.3.

After deployment (seen in Figure 15.3), the scenario execution continues by all the opponent missiles being fired at the same time. Each missile fired is aimed at the opposing cell of the defense side. Defense SAM sites detect the opponent missiles through their sensors and fire their defense missile toward the targets detected. The agent collects missile detection points, missile hit points, and the cells destroyed by opponent missiles during the execution to calculate the defense success. The scenario's termination condition is defined with regard to all the opponent missiles destroyed or whether they met their targets (hit their target cells). After the competition of a scenario execution, a new replication is started with a new configuration. All the configurations are generated automatically and the analyst may ignore any of them.

In a footprint parallel analysis scenario, the agent executes and repeats the scenario by changing the scenario parameters. The simplest one is the scenario under the assumption of the world being flat with no mountains, meaning that everything will be within the line of sight, and perfect communication and data sharing being available and the launch on remote data being allowed.

The second phase of the footprint analysis is to analyze nonparallel missile deployment. In this type of analysis, the opponent's missiles are fired from the

Table 15.3 The scenario assets.

Entity type		Side	# of entity
Fighters (F16)	F16-1, F16-2, F16-3, F16-4	Red	4
Choppers	Hel-1, Hel-2	Red	2
Defense sites (long range)	FFK-1, FFK-2, FFK-3	Blue	3
Defense sites (medium range)	KMR-1, KMR-2, KMR-3, KMR-4	Blue	4
Defense sites (short range)	AAA-1, AAA-2	Blue	2
Missiles	Guided-thrusted	Blue	28
Radars	Search radar	Blue	4
Seekers	Missile seekers	Blue	28
Warheads	Missile warheads	Blue	28

points designated in scenario design phase. The statistics collected during the scenario's execution phase and the scenario replication configurations are the same as those of the parallel end game analysis. As an intelligent part of simulation design, the C2 model applies target evaluation, weapon assignment, and fire doctrine rules.

In this analysis, 5×6 defense cells are created and 10 km distance and the missile grid parallel to the east direction are determined. The footprint parallel missile analysis is initiated by firing offensive missiles toward all the defense cells, with all the missiles being fired at each cell simultaneously Figure 15.3). The defense unit is defended according to the following different configurations, the

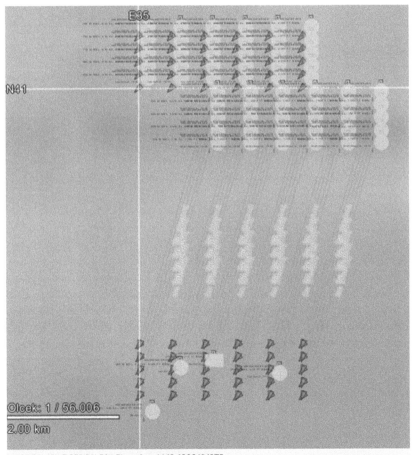

N 40° 56′ 40″ E 35° 01′ 58″ Elevation 1142,1396484375

Figure 15.3 Parallel footprint scenario deployment.

execution is repeated for each configuration, and the statistics were calculated for each execution.

- Scenario 1: The assumption that excellent communication will be available among the units, as well as a single radar, and with remote data engagement not being possible.
- Scenario 2: The assumption that excellent communication will be available among the units, as well as further point radars, and with the launch on remote data being possible.

After the scenario operation is completed, a nonparallel footprint analysis is carried out in which all the missiles are fired from the midpoint of the entry. The scenario conditions are also determined in the same way. The same statistics are performed for the nonparallel execution.

Operating Area

This view is used to answer the question of where a defense system should be located to provide engagement capability. This is used most often by the Navy systems, which have more of a challenge in positioning for an engagement to satisfy both surveillance and intercept functions.

In addition to the traditional measures of effectiveness (probability of negation and the number of engagements), in operating area analysis, the results for five MOEs (time in track, time in track post-boost, latest track loss time, earliest detection time, and earliest intercept time) are computed [36].

In the operating area analysis, the agent tries to deploy defense ships/land platforms (as a movable entity) to defend assets in the most effective way. Additionally, in the analysis, grid cell selection probability is determined, depending on the opponent missile trajectories aimed to the assets.

Analysts locate a ground-to-ground opponent missile site and a defended asset. A region on the opponent missile trajectory is determined and a movable sea or land platform is deployed on a grid drawn on the region. The movable entity has a defense system with sensors. It aims to hit opponent missiles targeting the defended assets. The end game agent changes the movable entity location for each scenario replication and collects statistics. The deployment strategy of the movable entity is selected by analysts, and it is possible to prepare a new strategy in the form of a plugin. Currently, the end game agent can change the location of the movable entity on the grid by using one of the algorithms given below:

- Changing the location from the North-West down to South-East one by one; the entity is deployed on North-West grid cell at first and then localized row by row and cell by cell down to the last South-East cell.
- In random order: The cell that is the entity to be deployed is selected randomly.

Each deployment is considered a new simulation replication. The end game agent compares the simulation replications to determine the location to which the movable entity must be deployed to give optimum defense success in run time. In the operating area analysis, the scenario is developed by using a ship as a movable platform. The ship is deployed on a gridded area between the area being defended and the opponent missile sites. The cell that the ship is deployed in is selected randomly during the first simulation run, and it is assumed that it has the number of missiles necessary for a successful defense to be executed. For each repeated execution, the ship's position is changed, and statistics are collected to measure the difference in success between these scenarios. Defeating all the opponent missiles or being unsuccessful in defending the area are defined as simulation termination conditions. After executing a scenario for all the deployment possibilities of the ship, a performance analysis is done for each scenario and they compared with each other statistically to identify the best defense configuration. In Figure 15.4, the operating area scenario's initial case deployment is depicted.

A ship with an air defense system is deployed between the defense side grid and the opponent side grid, at each simulation replication, the ship position is perturbated to find a perfect position to deploy.

Defended Area Analysis

This view is probably the one most used by operators and analysts. A defended area answers the question of whether a given defense architecture laydown can protect specified critical assets against the shown Intelligence Preparation of the Battlespace (IPB). In this analysis, there are multiple defense assets, land-based defense systems for those assets, and opponent missile systems. After deploying the assets to be defended, defense units, and opponent units on the map, a grid is drawn on the region to be measured so that it covers all the assets and the sub-areas surrounding those assets. As is in done in all the other analyses, the most atomic analysis region is determined to be a grid cell.

After the preparation of the scenario, the execution of the scenario is initiated, and the sensor and communication network, warfare choices such as command and control type, launch on remote data, sensor cueing, fire doctrines, etc. are set in place. The locations of the weapon and sensory systems are determined at the beginning of the replication. For each replication, statistical data is collected and compared with other replications to find the best deployment. Random deployment and genetic algorithm-based strategy are applied for sensory and weapon system permutations. As a displacement rule, it is said that a sensory system and weapon system must be together on the same cell for navy systems, but can be deployed separately for land systems. When the random deployment strategy is used, sensors and launchers are deployed on randomly selected grids.

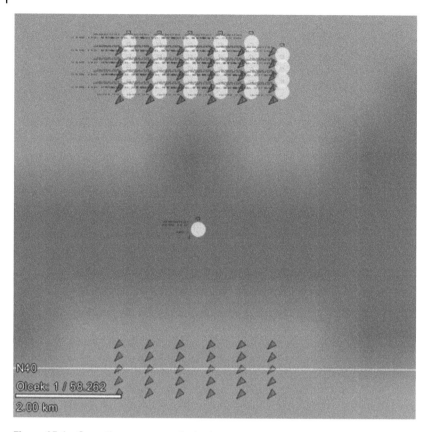

Figure 15.4 Operating area scenario deployment.

Each deployment results in a new scenario replication and the replications are created by the end game agent.

In the defended area scenario, a set of valuable assets such as an airport tower, a runway, and an airport building are chosen to defend. The end game agent sends the commander units information concerning the assets that are to be defended (the information consists of the runway, the tower, and the building in this example). If an entity is not listed as an asset to be defended, the commander units do not take any action to defend it and it is not considered in performance analysis. The importance level of an asset is defined on the basis of C2 units and may vary from the C2 unit to the C2 unit. The scenario deployment and execution are seen in Figure 15.5. The area to be defended surrounded by a polygon and the whole area is defended successfully. Defense is achieved by the air defense systems situated in the polygon and deployed on the ship.

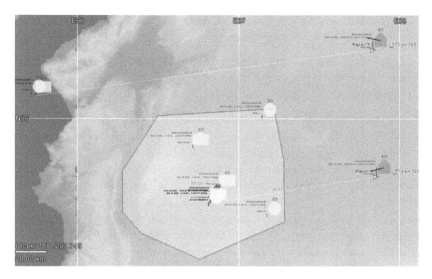

Figure 15.5   Defended area scenario deployment.

Scenario View

The last view (scenario) is used to determine the defense performance over multiple days or months, how well the defense design under consideration will perform against a specific raid or campaign, and to understand the impact of firing doctrine on inventory usage. Analysts manually create a scenario of BM/ABT threats. This view introduces the time component to the analysis and is particularly helpful in making clearer issues such as inventory coupled with firing doctrine, reconstitution, and re-supply.

The scenario view is particularly useful in gaining a better understanding of inventory usage, issues of time-phased deployment, threat timing, saturation, and reconstitution. In addition, since the campaign can encompass thousands of threats, statistics on defense performance can be generated for multiple Monte Carlo realizations.

There are many additional diagnostics that involve the use of a scenario tool, such as intercept and BM leakers, track histories, engagement flyouts, ground effects, and cruise missile threat profiles.

Online Analysis and Scenario Replication Design

The end game agent collects statistics in execution time. These are the defense levels of each grid cell (whether or not it is defended), opponent missile prevention points, prevention distance, detection distance, etc. The agent has a rule set to change the defense system.

- Weapon system (launcher and related equipment) placement.
- Sensor placements, types, and numbers.
- Communication device types between C2 and sensors.

The agent computes statistics using the data collected during simulation execution and changes the scenario so that it yields a better defense ratio and executes it repeatedly to attain an ideal configuration. In the case of the existence of never-defended cells, the agent adds new defense sites to the scenario to increase the cell's defense level. If the newly inserted defense sites do not increase the defense level statistically or in a meaningful way in general, the defense site is removed from the scenario. The decision of removing the defense site is made after a series of defense system placement perturbations. To be able to make choices that will lead to optimum deployment, defense systems are deployed to the randomly selected cells. The selection is done by dividing the grid into 2^n pieces. During the initial execution, each cell has the same selection probability and n is set to 2. After the first selection, cells in the piece resulting in a better defense ratio have a higher selection probability. To get more accurate result, the value n is increased during analysis for bigger grids.

The agent keeps each simulation execution's statistics during the configured scenario replications to be able to analyze and compare the defense system configurations. The analysis consists of statistically significant tests. In this sense, the agent configures and designs new scenarios, executes them sequentially, keeps their statistics, and uses them for a new scenario design.

An Air Defense Scenario: Scenario View

In the scenario displayed in Figure 15.6 and its assets in Table 15.3, the red fighters aim the blue air defense systems. Both blue air defense systems have launchers, missiles, and radars. Long-range and medium-range missiles have seekers and warheads. In this sense, they are guided, thrusted, and multiphase missiles. There is a perfect communication assumption between all the units.

The red fighters from three different directions and the red choppers from two opposite directions begin to follow their routes and enter the blue defense system control zone. The fighters depart to their destination almost two minutes earlier than the choppers. The red forces head to directly to the blue defense unit located at the center of the defense zone. The defense system radars detect the red fighters and send their detections to the command and control system they are related to. The relation "Sends Detections" is

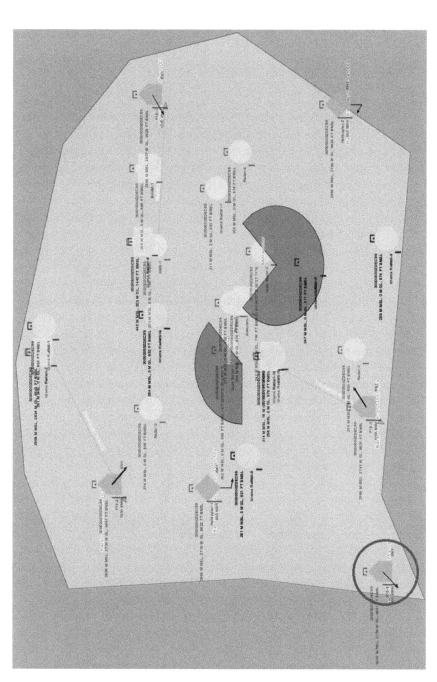

Figure 15.6 Air defense scenario.

established between the sensors and the commanders, and the relation "Fires" is established between the command and control system and the launchers. Each launcher is connected to a C2 system. Until the red forces enter to the defense zone, the C2 agent does not consider them to be targets. The C2 agent creates weapon target pairs, and evaluates and gives a score for each of them. After the scoring, it chooses the pair with the highest score for each defense site. There is no engagement started with the platform shown in the circle. Since it is moving further away, it is not evaluated as a target.

Since the AAALauncher-1 is closer to Chopper-1 than the AAALauncher-2, it has a higher score and is chosen to engage. Similarly, AAALauncher-2 is chosen for Chopper-2. For each target, the most apt missile is selected using the target–weapon pair scores. In the selection process, the round wheel algorithm that the higher score gives the higher selection probability is used. The function is defined as a plugin which can be interchanged with any other function, depending on the conditions defined.

The fired missiles are seen in Figure 15.6. A yellow line from a missile to a target shows which target is aimed at by which missile. A layered defense architecture is applied. The long-range missiles are deployed at the outer layer. The middle-range missiles are deployed following that of the long ranges. The short-range anti-aircraft artilleries (AAA) surround the command and control center. Because of the fire doctrine rules being applied in the scenario, it is not permitted to engage the choppers with the missiles. The only way possible to engage them is through AAAs and for this same reason, the single-shot fire doctrine is applied for the F16s.

The scenario is run for two different configurations. In the first configuration, the red forces fly at higher altitude (3000 m) and in the second, they follow the lower routes (~2000 m). At the end of the scenario, the blue land forces defend their area successfully in both cases. The duration of the red fighters' flight in both the scenarios is displayed in Table 15.4.

According to the chi-square test, the difference between fly duration in the two scenarios are randomly different at 95% level. The P value is equal to ~1.

According to the chi-square test results, the target engagement times are found randomly different. The calculated P value is 0.99887.

In Table 15.6, the target elimination distances and the explosion distances between the target red air platform and the missiles aimed them are shown. In the collision calculation, the physical dimensions of the entities are taken into account. In this sense, it is seen that the targets are eliminated through collision.

According to the chi-square test results, the target explosion distances are randomly different with P value 0.7248 and the target elimination times are not randomly different with P value ~0.0.

The detections of the sensor systems are given in Table 15.7. The table consists of the first detection times and the number of detections.

Table 15.4 Fly durations.

Red fighters and blue missiles	High-altitude scenario	Low-altitude scenario
FFKMissile-1	41 359	39 519
FFKMissile-3	35 433	34 347
FFKMissile-5	41 182	39 130
F16-1	79 518	72 932
F16-2	57 444	50 465
F16-3	86 353	80 215
Chopper-1	43 219	37 001
Chopper-2	176 238	172 506
Mean	7 009 325	6 576 438

Table 15.5 Target engagement times.

	High-altitude scenario		Low-altitude scenario	
Defense site	Target	Engagement time	Target	Engagement time
FFK-1	F16-2	136 086	F16-2	130 947
FFK-2	F16-1	164 086	F16-1	158 086
FFK-3	F16-3	165 172	F16-3	161 086
AAA-1	Chopper-1	82 204	Chopper-1	76 204
AAA-2	Chopper-2	94 095	Chopper-2	88 095
Mean		1 283 286		1 228 836

Table 15.6 Target elimination distances.

	High-altitude scenario		Low-altitude scenario (unit meter)	
Missiles	Explosion distance	Engagement distance	Explosion distance	Engagement distance
FFKMissile-1	207 444	163 075	120 526	165 345
FFKMissile-3	102 273	190 132	845 495	194 807
FFKMissile-5	205 852	12 007	172 139	119 381

Table 15.7 Target detection times and detection numbers.

	High-altitude scenario		Low-altitude scenario	
	First detection time	Number of detection	First detection time	Number of detection
Chopper-1	82 000	8	76 000	8
Chopper-2	97 000	15	91 895	10
F16-1	158 000	54	154 000	54
F16-2	161 000	10	130 861	15
F16-3	165 086	57	161 000	52
Mean		28.8		27.8

According to the statistics test, the differences between the two scenarios are random ($P = 0.794\,843$ and $P = 0.709\,718$, respectively).

Discussions

An end game analysis starts from the most ideal cases and ends with very complex scenarios that have many alternative cases. Using simple to complex scenarios gives an interval of the air defense success level that is possible for an air defense system to reach. Having an interval gives the modelers a limit that is possible to get closer to, and not able to reach in practice in an air defense scenario. In other words, the interval also gives a measurement for a marginal defense success for a new investment such as installing a new radar or a launcher etc.

Agent-based modeling makes scenario definition, scenario execution, and analysis automated and also presents an analysis flow that is easy to customize. The agent modeling and execution language AdSiF provides a flexible, reusable, and extendable programming environment. The ontological view that AdSiF has, it's software engineering criteria, and the programming paradigms it supports are displayed in [37]. Having an ontological commitment provides the modelers with the ability to define an entity from scratch and having multiprogramming support gives the modelers an enriched worldview. Because of its flexibility and script-based programming, it is possible to extend the agent trace parameters using simple declarations. This allows modelers and analysts to enrich their analysis by adding new features.

Agent-based modeling not only gives an automatized scenario design and execution but also provides a run-time analysis and scenario drive opportunities based on the results of the analysis. It is possible to define a rule set to drive the scenario being run and/or schedule new replications. For example, a rule may be defined to start a new replication that 60% of airplanes are lost, at any time. In this sense, an agent in a simulation scenario is an intelligent component that plans missions, measures performance criteria, interacts with simulation entities, and drives the scenario being executed. Being capable of managing a scenario also allows modelers to design meta-level scenario rules to guide the scenario entities to optimize a given goal. In our example, the agent determines reachable performance intervals by designing and measuring the end game analysis and provides a dynamic scenario view by handling the scenario management rules. Being able to keep detections and actions in the knowledge base as time-tagged information also provide modeling capability for time-delayed systems and information to use generate estimations about next situations to be lived in the close future.

References

M. J. Liebhaber and B. Feher, *Air Threat Assessment: Research, Model, and Display Guidelines*. 2002.

Liu, H., Ma, Z., Deng, X., and Jiang, W. (2018). A new method to air target threat evaluation based on Dempster–Shafer evidence theory. *2018 Chinese Control And Decision Conference (CCDC)* 0(4):2504–2508.

Brancalion, J.F.B. and Kienitz, K.H. (2018). Threat evaluation of aerial targets in an air defense system using Bayesian networks. *Proc. – 2017 IEEE 15th Int. Conf. Dependable, Auton. Secur. Comput. 2017 IEEE 15th Int. Conf. Pervasive Intell. Comput. 2017 IEEE 3rd Int. Conf. Big Data Intell. Comput. 2017 IEEE 3rd Int. Conf. Big Data Intell. Comput.* 2018(1): 897–900.

Wu, H. (2002). Sensor fusion using Dempster–Shafer theory. *IEEE Instrumentation and Measurement Technology Conference, 2002.*

Louvieris, P.P., Gregoriades, A., and Garn, W. (2010). Assessing critical success factors for military decision support. *Exp. Syst. Appl.* 37: 8229–8241.

Hocaoğlu, M.F. (2019). Weapon target assignment optimization for land based multi-air defense systems: A goal programming approach. *Comput. Ind. Eng.* 128: 681–689.

Wang, J., Luo, P., Hu, X., and Zhang, X. (2018). A hybrid discrete grey wolf optimizer to solve. *Discret. Dyn. Nat. Soc.* 2018: 1–17.

Ahuja, R.K., Kumar, A., Jha, K.C., and Orlin, J.B. (2007). Exact and heuristic algorithms for the weapon-target assignment problem. *Oper. Res.* 55 (6): 1136–1146.

Kline, A., Ahner, D., and Hill, R. (2019). The weapon-target assignment problem. *Comput. Oper. Res.* 105: 226–236.

Lai, C. and Wu, T. (2019). Simplified swarm optimization with initialization scheme for dynamic weapon – target assignment problem. *Appl. Soft Comput. J.* 82: 105542.

Lu, R.H.H., Zhang, H., and Zhang, X. (2006). An improved genetic algorithm for target assignment optimization of naval fleet air defense. *6th World Cong. Intell. Contr. Autom.*: 3401–3405.

Malhotra, A. and Jain, R.K. (2001). Genetic algorithm for optimal weapon allocation in multila. *Def. Sci. J.* 51 (3): 285–293.

Li, Y., Kou, Y., Li, Z. et al. (2017). A modified pareto ant colony optimization approach to solve biobjective weapon-target assignment problem. *Int. J. Aerosp. Eng.* 2017: 1–14.

Lee, S.F., Lee, Z.J., and Su, C.Y. (2002). An immunity based ant colony optimization algorithm for solving weapon-target assignment problem. *Appl. Soft Comput.* 2: 39–47.

Duan, J., Zhang, Y., Ma, W., et al. (2010). Analysis of aerial targets threat degree in air defense system. *2010 International Conference on Computational Intelligence and Software Engineering*, 1–4.

Jackson, B.A. and LaTourrette, T. (2015). Assessing the effectiveness of layered security for protecting the aviation system against adaptive adversaries. *J. Air Transp. Manag.* 48: 26–33.

Hu, X., Luo, P., Zhang, X., and Wang, J. (2018). Improved ant colony optimization for weapon-target assignment. *Math. Probl. Eng.* 2018: 1–14.

Chang, Y., Li, Z., Kou, Y. et al. (2017). A new approach to weapon-target assignment in cooperative air combat. *Math. Probl. Eng.* 2017: 1–17.

Das, S.K. (2014). Modeling intelligent decision-making command and control agents: an application to air defense. *IEEE Intell. Syst.* 29 (5): 22–29.

Ling, Y. and Zhang, Y. (2016). *Modeling of Air Defense and Antimissile System Based on FUML/FPN*, 215–218.

Chen, X. and Shi, H. (2017). Multi-platform air defence scheduling based on 0–1 integer linear programming. *Proc. – 4th Int. Conf. Inf. Sci. Control Eng. ICISCE 2017*, 480–483

Zhang, J., Wang, G., Yao, X., Song, Y., and Zhao, F. (2019). Research on task assignment optimization algorithm based on multi-agent. *Proc. 2018 Chinese Autom. Congr. CAC 2018*, 2179–2183.

Zhang, H.F., Liu, X.M., Liu, Q., and Wu, Z.W. (2019). Simulation design of air defense forces operation action supported by space information. *Proc. – 2018 5th Int. Conf. Inf. Sci. Control Eng. ICISCE 2018*, 1249–1252.

Garcia, R.J.B. and Von Winterfeldt, D. (2016). Defender – attacker decision tree analysis to combat terrorism. *Risk Anal.* 36 (12): 2258–2271.

Niznik, C. (2013). Games of timing theoretical protocol development and performance analysis for missile defense. *22nd International Conference on Computer Communication and Networks (ICCCN)*, 1–6.

Liu, K. and Tian, X. (2012). Portable air defence missile system reliability analysis and design. *International Conference on Quality, Reliability, Risk, Maintenance, and Safety Engineering*, 1171–1174.

Wooldridge, M. (2002). *An Introduction to MultiAgent Systems*. Wiley.

Firby, W.F.J. (2000). *The RAP System Language Manual, Version 2.0*. Evanston, IL: Neodesic Corporation.

M. W. R. H. Bordini, J. F. Hübner, *Programming Multi-agent Systems in AgentSpeak Using Jason*. Wiley Series in Agent Technology, 2007.

Zeigler, T.G.K.B.P. and Praehofer, H. (2000). *Theory of Modeling and Simulation*. Florida: Academic Press.

Sarjoughian, V.E.H. (2009). CoSMoS: a visual environment for component-based modeling, experimental design, and simulation. In: *SIMUTools 2009*.

Dutta, D. (2014). Probabilistic analysis of anti-ship missile defence effectiveness. *Def. Sci. J.* 64 (2): 123–129.

Haghighat, M., Abdel-Mottaleb, M., and Alhalabi, W. (2016). Discriminant correlation analysis: real-time feature level fusion for multimodal biometric recognition. *IEEE Trans. Inf. Forensics Secur.* 11 (9): 1984–1996.

Gray, M.D. and Jones, C.K. (1994). Realtime data analysis for the joint theater missile defense simulation network (JTMDSN). *Proc. 5th Annu. Conf. AI, Simulation, Plan. High Auton. Syst. Distrib. Interact. Simul. Environ. AIHAS*, 77–81.

Tombach, H. (1961). Critique of air defense measures of effectiveness. *Manag. Technol.* 1 (3): 52–62.

Sparta Inc. (2004). *Commander's Analysis and Planning Simulation (CAPS). v 8.0 User's Manual*, San Diego, CA.

Hocaoğlu, M.F. (2018). AdSiF: agent driven simulation framework paradigm and ontological view. *Sci. Comput. Program.* 167: 70–90.

Epilogue
Andreas Tolk

In the prologue, we formulated the challenge to bring human creativity and intuition of wargames closer together utilizing the power of computation provided by simulation systems. Did we get there? Surely, the contributions in this volume provide some impressive visions, from the use of latest technology, such as visualization tools for immersive presentations of simulation as well as resulting data, to providing common ontological representations allowing the application of artificial intelligence techniques, to the increased use of analysis and latest simulation methods to support them. There are examples of successful applications in several domains, demonstrating the new dimension of warfare. But are we truly growing closer together? Or were simulation and wargaming never divided in the first place? Is it perhaps simply a matter of perspective?

Wargamers always were interested in getting better support. From the introduction of movement and attrition tables in the beginning to the use of physics-based simulation models to compute effects, wargamers accepted help, but as Seth Bonder presented in his address on Military Operations Research, Science, and Models: Some Lessons Learned during the 68th Military Operations Research Society meeting in Colorado Springs in June 2000: "[It is] not models [that] produce new insights, but the analysts with ten to fifteen years of experience in the respective domains."

Computer scientists know this very well, and so should simulationists. Simulation systems are computer programs that operate under the general constraints on computer science. One of these constraints is that computers cannot be creative. They use computable functions to transform input parameters into output parameters. In other words, all information that can be extracted from a computer program either is encoded into the algorithms or is part of the input data. Hidden patterns and information can be discovered by visualization and presentation in different modes, but the computational process, even when it has

Simulation and Wargaming, First Edition. Edited by Charles Turnitsa, Curtis Blais, and Andreas Tolk.
© 2022 John Wiley & Sons, Inc. Published 2022 by John Wiley & Sons, Inc.

stochastic elements, is purely transformative, not creative. However, computer programs can mimic various forms of creativity, which are combinational, exploratory, and transformational. With the immense computing power we have available today, we can easily combine pieces of knowledge in many ways to find new solutions using known components in new and unfamiliar ways. The same is true for exploration of very large solution spaces under constraints, and modifying some of these constraints allows for transformational results. In his 2004 book on *Extending Ourselves: Computational Science, Empiricism, and Scientific Method* (Oxford University Press), Humphreys observes that "What cannot be done in principle, can't be done in practice either. . . . Yet when one is concerned with positive rather than negative results, the situation is reversed – what can be done in principle is frequently impossible to carry out in practice." That is the power of computer simulations: They are continuously pushing the border of what is practically possible, and the visionary contributions of this compendium provide examples of how it is done.

Wargamers need this new form of support to address the complexity of modern warfare challenges. Such challenges of complexity and emergence have been addressed in language and constructs of their times by authors from Sun Tzu to Clausewitz. However, due to jointness, networking, long-range fires, precision weapons, special operations, gray-area operations, and other developments, the complexity of military planning and operations has increased substantially. Traditionally separated domains merge into each other on the modern battleground, including the new elements of space and cyber. This requires true creativity of wargamers, but also the support by simulation in two main categories: (1) providing a decision space with the necessary complexity to be representative of the real-world scenario, and (2) allowing for exploratory analysis of situations.

The first category deals with providing a realistic environment embedding the wargame. We often use tabletop exercises to experiment with new ideas, but they are inherently linear in nature, and our decision spaces are no longer linear. The current environment is complex, highly nonlinear, and loaded with overlapping effects from modern weapon systems. Effect- and kill-chains are resulting in multidimensional effect- and kill-webs with many alternatives, feedback loops, and often delays in observable effects, to include higher order effects. Even small variations can lead to significant changes in the outcome, and some regions may even expose chaotic behavior. This is the reality of the modern warfighter, so it needs to be the reality for the modern wargamer as well. Multidomain and joint all-domain operations may be just the beginning of allowing for more local self-organization to provide the necessary operational agility to cope with complexity. Command and control and high-speed interoperability of all systems will become a focal point of future research. The introduction of intelligent autonomous systems that need to be integrated in the operations will add more challenges. Such a realistic representation of the decision space and resulting effects is a major challenge for simulation systems and supporting visualization methods.

The second category is directly related to the first one, as it provides the analytic tools needed to cope with the complex situation. Data science, visualization, and other means are needed in real operations, and so they are for the wargamers to support their decision process. The classic use of simulation systems for optimization is based on high-resolution presentations of known systems to be fine-tuned to look for the optimal outcome, confirmative in nature. Such simulation systems often use hundreds or thousands of input parameters and require significant effort to initialize a scenario described by all these data. But the technical challenge of preparation is not even the main challenge, as new scenarios are characterized by deep uncertainty about technical details and even more about the doctrinal challenges. Instead of optimizing for a well-defined but extremely speculative point-case future, these challenges require us to better understand the possibility space and design solutions accordingly. Decision-makers and wargamers must be enabled to detect, comprehend, visualize, and manage possibilities in the complex battlespace, including anticipating and dealing with emergent phenomena, requiring moving from confirmative to exploratory analysis. Command and control structures need to be configurable quickly to evaluate new, unconventional options. The focus is more on understanding the options than jumping to optimization tasks prematurely. In the complex decision space, it is often as important to exclude bad options early as it is to identify areas to focus on.

There are many open questions to be addressed in the future before we can expect the full support for wargames envisioned here. One example are megacities in coastal regions that are growing and likely to become a place for future conflicts. This will result in a new form of conflict that is no longer characterized by two groups of combatants facing each other on the battlefield, but operations will be entangled with noncombatant activities, increased focus on avoiding collateral damage, and the reactions of the population. The latter requires new insights of human, social, and cultural behavior. Such challenges need to be addressed and require experts beyond the physics-based models we grew so fond of. They are necessary, but no longer sufficient.

Coming back to our original question: are we growing closer together? Considering the contributions to this volume I believe we can be optimistic. Recent research in complexity science, artificial intelligence, and operations research helped to understand the potential and constraints of simulation systems better, and the complexity of the modern battlespace requires increased computational support. New wargaming centers will have to take this into account and provide for the necessary infrastructure for simulation support as required by the wargamers as well as for the supporting cells.

Index

Simulation and Wargaming, First Edition. Edited by Charles Turnitsa, Curtis Blais, and Andreas Tolk.
© 2022 John Wiley & Sons, Inc. Published 2022 by John Wiley & Sons, Inc.